Y0-BSD-047

Koninklijke Nederlandse Akademie van Wetenschappen
Verhandelingen, Afd. Natuurkunde, Tweede Reeks, deel 103

Functioning of microphytobenthos in estuaries

Edited by Jacco C. Kromkamp, Jody F.C. de Brouwer,
Gérard F. Blanchard, Rodney M. Forster and Véronique Créach

Amsterdam, 2006

P.O. Box 19121, 1000 GC Amsterdam, the Netherlands
T +31 20 551 07 00
F +31 20 620 49 41
E edita@bureau.knaw.nl
www.knaw.nl

ISBN 90-6984-453-2

The paper in this publication meets the requirements of ♾ ISO-norm 9706 (1994) for permanence.

Table of Contents

Preface and acknowledgements

Intertidal mudflats, sandbanks and shallow subtidal sediments are an important component of estuaries and they fringe large areas of the European coastline. Together with salt marshes and mangroves, they form natural barriers which protect coastal areas from the sea by reducing erosion and attenuating the energy of storm surges. The removal of these natural coastal defences can have devastating consequences, as shown in August 2005 after the passage of hurricane Katrina over the Gulf coast of the United States.

Intertidal and subtidal sediments are also biologically highly productive areas, representing important nursery and feeding grounds for higher organisms including birds, fish and shellfish. Dense concentrations of wading birds in particular are attracted by the rich supply of food, chiefly sediment-dwelling invertebrates, which in turn depend upon the presence at the sediment surface of a diverse assemblage of phototrophic microalgae (*microphytobenthos*[1]). Microphytobenthos supply organic matter to the ecosystem by the process of photosynthesis, and also influence nutrient fluxes in and out of the sediment. The microphytobenthic assemblages are most frequently dominated by diatoms, with other micro-organisms such as cyanobacteria and euglenoids found in lesser numbers. Up to 50% of primary production in estuaries can take place on sediments, despite the fact that the microalgal cells are only active in a thin surface layer of the sediment subject to rapid fluctuations in light, temperature and salinity. Although microphytobenthos activity is restricted to the uppermost surface sediment layer, they have a large impact on whole estuary morphology and ecosystem functioning. In addition to their impact on sediment chemistry and bentho-pelagic nutrient exchange, diatoms together with their associated communities of bacteria and animals can produce copious amounts of extracellular polysaccharides. These sticky substances can bind and trap sediment particles and cause the formation of cohesive surface biofilms that eventually become strong enough to stabilise the mudflats against erosion. In this way microphytobenthos can influence the morphology of coastal zones, and play an important role in the sediment balance of estuaries. Thus, microphytobenthos exerts its effect over a range of different spatial scales going from the hundred micron thick photic zone where photosynthesis occurs, to the scale of whole estuaries of many square kilometers where biofilms are able to influence the shape of the coastline.

[1] commonly called *microphytobenthos* (*MPB*) in Europe, and *benthic microalgae* (*BMA*) in the United States and elsewhere

In spite of the important roles that microphytobenthos play in estuarine environments, research on the physiology and ecology of these organisms is still at an early stage. For example it is apparent that benthic diatom species differ from phytoplanktonic diatoms in their photosynthetic responses, ability to move, and in a greatly increased resistance to environmental stressors such as ultraviolet radiation. However, the biochemical and genetic mechanisms underlying these differences are not known. At the ecosystem level, the contribution of microphytobenthos to the dynamics of estuaries is apparent, but integration of biological processes with hydrodynamic models of sediment movement has rarely been attempted. This is partly because of the highly dynamic nature of estuarine ecosystems and partly because of the extremely complex interplay between biological, physical and sedimentological processes. An interdisciplinary approach is clearly needed in order to come to a more thorough understanding of these processes. Almost all estuaries now have multiple, conflicting uses: important transport hubs such as container ports are located next to sites for migratory birds to rest and forage, excess nutrients from the land are deposited in the same waters that commercial fish stocks use as spawning grounds. Our ability to monitor the status of these ecosystems and to manage the multiple functions of estuaries in a sustainable manner will definitely require a more holistic approach. Therefore to facilitate exchange of ideas in the diverse fields of evolutionary ecology, ecophysiology, benthic ecology, sedimentology, and coastal zone management a colloquium was organised in Amsterdam in August 2003. Fifty leading scientists attended, and the contributions of many of the speakers can be found in this book. The meeting was financed by the Royal Netherlands Academy of Arts and Sciences (KNAW) and we are very grateful for their sponsorship and logistical support.

Contents

The symposium was organised around eight central themes, each with three speakers and followed by a discussion session. Additional round-table discussions were held for the exchange of views on several 'hot topic' themes in which there is currently division of opinions, or in which the introduction of new methods has caused a rethink of old ideas. These themes were 'Biodiversity of microphytobenthos' in which the classical and modern taxonomic methods were discussed. 'Methods for estimating microphytobenthic biomass and primary production' debated the wide variety of methods developed for sampling in a spatially heterogenous environment, and the difficulties of comparing photosynthetic measurements between sites and studies, and participants in 'Quantification and definitions of extracellular polymeric substances' discussed the problems of extraction and nomenclature of different extracellular fractions produced by microphytobenthos. The submitted papers are arranged in this volume in the original sequence of the symposium, beginning with studies of microalgal genetics and ecophysiology, through to holistic works on ecosystem modelling and socioeconomic perspectives.

(I) Taxonomy, evolution and biodiversity of microphytobenthos
Diatoms are the most important functional group within the microphytobenthos, but the taxonomy of this group is currently in a state of flux since the introduction of molecular classification methods in the 1990s. The paper by Linda Medlin describes the evolutionary history of diatoms, explains the most recent classification system for these algae based on DNA sequences and morphological studies, and describes the relationships between benthic diatoms and other diatom clades.

(II) Photosynthesis in marine diatom assemblages
Two papers are presented in this section: the paper by Jacco Kromkamp and Rodney Forster gives an overview of the methodologies used to measure biomass, photosynthesis and primary production of microphytobenthos (MPB). Photosynthetic physiology is described and the most frequently used variants of oxygen evolution and radiocarbon-based methods are described. The use of variable fluorescence techniques is covered in detail, and procedures (and precautions) are given for conversion from fluorescence-based measurements of electron flow to more conventional units of carbon fixation. The paper by Ronnie Glud describes the use of microsensor techniques to measure photosynthesis, particularly in subtidal sediments. Also the recent application of planar optodes is discussed which gives unparalleled 2D-information on O_2-fluxes.

(III) Extracellular Polymeric Substances: function and role in mudflats
In this section Jody de Brouwer et al. review the origin, structure and function of extracellular polymeric substances. The composition of EPS extracted from laboratory cultures of diatoms is compared to material from natural sediments. The focus lies on the difficulties of obtaining good samples, understanding the complexity of EPS and the need for better definitions.

(IV) Use of stable isotopes in foodweb research
It is well known that microphytobenthos can be an important foodsource for meiofauna. The paper by Kevin Carman *et al.* explores the role of MPB as foodsources in relation to temporal variation in MPB availability for different copepods using stable isotope signatures of both C and N.

(V) Upscaling primary production
There is a large mismatch between the scale and frequency at which samples are taken during field measurements, and the need for results at much greater scales. For example, photosynthetic rates measured at only one time during the day are frequently used to predict rates of primary production for the whole of that day. In this section, Gerard Blanchard *et al.* investigate short term dynamics in MPB biomass, and modelling results indicate that loss rates will balance productivity. Lawrence Cahoon reviews (the lack) and uncertainty of regional and global estimates of primary production. Finally, Rodney Forster and Jacco Kromkamp use different approaches to model net primary production and propose a method using remote sensing data to map primary production at the whole estuary scale.

(VI) Microphytobenthos and benthic-pelagic exchange
Two papers are presented in this section: Graham Underwood and Mark Barnett use a statistical approach to investigate the relationship between nutrient availability and other abiotic variables and species composition. Jay Pickney takes an

ecosystem approach to study nutrient dynamics in order to understand interannual variation in nutrient concentrations in Galveston Bay, Texas.

(VII) Mudflat ecosystem models

Two very different contributions are presented in this section. The paper by Laurent Seuront and Céline Leterme investigates the mechanisms and scale relationships behind the patchiness in MPB biomass distribution. The paper by Jean-Marc Guarini *et al.* proposes an integrated model which simulates local dynamics in MPB biomass and primary productivity based on physiological behaviour and changes in the abiotic environment.

(VIII) Mudflats and socioeconomics

Two papers are presented in this section: eutrophication of estuaries by agriculture is a well known phenomenon, usually with deleterious consequences. Increased nutrient inputs may increase turbidity in the water column, and cause blooms of opportunistic macroalgal species, with resulting negative effects on faunal biodiversity. However, the link between processes on land, and biological processes in the estuary has rarely been researched. The role of farming strategies and awareness of farmers of their role in the context of 'nitrate vulnerable zones' is investigated in the paper by Colin Macgregor. Carsten Brockmann *et al.* investigate the use and its implementation of remote sensing as a tool for monitoring and managing of estuarine ecosystems.

We hope you enjoy this special book about the role of microphytobenthos in estuarine ecology.

Jacco C. Kromkamp
Jody F.C. de Brouwer
Gérard F. Blanchard
Rodney M. Forster
Véronique Créach

I. Taxonomy and systematics of microphytobenthos

Linda K. Medlin

MINI REVIEW: The evolution of the diatoms and a report on the current status of their classification

Abstract

This mini review describes the newest classification of diatoms based on their evolution, which was obtained from molecular data. Because centric forms were found earlier in the geological record it is assumed that the pennate diatoms evolved from the centric forms and 3 classes were described: the centric diatoms, the araphid pennate and the raphid pennate diatoms. However, molecular data showed that the centric diatoms are most likely paraphyletic and the diatoms are now divided into two groups: Clade 1 contains the radial centric diatoms and Clade 2 contains two sub-Clades: the first sub-Clade contains the bipolar centrics and the Thalassiosirales and the second sub-Clade contains the pennate diatoms to which many of the microphytobenthos species belong. These Clades and additional morphological support for this new taxonomy is discussed in this mini review.

The diatoms are one of the most easily recognisable groups of major eukaryotic algae, because of their unique silicified cell wall, which consists of two overlapping thecae, each in turn consisting of a valve plus a number of hoop-like or segmented girdle bands. Such structures are present in all living diatoms (except following secondary loss, e.g., in the case of the endosymbiotic diatoms living in foraminifera), and also in early, well-preserved fossil diatoms from the early Albian (Lower Cretaceous) of what is now the Weddell Sea, Antarctica (Gersonde and Harwood, 1990). Molecular sequence data have consistently shown that diatoms belong to the heterokont algae. These chlorophyll *a*+*c*–containing algae typically have motile cells with two heterodynamic flagella, one covered with tripartite mastigonemes and the other smooth. In diatoms, the flagellar apparatus is reduced or absent; in fact, only the spermatozoids of the oogamous 'centric' diatoms are flagellated (Manton and von Stosch, 1966) and these are uniflagellate, lacking all trace of a smooth posterior flagellum or basal body. Nevertheless, the characteristic heterokont mastigonemes are present on the single flagellum and diatoms also possess similar plastid ultrastructure (with four bounding membranes, lamellae of three thylakoids, and usually a peripheral ring nucleoid) and pigment composition to e.g., the brown algae.

Diatom origin has been speculated upon by several workers. Diatoms may be derived from a spherical uniformly scaled monad (Round, 1981; Round and Crawford, 1981, 1984) with an anterior flagellum (Cavalier-Smith, 1986), or be derived from a cyst-like form like the extant Parmales in the chrysophyte algae (Mann and Marchant, 1989). A scaly ancestor likely existed at some point in their phylogeny because of the presence of scales on reproductive cells of diatoms and scales on the reproductive stage of the Labyrinthuloides, an earlier divergence in the heterokont lineage (Medlin *et al.* 1997a). Recent phylogenies constructed from nuclear-encoded small-subunit ribosomal RNAs place the diatoms within the pigmented heterokont algal lineages (Bhattacharya *et al.*, 1992; Leipe *et al.*, 1994, Medlin *et al.*, 1997b), most closely related to the new algal class, the Bolidophyceae, which are picoplanktonic algae with a simplified cellular organization (Guillou *et al.* 1999). Both diatoms and bolidomonads commonly possess similar pigments and two transverse plates at the base of each flagellum. A molecular clock constructed from 4 genes has placed the average age of the diatoms ca. 135 Ma ago with their earliest possible age being no earlier than 240 Ma ago (Medlin *et al.* 1997a, 2000).

Most diatomists have long assumed that the diatoms contain two groups: the centrics and the pennates, which can be distinguished by their type of sexual reproduction, pattern centers or symmetry, and plastid number and structure (Figure 1A, Round *et al.* 1990). The oogamous centric diatoms with radially valve symmetrical ornamentation and with numerous discoid plastids are distinct from the isogamous pennate diatoms with bilaterally symmetrical pattern centres and with fewer plate-like plastids. Both groups are known to most aquatic and cell biologists under these terms. Historically, the centric and pennate diatoms have been classified into two distinct classes based on these characters. Pennate diatoms undoubtedly evolved from the centric forms because they first appear later in the geological record. Coscinodiscophyceae (centric diatoms), Fragilariophyceae (araphid pennate diatoms), and Bacillariophyceae (raphid pennate diatoms) are the three classes presently recognized in Round *et al.* (1990) (Figure 1A) because the raphid pennate diatoms (those with a slit opening [raphe] in the cell wall for movement) were given equal taxonomic ranking with the araphid pennate diatoms (those without this slit). An alternative classification based on molecular data supported by different cellular features has been presented by Medlin and Kaczmarska (2004) and in Figure 1B.

The first molecular evidence that clearly demonstrated that the centric diatoms were paraphyletic was presented by Medlin *et al.* (1993). In the same study, araphid pennate diatoms were also shown to be paraphyletic. Among the taxa studied in this first paper, the centric diatom, *Skeletonema costatum* was most closely related to the pennate diatoms, with high bootstrap support in molecular analyses. Additionally, Sörhannus *et al.* (1995) showed that centric and araphid taxa were paraphyletic using an analysis of partial sequences from the 28S large-subunit (LSU) rRNA coding region from eight diatoms. These initial data suggested that presently used higher level diatom systematics do not reflect their evolutionary history. All subsequent analyses from three more genes have supported this finding (Medlin *et al.* 1996, Medlin *et al.*, 2000, Ehara *et al.* 2000), and the diatoms have been consistently divided into 2 groups: Clade 1 contains the radial centrics and Clade 2 can be subdivided into two sub-Clades; the first of which contains the bipolar centrics and the radial Thalassiosirales (Clade 2a), and the

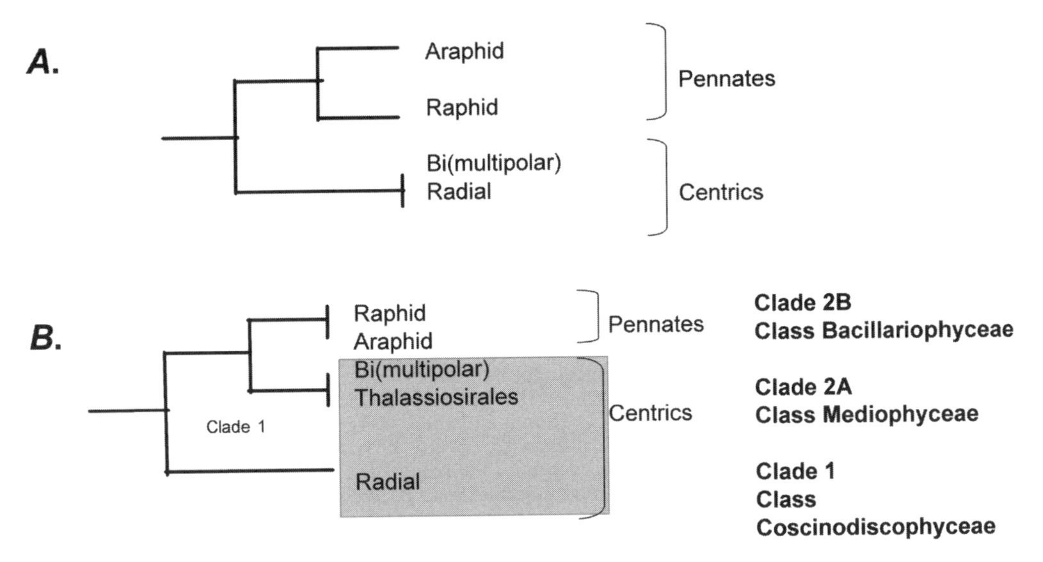

Figure 1. Stylised tree depicting current classification of the diatoms (A) based on morphological data and that proposed by Medlin and Kazcmarska 2004 based on molecular data (B).

second, the pennates (Clade 2b). Morphological and cytological support for the two Clades was reviewed in Medlin *et al.* (2000) and Medlin and Kaczmarska (2004).

Clade 1 and Clade 2, are recognised now at the subdivision level, as the Coscinophytina and Bacillariophytina, respectively, and Clades 1, 2a and 2b are recognised now at the class level: Class Coscinodiscophyceae, Mediophyceae and Bacillariophyceae (Medlin and Kaczmarska (2004). Primary support for two subdivisions comes from the Golgi arrangement, which is in Clade 1 primarily the G–ER–M unit and in Clade 2, a perinuclear arrangement. Weaker support for the two subdivisions comes from the sperm/chloroplast arrangement, which is in Clade 1 primarily merogenous and in Clade 2 primarily hologenous.

The primary support for the three classes comes from the auxospore structure. Isodiametric auxospores with scales are characters of Clade 1, anisodiametric auxospores with scales and hoops or bands (a properizonium) are noted in Clade 2a, and anisodiametric auxospores that form a complex tubular perizonium, usually consisting of transverse hoops and longitudinal bands, are only in Clade 2b. Weaker support for the three classes comes from the pyrenoid structure, which has a single thylakoid crossing the pyrenoid that is not connected to the plastid thylakoids in Clade 1, is usually without a crossing thylakoid in Clade 2a (or if present, is similar to Clade 1 or just lies along the periphery of the pyrenoid centre), or has a single thylakoid crossing the pyrenoid that is connected to the plastid thylakoids in Clade 2b. Fossil support from the earliest best preserved fossil deposit (Gersonde and Harwood (199)) suggests that diatoms lacking any structure in the valve centre and having complex linking structure were the likely ancestors of Clade 1 diatom, whereas those with a central tube structure in the valve and less complicated linking structures likely gave rise to Clade 2 diatoms.

References

BHATTACHARYA, D., L. MEDLIN, P.O. WAINWRIGHT, E.V. ARIZTIA, C. BIBEAU, S.K.STICKEL, and M.L. SOGIN. 1992. Algae containing chlorophylls *a* + *c* are paraphyletic: molecular evolutionary analysis of the Chromophyta. Evolution **46**: 1808-1817. Errata *Evolution* 47: 98

EHARA M, Y. INAGAKI, K.I. WATANABE and T. OHAMA. 2000. Phylogenetic analysis of diatom coxI genes and implications of a fluctuating GC content on mitochondrial genetic code evolution. Current Genetics **37**: 29-33.

CAVALIER-SMITH T. 1986. The Kingdom Chromista: Origin and systematics. Progress In Phycological Research **4**: 319-358.

GERSONDE R, and D.M. HARWOOD. 1990. Lower Cretaceous diatoms from ODP Leg 113 site 693 (Weddell Sea) Part 1: Vegetative cells Proc Ocean Drilling Program, Scientific Results **113**: 365–402.

GUILLOU L, M-J. CHRÉTIENNOT-DINET, L.K. MEDLIN, H. CLAUSTRE, S. LOISEAUX-DE GOËR S, and D. VAULOT. 1999. *Bolidomonas*: a new genus with two species belonging to a new algal class, the Bolidophyceae (Heterokonta). J. Phycol. **35**: 368–381

LEIPE, D.D., P.O. WAINWRIGHT, J.H. GUNDERSON, D. PORTER, D. PATTERSON, F. VALOIS, S. HIMMERICH, and M.L. SOGIN. 1994. The stramenopiles from a molecular perspective: 16S-like rRNA sequences from *Labyrinthuloides minuta* and *Cafeteria roenbergensis.* Phycologia **33**: 369-377

MANN, D.G., and H. MARCHANT. 1989. The origins of the diatom and its life cycle. In JC Green, BSC Leadbeater and WL Diver (eds).The Chromophyte Algae: Problems and Perspectives (Systematics Association Special Volume 38), 305-321 Clarendon Press, Oxford.

MANTON, I., and H.A. VON STOSCH. 1965. Observations on the fine structure of the male gamete of the marine centric diatom *Lithodesmium undulatum* – J. Roy. Microsc. Soc. **8**: 119–134.

MEDLIN, L.K., I. KACZMARSKA. 2004 Evolution of the diatoms: V Morphological and Cytological Support for the Major Clades and a Taxonomic Revision Phycologia *Phycologia* 43: 245-270

MEDLIN, L.K., R. GERSONDE, R. KOOISTRA, W.H.C.F, and U. WELLBROCK. 1996. Evolution of the diatoms (Bacillariophyta) II Nuclear-encoded small-subunit rRNA sequence comparisons confirm a paraphyletic origin for the centric diatoms Mol. Biol. Evol. **13**: 67-75

MEDLIN, L.K., D.M. WILLIAMS, and P.A. SIMS. 1993. The evolution of the diatoms (Bacillariophyta) I Origin of the group and assessment of the monophyly of its major divisions Eur. J. Phycol. **28**: 261–275

MEDLIN, L.K., W.H.C.F. KOOISTRA, R. GERSONDE, P.A. SIMS, and U. WELLBROCK. 1997a. Is the origin of the diatoms related to the end-Permian mass extinction? Nova Hedwigia 65: 1–11

MEDLIN, L.K., W.H.C.F. KOOISTRA, D. POTTER, G.W. SAUNDERS, and R.A. ANDERSON. 1997b. Phylogenetic relationships of the 'golden algae' (hepatophytes, heterokont chrysophytes) and their plastids *Plant Systematics and Evolution (Supplement)* 11: 187-210

MEDLIN, L.K. W.H.C.F. KOOISTRA, and AM-M. SCHMID. 2000. A review of the evolution of the diatoms – a total approach using molecules, morphology and geology In *The origin and early evolution of the diatoms: fossil, molecular and biogeographical approaches* (Ed by A Witkowski & J Sieminska) pp13-35 Szafer Institute of Botany, Polish Academy of Science, Cracow, Poland

ROUND, F.E. 1981. Some aspects of the origin of diatoms and their subsequent evolution Biosystematics **14**: 486-486

ROUND, F E, and R.M.Y,.1981. The lines of evolution of the Bacillariophyta I Origin. Proc. Royal Soc. London *B* 211: 2 37-260

ROUND, F.E., and R.M. CRAWFORD. 1984. The lines of evolution of the Bacillariophyta II The centric series *Proceedings of the Royal Society of London B* 221: 169-188

ROUND, F.E., and D.G. MANN. 1990. The diatoms Biology and morphology of the genera Cambridge University Press, Cambridge.

SORHANNUS, U., F. GASSE, R. PERASSO, and A. BAROIN-TOURANCHEAU. 1995. A preliminary phylogeny of diatoms based on 28S ribosomal RNA sequence data. Phycologia **34**: 65-73.

II. Photosynthesis in marine diatom assemblages

Jacco C. Kromkamp and Rodney M. Forster

Developments in microphytobenthos primary productivity studies

Abstract

Primary production by microphytobenthos can be a significant fraction of the total estuarine primary production. The productivity can be measured using a variety of techniques, which are reviewed in this contribution. Photosynthesis of intertidal benthic microalgae is increasingly measured using the pulse amplitude modulated (PAM) fluorescence technique and we will discuss this technique in detail, putting emphasis on a number of inherent assumption. PAM fluorescence is rapid, non-intrusive and can be used in situ. These measurements assume that the chlorophyll fluorescence is proportional to the inherent photosynthetic properties of the algae. We will demonstrate that algal fluorescence from below the surface ('deep layer' fluorescence) can contribute significantly to the fluorescence measured at the sediment surface. This will cause overestimates of the truc apparent quantum efficiency and thus of the estimated rate of photosynthesis. This error, which is influenced by the pattern in depth distribution in algal biomass, is significant at higher irradiances and in most cases ranges between 20 to 40%. Most estuarine sediments show that microphytobenthos biomass decreases exponentially with depth during low tide. For these sediments calculated total production and modeled production showed a linear relationship, demonstrating that under these circumstances PAM fluorescence techniques can be used.

Introduction

Primary production by pelagic phytoplankton and by benthic algae make estuaries belong to the most productive ecosystems of the world. Estimates for microphytobenthos primary production range from 27 to 234 gC m^{-2} $year^{-1}$ (Heip *et al.* 1995; Macintyre *et al.* 1996; Underwood and Kromkamp 1999) while those for phytoplankton range from 7 to 875 gC m^{-2} $year^{-1}$ (Boynton *et al.* 1982; Underwood and Kromkamp 1999). This primary production drives a rich foodweb, making estuaries ecologically important areas and of crucial importance for shorebirds and fish(eries). However, microphytobenthos are not only important as primary foodsource for higher trophic levels; they also play a very important role in sediment dynamics as they are important stabilizers of estuarine sediments through the excretion of extracellular polymeric substances (Blanchard *et al.* 2000; De Brouwer *et al.* 2000; Paterson 1989; Paterson and Black 1999).

Because estuaries are not steady state systems they change slowly over time. To follow these changes with respect to primary productivity good quantitative approaches are needed and this is not an easy task with respect to microphytobenthos biomass and production because of the complex interaction between physical, chemical and biological processes. As estuaries are often manipulated by man, discriminating the effects of 'normal' change and/or variability from year to year and anthropogenic induced changes is even more difficult, requiring good methods and avoiding the ever present danger of undersampling.

Microphytobenthos (MPB) primary productivity is a function of the biomass present, the photosynthetic characteristics, and the sediment characteristics (nutrient availability and sediment optics). However, total primary production is not only dependent on the productivity, but also in the case of intertidal MPB on the exposure time, water transparency (in case of submerged photosynthesis) and possible vertical migration (Paterson 1986; Pinckney and Zingmark 1993). These latter components will not be treated in this overview, but the reader is referred to the chapter from Forster & Kromkamp (this book).

Measurements of MPB biomass

Although this overview will deal with MPB productivity measurements, some attention has to be paid to the different methods to determine MPB biomass. Routinely, samples are taken by using 'syringe like' corers of different type, and the sampling depth varies generally between 2 and 10 mm (see MacIntyre *et al.* 1997). In the more cohesive sediments most of the MPB biomass of intertidal mudflats is concentrated near the surface (De Brouwer and Stal 2001b; Kelly *et al.* 2001), and very little MPB is present below 2 mm depth (see Figure 1). In this case, the sampling depth is not so important as long as the chlorophyll concentration is expressed per unit surface. If it is expressed per weight, then adding more depth will 'dilute out' the chlorophyll content, making comparison difficult. In sandy sediments the chlorophyll is normally more evenly distributed with depth. In this case the chlorophyll content per unit weight is relatively independent of the sampling depth, but the chlorophyll concentration per surface area will vary with sampling depth. As significant compaction can occur during low tide as the water evaporates, this can add to the error in the estimate of the chlorophyll content of estuarine sediments (Perkins *et al.* 2003).

Estimating MPB biomass is also difficult because of the patchy nature of the algae. We investigated this briefly on several sites in the Eden estuary in Scotland, and found that the standard error ranged from 6-22% on samples (20 cm^2) spaced 20 cm apart, from 6-53% on samples spaced 100 cm apart and from 6 to 145% on samples spaced 500 cm apart. More information on spatial distribution of MPB can be found in Seuront *et al.* (this book) and References therein.

Another factor complicating MPB biomass estimation is the occurrence of vertical migration (Kromkamp *et al.* 1998; Paterson 1986; Paterson *et al.* 1998; Serôdio *et al.* 1997). Pinckney and Zingmark (1991) developed a model, based on sun angle

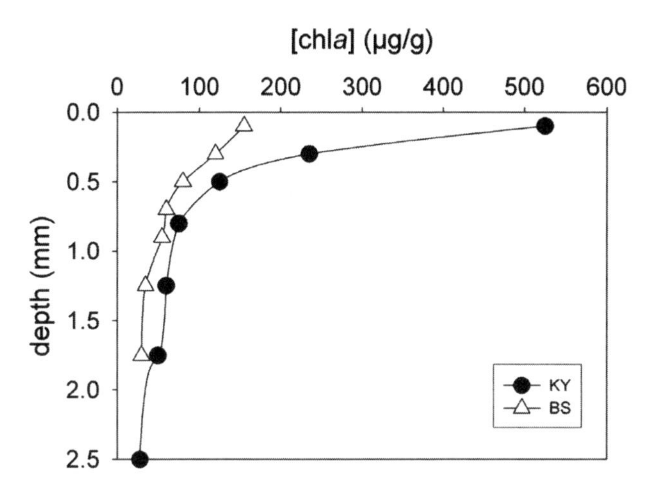

Figure 1. Two examples of vertical distribution of chlorophyll *a*. Profile KY is taken from (Kelly *et al.* 2001) and profile BS is taken from De Brouwer and Stal (2001).

and tidal stage, to describe vertical migration patterns. This model accurately predicted vertical migration of epipelic diatoms on an intertidal mudflat in the Westerschelde estuary (Biezelingsche Ham, SW Netherlands, R.M. Forster pers. comm.). A preliminary conclusion from this model is that most of the migration takes places within 30-60 minutes after the tide leaves the mudflat, and that the algae start to migrate down 30-60 min before the tide returns. This leaves a sampling window open, centered around low tide, in which the vertical depth profiles are probably rather stable. Nevertheless caution is necessary, as the migration pattern, which seems to be driven by internal clock(s) as it persists for several days in darkness without tidal influence (Serôdio *et al.* 1997, own observation), can be modulated by environmental conditions. Perkins *et al.* (2002) observed that high light may drive MPB downwards, in contrast to the situation observed by Kromkamp *et al.* (1998) who found evidence that in high light the biomass stayed at the surface, but that a 'micro-migration' prevented MPB from being exposed to deleterious conditions too long.

Measuring photosynthesis

Photosynthesis starts with the absorption of light by the light harvesting antenna complexes, which are phycobilosomes in cyanobacteria or red algae, and complexes intrinsic in the thylakoid membrane containing chl *a/b*- or chl *a/c*-binding proteins in eukaryotic algae. Absorbed light energy is passed on via several different light harvesting protein complexes to the reaction centra (RC) where it is used for charge separation. The charge separation occurring in the reaction centres of photosystem II (PSII), leading to the formation of oxygen by the splitting of water, can be measured using variable fluorescence techniques like the pulse amplitude modulated (PAM) techniques (Schreiber *et al.* 1986), the pump and probe technique (Falkowski *et al.* 1986; Kolber and Falkowski 1993) and its successor the fast repetition rate fluorometer (FRRF, Kolber *et al.* 1998). These variable fluorescence techniques thus measure the initial steps in photosynthesis and can be used to estimate photosynthetic electron transport (ETR). Because PSII charge separation leads to the formation of O_2 it might be expected that rates of photosynthetic electron transport and rates of oxygen evolution are closely coupled (Kolber and Falkowski 1993). This is not necessarily the case though. Normal methods (Winkler titration, oxygen electrodes) measure the net oxygen exchange (NPP), i.e. the result of oxygen production (i.e. gross photosynthesis, GPP), minus the rate of oxygen consumption (R). In the dark, most of the oxygen consumption will be caused by respiratory electron transport (mitochondrial respiration), but in the light there may be other oxygen consuming processes like photorespiration (caused by the oxygenase activity of RUBISCO) and the Mehler reaction (i.e. O_2 uptake by PSI). In the latter case the oxygen produced from water by PSII is later consumed by PSI, producing water, (via O_2^- and H_2O_2) and this complex series of reactions is therefore also called the water-water cycle (see Asada 2000 for a recent review). Thus, although the Mehler reaction gives rise to ETR, there is no net O_2-production. The Mehler reaction is likely to play a role in photoprotection: it keeps the redox state of the PQ-pool lower and increases the ΔpH, which might initiate the xanthophyll cycle (Neubauer and Yamamoto 1994; Schreiber and Neubauer 1990) which serves to protect PSII by over excitation because it dissipates absorbed irradiance as heat (for a recent review see Horton *et al.* 2000).

Photosynthesis can also be measured by the incorporation of CO_2. In terrestrial situations CO_2 uptake is often measured directly by measuring changes in CO_2 concentration with an infrared gas analyzer (IRGA). In aquatic systems CO_2 fixation is measured indirectly by measuring the uptake of $NaH^{14}CO_3$ into organic material, either in particulate form (i.e. the sample is filtered and the material on the filter is counted after acidification to remove any remaining $NaH^{14}CO_3$), or both the particulate and possible excreted DOC are measured after acidifying the solution.

Photosynthesis can be described by the following, well known equation:

$$6CO_2 + 6H_2O \rightarrow C_6H_{12}O_6 + 6O_2$$

This suggests that both methods are interchangeable. This is not true, because it depends also on the nitrogen component used for growth (Williams 1993a): for growth on ammonia, and assuming a C: N-ratio of 6:

$$6CO_2 + 6H_2O + NH_3 \rightarrow C_6H_{12}O_6NH_3 + 6O_2$$

whereas for growth on nitrate the situation is different:

$$6CO_2 + 6H_2O + HNO_3 \rightarrow C_6H_{12}O_6NH_3 + 7.5O_2$$

Thus, if the rate of C-fixation is to be calculated the photosynthetic quotient ($PQ=CO_2/O_2$) has to be known. For growth on ammonia a PQ of 1.1 is often observed, whereas for growth on nitrate a PQ of 1.4 seems the best value. However, a larger range in PQ values has been observed in field observations, pointing to possible technical errors or to different regulation patterns in the N-metabolism leading to different C: N-ratios (Williams and Robertson 1991; Williams 1993b).

Thus, to measure photosynthesis and primary productivity in MPB, three principle techniques exist: Oxygen evolution measurements, C-fixation measurements, and measurements of variable fluorescence. For every principle method, a number of different approaches exist which we will very briefly review here. For all cases it can be said that comparisons are difficult, even if the same techniques are used, because different researchers use different incubation times

Oxygen-based techniques

1. One of the older techniques is the bell-jar. With this technique a benthic chamber is pushed in the sediment, and the rate of change in oxygen in the overlying water over a specified period is measured (Lindeboom and De Bree 1982; Pomeroy 1959). A big advantage of this technique, assuming a larger size bell jar, is that it is less sensitive to small scale variation in patchiness. Another advantage of this technique is that it leaves the vertical chemical and physical gradients intact. This is important as these physical and chemical gradients might influence the rate of photosynthesis. Because the bell-jar technique measures net oxygen exchange, it measures net community production, which is not equal to net photosynthesis. Adding the respiration, measured in the dark, gives an estimate of gross primary production, but this assumes that the total rate of oxygen consumption is not influenced by light, an assumption which does not hold. Another disadvantage of this technique is that the measured rates are underestimates of the true rate of net oxygen production, because only the upward oxygen flux is measured, whereas part of the oxygen produced is consumed at greater depths.

2. Oxygen microelectrode techniques. These electrodes measure gross photosynthesis. The light-dark shift technique (Revsbech and Jørgensen 1983) measures the initial rate of oxygen consumption after darkening a sample, which equals the gross rate of photosynthesis. This can be done at several depths, so that a complete photosynthesis-depth profile can be measured. This technique is widely used (Glud *et al.* 1992; Kromkamp *et al.* 1995; Kühl *et al.* 1994; Yallop *et al.* 1994). The advantage of the light-dark shift technique is that it measures gross photosynthesis without destroying the chemical and physical gradients in the sediment. The technique can even be used *in situ*. When production profiles at different irradiances have to be measured it takes quite some time to finish the measurements because of the time needed for the oxygen gradients to reach steady state. As a result, the rates might be considerably influenced by vertical migration, if occurring during the measurements. Another disadvantage is that the method is very sensitive to patchiness, making it less suitable as a method to determine primary production at larger space scales, but very suitable to measure the occurrence of it.

Production rates can also be calculated from modeled oxygen depth profiles at fixed irradiances (Epping and Jorgensen 1996; Wiltshire *et al.* 1996). This method is quicker than the light-dark shift technique and is therefore sometimes chosen when vertical migration might play a role, but it still takes several hours to complete a number of profiles at different irradiances. An additional disadvantages of the oxygen profile method are that several parameters have to be known or assumed (porosity, diffusivity and photic depth). An advantage of this method relative to the light-dark shift method is that a more robust electrode can be used because fast reaction times are not needed.

3. Optodes. Two different optodes techniques are described in the literature. Fiber-optic microprobes (Glud *et al.* 1999a; Holst *et al.* 1997) and planar optodes (Glud *et al.* 1999b; Glud *et al.* 1998; Glud *et al*, this book). Optodes detect oxygen with a special chromophore, which then emits light, which is detected by a sensor. An advantage of the optodes is that they do not consume oxygen, which Clark-type oxygen electrodes do. Microfiber optodes are commercially available and can be operated analogous to oxygen microprofiles. This technique has the same advantages and disadvantages as estimating primary production by modelling of vertical oxygen profiles. Because the optode is based on a glass microfiber, they are flexible and will break less easily than the glass based Clark-type oxygen electrodes. In coarse sands they cannot be used, because the chromophore comes off too easily (personal experience).

The planar optode is a very recent development and is not commercially available yet. The big advantage of a planar optode is that it quantifies the oxygen field across an area of 10-100 cm^2 . For this reason the planar optode is ideal for scaling up productivity estimates because estimates are not sensitive to micro-small scale patchiness. For more information on optodes, especially planar optodes see Glud *et al.*, this book.

^{14}C based techniques

1. Uptake of radioactive labeled bicarbonate can also be measured using the bell-jar technique, where the overlying water is labeled and the rate of uptake is measured (Colijn and De Jonge 1984; Rasmussen *et al.* 1983). A disadvantage of the technique is that is difficult to quantify the rates, because the interstitial DIC in the thin MPB layer is not known, leading to underestimates of the true rate (Heip *et al.* 1995; Smith and Underwood 1998). This technique seems hardly used nowadays.

When ^{14}C-labeled bicarbonate is added to the overlying waters in a bell-jar, the algae in the sediment are exposed to varying concentrations of the tracer. To overcome this problem Jönsson (1991) developed the percolation technique, allowing the ^{14}C-labelled bicarbonate to percolate through the sediment. He observed that the percolation tube gave 1-16x higher rates than obtained using a bell jar-technique, but the rates were still 3-8 times lower than observed when using slurries (see below). Percolation on porous, more sandy substrates is possible, but the technique does not work with very cohesive sediments (Sundback *et al.* 1996; Underwood and Kromkamp 1999). Because of difficulties with the specific activity, this method is not reliable for obtaining good estimates of photosynthetic activities.

2. Slurry technique. For this technique, the upper slices of sediment cores are used, often pooled, and diluted with filtered seawater. The pooled samples are then

incubated in a photosynthetron with a known concentration of ^{14}C-labeled bicarbonate after which the samples are counted. This technique, employed by many authors (Barranguet and Kromkamp 2000b; Blanchard and Cariou-Le Gall 1994; Blanchard and Montagna 1992; Macintyre and Cullen 1998), is a good way to obtain reliable photosynthetic parameters. In principle the rate of oxygen evolution can also be measured with slurries (Wolfstein *et al.* 2000), but because the oxygen electrode is not so sensitive thick slurries are needed, causing a steep light gradient in the cuvet, thus giving rise to a poorly defined light climate. The drawback of the slurry technique is that is measures potential production, because the existing gradients in the sediments are destroyed. However, if no nutrients are limiting, especially CO_2/HCO_3^-, the measured potential photosynthetic rates might reflect real rates (Barranguet *et al.* 1998).

Variable fluorescence techniques

Recently, photosynthetic activity of microphytobenthos is also measured with variable fluorescence techniques, either *in situ* (Barranguet and Kromkamp 2000a; Kromkamp *et al.* 1998; Perkins *et al.* 2002; Serodio 2003), or in slurries (Wolfstein *et al.* 2000). An exciting new development is the measurement of photosynthesis using variable fluorescence on single cells (Gorbunov *et al.* 1999), either using fluorescence microscopy (Snel and Dassen 2000) or in combination with imaging techniques (Baker *et al.* 2001; Oxborough *et al.* 2000; Perkins *et al.* 2002).

In the dark, when the primary electron acceptor of PSII, Q_A, is in its oxidized state, it is an efficient quencher of fluorescence. As a result, the measured minimal fluorescence (F_0) is a good proxy for the chlorophyll concentration. In the light, part of the Q_A becomes reduced, causing a rise in fluorescence. When a single turnover flash (STF) is given (i.e. a flash of less than 2-5 μs duration giving a reduction of Q_A only, thus without electron transport from Q_A^- to Q_B) the fluorescence rises rapidly to a fluorescence level called I_1 (Neubauer and Schreiber 1987; Schreiber *et al.* 1995) and which equals the level reached by the pump and probe and Fast Repetition Rate Fluorometer (Kolber and Falkowski 1993; Kolber *et al.* 1998; Kromkamp and Forster 2003; Samson *et al.* 1999; Schreiber *et al.* 1995). This rapid rise in fluorescence, called the photochemical phase, can be followed by a slower, thermal phase when a saturating multiple turnoverflash (MTF) is used. Unlike the rapid photochemical phase, this slower phase in the fluorescence induction curve disappears partially at temperatures below 0 °C and is completely abolished at temperatures below -35 °C (see Neubauer and Schreiber 1987). Although the origin of the thermal phase is still uncertain (see reviews by Lazar 1999, and Samson *et al.* 1999) it suffices to say here that the F_M level reached by a STF ($=F_{M\text{-}ST}$) is lower than the F_M level reached by a MTF ($=F_{M\text{-}MT}$), and that the F_M reached with a MT-flash equals the fluorescence yield after application of the herbicide DCMU. The ratio of the variable fluorescence F_V ($=F_M-F_0$) to F_M (i.e. F_V/F_M) equals the maximum quantum efficiency of PSII charge separation and this is a valuable parameter as it reveals whether the algae are experiencing stressful conditions, as damage to PSII (Melis 1999; Ohad *et al.* 1994; Park *et al.* 1996; Parkhill *et al.* 2001; Young and Beardall 2003), which might be caused by excess light or UV-radiation, as well as unbalanced growth conditions due to nutrient limitation (Green *et al.* 1994; Kolber *et al.* 1988; Kromkamp and Peene 1999; Lippemeier *et al.* 2001). All of these conditions will lead to a reduction in F_V/F_M.

The reaction centre of PSII is very susceptible to photoinhibitory damage, especially when Q_A is in the reduced (closed) state. It is therefore not surprising that several protection mechanisms exist. One of the prominent mechanisms is the activation of the xanthophyll cycle. At high rates of PSII electron transport, the lumen of the thylakoid membranes will acidify, causing a build up in the ΔpH. This will in turn activate an enzyme called a de-epoxidase. In higher plants and green algae the de-epoxidation will convert in a two step way the carotenoid violaxanthin (V) via anteraxanthin (A) to zeaxanthin (Z). The conversion of V→A→Z occurs within a few minutes, causing a conformational state in the light harvesting protein complexes of PSII (LHC II) (Bassi and Caffarri 2000; Horton *et al.* 2000). Under low light, the reverse reaction takes place, but the rate of epoxidation is slower (10-20 min) compared to the de-epoxidation reaction. In diatoms and haptophytes a different xanthophyll cycle occurs: the xanthophyll diadinoxanthin (Dd) is de-epoxidized to diatoxanthin (Dt) in a one step process (Figure 2) (Young and Frank 1996). Like the conversion from V→A→Z the epoxidation from Dd to Dt is stimulated by an increase in the ΔpH and occurs within a couple of minutes. Activation of the xanthophyll cycle induces loss of absorbed light as heat. This heat dissipation causes a decrease in the functional absorption cross section of PSII (σ_{PSII}) and an increase in the quenching of fluorescence. This energy quenching (q_E) is one of the components of non-photochemical quenching (q_N), together with fluorescence quenching caused by a transition of the high fluorescent state 1 to the low fluorescent state 2 transition (q_T) and quenching caused by photoinhibitory damage (q_I). State transitions can be observed

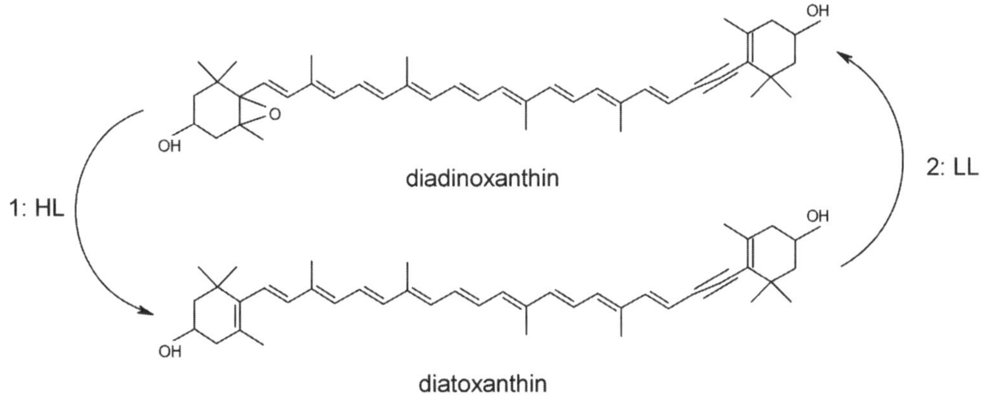

Figure 2. The xanthophylls cycle in diatoms and other chromophyte algae. In high light (HL) diadinoxanthin (DD) is de-epoxidized to diatoxanthin (DT), whereas in low light (LL) the reverse reaction takes place. In diatoms both ΔpH and diatoxanthin are necessary to induce NPQ (Lavaud *et al.* 2002).

in green algae and are especially important in cyanobacteria, whereby the phycobilisomes can move from PSII (state 1) to PSI (state 2) as the redox state of the PQ becomes more reduced in high light (see Allan and Pfannschmidt 2000 for a recent review on state transitions and Campbell *et al.* 1998 for a review on chlorophyll fluorescence and photosynthesis in cyanobacteria). Non-photochemical fluorescence quenching can be calculated as follows:

$$q_N = 1 - \frac{F'_M - F}{F'_M - F'_0} \qquad (1a)$$

or, in the absence of F_0' measurements as

$$NPQ = \frac{F_M - F'_M}{F'_M} \qquad (1b)$$

Photochemical fluorescence quenching as can be calculated as:

$$q_P = \frac{F'_M - F}{F'_M - F'_0} \qquad (2).$$

The photochemical quenching coefficient q_P is a proxy for the fraction of open reaction centres (Kolber and Falkowski 1993) although in the presence of excitation energy transfer between PSII units (connectivity) the relationship between q_P and the number of open RCII is not straightforward (Dau 1994; Lazar *et al.* 2001). Genty *et al.* (1989) demonstrated, using MT-flashes, that the effective quantum efficiency of linear PSII electron transport (ETR) equals the photochemical quenching coefficient and the efficiency of open PSII centres (F_V'/F_M'):

$$\Delta F/F_M' = qP \ x \ F_V'/F_M' \qquad (3)$$

Under optimal conditions $\Delta F/F_M$' often shows a linear relationship with the quantum efficiency of C-fixation (Genty *et al.* 1989). The rate of PSII electron transport can thus be calculated as:

$$ETR = E \ x \ a^*_{PSII} \ x \ \Delta F/F_M' \ x \ 0.25 \qquad (4)$$

E is the irradiance (PAR) and a^*_{PSII} is the optical absorption cross section (Å^2) of a PSII. The product of E and a^*_{PSII} is the quantity of light absorbed by a PSII. These equations assumes that (minimally) 4 absorbed photons are needed to produce an electron. If the quantum requirement does not change, the rate of C-fixation can be calculated using a MTF-protocol (i.e. using a PAM-like instrument) as follows:

$$P^B \ (mg \ C \ (mg \ chl)^{-1} \ h^{-1}) = E \ x \ a^* \ x \ 0.5 \ x \ \Delta F/F_M' \ x \ 0.25 \ x \ PQ^{-1} \qquad (5a)$$

Here is a* the optical absorption cross section per mg chl*a* (m^2 mg chl^{-1}). It is further assumed that 50% of the absorbed light is utilized by PSII. The photosynthetic quotient (PQ) is used so that the rate of oxygen production can be converted into a rate of C-fixation. Because the ratio of PSII to PSI can be variable, and the minimum quantum requirement of both PSI and PSII can be greater than the 4 absorbed photons per photosystem, it is perhaps better to replace rewrite eq. 5a as follows:

$$P^B = E\ x\ a^*\ x\ \Delta F/F_M'\ x\ \Phi_e\ x\ PQ^{-1} \qquad (5b)$$

Here Φ_e is the electron requirement for oxygen production (O_2/e^-). The minimal theoretical value according to the Z-scheme of photosynthesis is 0.125 O_2/e^- (which is equal to 0.5 x 0.25, see eq. 5a), although it must be stressed that the electron produced by PSII is used again during charge separation in PSI: so Φ_e is actually the reciprocal of the minimum quantum requirement of both PSI and PSII (i.e. the total number of charge separations needed by both PSI and PSII), and is closer to 0.1 than to 0.125 according to (Mauzerall 1986). The quantum requirement (Φ_e^{-1}) has been shown to vary often between 10 and 14 for a large variety of higher plants (Seaton and Walker 1990), and our own observations show that by using a value of 11-13 for charge separations of both PSI and PSII an accurate prediction of the rate of oxygen evolution can be predicted. An example is given for a benthic diatom in Figure 3.

For the pump and probe and FRRF fluorometers, which use a STF, a slightly different equation to predict the rate of C-fixation was developed (Kolber and Falkowski 1993):

$$P^B = E\ x\ \sigma_{PSII}\ x\ q_P\ x\ n_{PSII}\ x\ \Phi_e\ x\ PQ^{-1} \qquad (6).$$

Here, σ_{PSII} (Å^2) the functional absorption cross section (that fraction of the optical absorption cross section which is used to drive photochemistry) and n_{PSII} the number of the PSII units per mg chlorophyll *a* are required. For more information about the differences in the two methods see Kromkamp and Forster (2003).

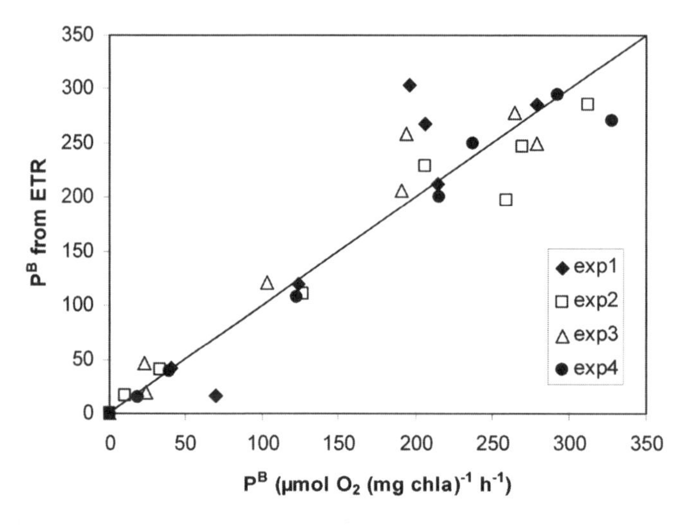

Figure 3. Rate of gross oxygen evolution measured using an oxygen electrode plotted against the gross rate of oxygen evolution calculated from ETR measurements for the diatom *Cylindrotheca closterium*.

Estimating photosynthesis and primary production from PAM measurements

As described above the standard techniques using either oxygen evolution or C-fixation all have their pros and cons. Because of its non intrusive nature and the relative ease with which a portable PAM[1] fluorometer can be used to measure photosynthetic activity this fluorometric technique has found its way in *in situ* microphytobenthos studies. The minimal fluorescence signal has be used to derive chlorophyll content (Honeywill *et al.* 2002; Serôdio *et al.* 2001), to follow vertical migration (Kromkamp *et al.* 1998; Serôdio *et al.* 1997) and variable fluorescence has been used to estimate photosynthesis (Barranguet and Kromkamp 2000a; Kromkamp *et al.* 1998; Perkins *et al.* 2002). The results of these studies are mixed: Barranguet and Kromkamp (2000a) found in general a good relationship between areal production measured with PAM fluorescence and C-fixation (measured with the slurry technique, Figure 4), despite the fact that in a number of cases PAM measurements could overestimate C-fixation at high irradiances. Serodio (2003) also showed that total primary production (ΣPP), estimated from oxygen microelectrode measurements could be adequately predicted by both biomass and photosynthetic activity: $\Sigma PP = F_0/F_{0,sed} \times E \times \Delta F/F_M'$ ($F_{0,sed}$ is the background fluorescence of sediments without chl). However, Perkins *et al.* (2002) found a poor relationship between ETR and C-fixation (measured with the percolation technique), which they attributed to downward migration of the epipelic

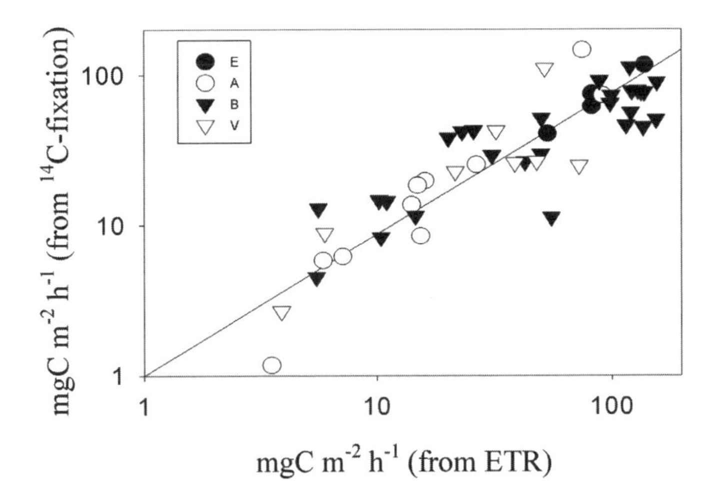

Figure 4. Relationship between C-fixation estimated using PAM surface measurements (ETR) and C-fixation measured using slurries in a photosynthetron. The symbols E,A,B are stations located on an intertidal flat in the Western Scheldt estuary (Molenplaat), whereas sampling location V was located on an intertidal flat in the central Eastern Scheldt. Both estuaries are located in the Netherlands. From Barranguet and Kromkamp (2000).

[1] There are several different manufacturers selling Pulse Amplitude Modulated (PAM) Fluorometers, and most of them use a MT-flash to measure F_M.

diatoms during the measurements of the PE-curve at irradiances exceeding 1200 μmol photons m^{-2} s^{-1}.

It is important here to emphasize that there are a number of difficulties associated with working with a PAM fluorometer on benthic biofilms which might explain the varying results described in the literature.

- The signal measured by a PAM is depending on the sediment optics. The 'measuring depth' of the modulating light will depend on the (absorption by the) algal biomass, its vertical distribution and sediment characteristics such as grain size, water content, and organic matter content. For muddy sediments it is estimated that the measuring depth is approximately 150-200 μm. Therefore, the total F_0 signal measured depends not only on the algal biomass in the top layer of the sediment, but also by its vertical distribution. This is nicely demonstrated by Honeywill *et al.* (2002) who, using cryolanders, showed that the regression coefficient of the relationship between F_0 and [chl*a*] varied with the sampling depth.
- In high light, quenching of fluorescence can decrease the fluorescence yields, and this can falsely be mistaken for downward migration.
- For accurate measurements of ETR, the absorption in each layer of the sediment should be known. This requires accurate knowledge of the sediment light field, which is a function of the biomass density and depth distribution, algal absorption characteristics, sediment characteristics, and the incident irradiance and sun angle.
- It is assumed that the photosynthetic parameters describing the P/E-curve do not change with depth.
- If C-fixation of oxygen evolution is to be calculated from PAM measurements, the relationship ***between*** P^B and ETR should not be influenced by temperature.
- It is furthermore assumed that no significant depth integration of the fluorescence signal occurs, i.e. that $\Delta F/F_M$' measured at the surface reflects the effective quantum efficiency from the cells at the surface and the signal is not influenced by variable fluorescence signals from below.

The latter two assumptions will be now investigated.

Is the relationship between ETR and P^B temperature dependent?

This is an important question, as during sunny days the temperature on mudflats can easily rise by 10 °C during low tide. Harrison (1985) observed that temperatures can rise as quickly as 4 °C h^{-1}. At low light the rate of photosynthesis is limited by the rate of absorption of light, thus the light-limited region of the P/E-curve is hardly influenced by temperature (Davison 1991; Post *et al.* 1985), and light limited growth rates are therefore also hardly influenced by temperatures (Post *et al.* 1985). On the other hand, light saturated rates of photosynthesis are determined by enzymatic reactions and are therefore highly sensitive to changes in temperature. Generally $P^B{}_{max}$-values increase from low towards maximal values at an optimal temperature, after which it declines rapidly (Blanchard *et al.* 1996; Davison 1991). This temperature optimum can be quite wide, and be higher than the growth temperature (Morris and Kromkamp 2003). Q_{10} values for $P^B{}_{max}$ can vary widely (Davison 1991) and be dependent on

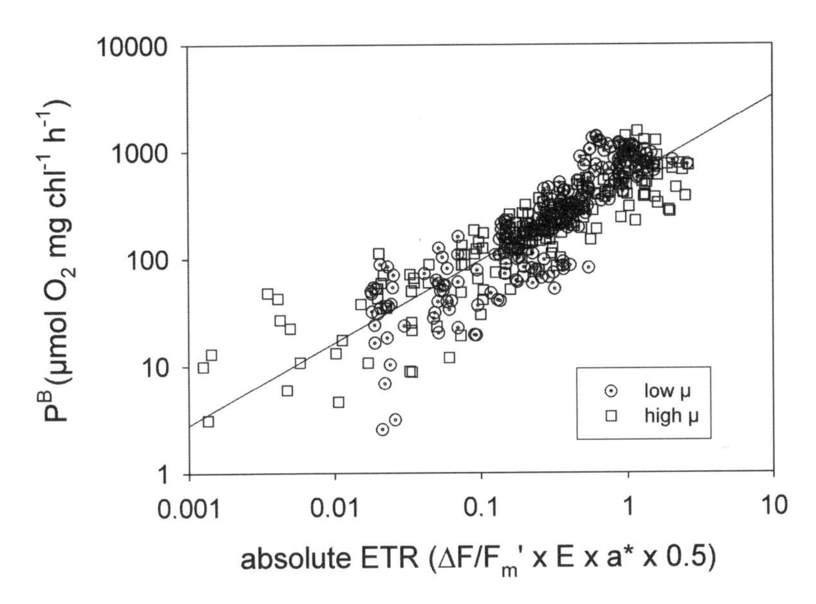

Figure 5. Relationship between ETR and oxygen evolution of light limited steady state grown *Cylindrotheca closterium* cells at a low and high irradiance. Photosynthesis was measured for each growth rate (μ) at temperature varying between 5 and 35 °C. From Morris and Kromkamp (2003)

growth conditions (Kromkamp *et al.* 1988). Morris and Kromkamp (2003) studied the short term acclimation in photosynthesis to different temperatures by simultaneously measuring the rate of oxygen evolution and ETR of the benthic diatom *Cylindrotheca closterium,* which was grown at two different light limited growth rates. They observed that the photosynthetic efficiency α measured with both methods was hardly influenced by photosynthesis. The optimal quantum efficiency of PSII, F_V/F_M, showed a very small, but significant decrease at lower temperatures. $P^B{}_{max}$ and ETR_{max} was very sensitive to changes in temperature, and the optimal temperature was 10 °C higher than the growth temperature of 20 °C. More interesting, the relationship between ETR and P^B was very robust (Figure 5) and only affected at the most extreme temperatures used and was not significantly affected by the growth rate. This indicates that temperature changes during a low tide will not influence microphytobenthos C-fixation estimates which are based on ETR measurements.

Does fluorescence from subsurface influence the fluorescence measured at the surface of an intact microphytobenthos community?

When performing fluorescence measurements on intact microphytobenthos communities it is implicitly assumed that upwelling fluorescence from algae below the surface does not distort the signal measured at the surface, and that the fluorescence values measured at the surface truly reflect the fluorescence yield under the ambient environmental conditions. However, light is attenuated steeply within microphytobenthos communities, and as a result levels of F and F_M' may change rapidly with depth, and if part of this fluorescence will reach the sediment surface it will be part of the overall fluorescence measured. The effect of this was recently investigated by Forster and Kromkamp (2004) in a mathematical model with varying pigment depth profiles and a realistic simulation of the sediment light field. Four different situations were investigated: a) assuming an uniform distribution with depth; b), assuming an exponential decrease with depth, using 2 published profiles by Kelly *et al.* (2001) and De Brouwer and Stal (2001b) (resp. KY & BS in Figure 1); c), assuming a surface distribution only, with either a single surface monolayer (SSL) or a multilayer surface layer assemblage (MSL) and d), a multilayer subsurface assemblage (SUB). The first situation can bc found in environments where the sediments are regularly stirred by the tides, normally the sandier sediments or in situation with high tide in the night. The profiles KY and BS are typical for epipelic diatoms in the more silty sediments during daytime low tide, where, according to low temperature scanning electron microscopy diatoms can also be present only at the surface in a single or multilayer organization (Paterson 1986). A subsurface situation is seldom encountered, but might arise when the epipelic pennate diatoms migrate away when the incoming tide approaches. Figure 6 shows what happens if fluorescence originating at increasing depth layers contributes via upwelling to the signal measured at the surface. In this case the vertical profile 'BS' was used as an example (see Figure 2). It can be seen that if the contributing depth increases, the measured values for F, F_M' and $\Delta F/F_M$'

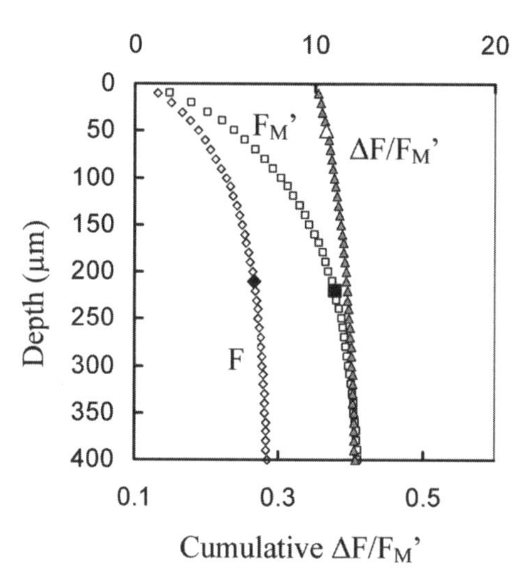

Figure 6. Cumulative $\Delta F/F_M$' measured at the surface with increasing depths contributing to the signals measured at the sediment surface. The filled symbols represent the depth where 90% of the final value at the surface is measured. Surface irradiance 500 µmol photons m^{-2} s^{-1}. Data from Forster and Kromkamp (2004).

increase asymptotically: approximately 90% of the maximum values is derived from the upper 210 μm. Whereas the $\Delta F/F_M$' of the cells in the upper 10 μm is 0.355, the cumulated $\Delta F/F_M$' measured at the surface reaches a value of 0.406: thus, in the light, especially in strong light, upwelling fluorescence from deeper layers cause an overestimation of the true $\Delta F/F_M$' of the cells at the surface. Similar results were obtained by Serôdio (2004). At an irradiance of 2000 μmol photons m^{-2} s^{-1}, this overestimation of the true $\Delta F/F_M$' can be as large as 46%. In a dark adapted core contribution of fluorescence below also contributes to the F_0 and F_M values measured at the surface, but the value of F_v/F_M does not change. Next these authors investigated the effect chlorophyll distribution pattern on changes in $\Delta F/F_M$' and rETR at different irradiances (Figure 7). As can be seen, the effect of the profile is quite limited: a uniform profile or a profile with monotonous increase in biomass towards the surface as shown by the BS, KY profiles hardly effect the shape of the $\Delta F/F_M$'-irradiance curves, and all overestimate the true ETR by approx. 40-50% at the highest irradiance. The biggest overestimate in true $\Delta F/F_M$' occurs in a subsurface layer, which is caused by the fact that the true irradiance experienced by the algae below the surface is of course lower than the surface irradiance, which is used to calculate the electron transport rates. Interesting are the overestimates in the single and multilayered surface assemblages, both of which had a thickness of 50 μm. It might be expected that a single cell surface layer might behave very similar to a leaf, but even here due to light attenuation within the upper 50 um caused a strong decrease in average irradiance, leading to overestimation of the true $\Delta F/F_M$'. The overestimation is stronger in the multilayered surface layer (MSL), where light absorption is stronger than in the SSL because the chlorophyll content is higher in the same depth stratum. It is thus clear that if the optical conditions in even a small layer are unknown, considerable errors (overestimates) might be made in the estimates of the inherent $\Delta F/F_M$'. This also suggest that effective PSII quantum efficiencies made on leaves and macroalgae might suffer from the same artefact, although the error is probably minimized by the focussing effect of the upper palisade parenchyma cells on the underlying mesophyll cells which contain most of the chlorophyll. A striking feature of the contribution of 'deep layer fluorescence' is that the ETR-irradiance (ETR-E) curves do show a varying degree of saturation at high irradiances: especially the conditions with a high surface biomass content like the MSL or the subsurface layer do not show a maximum rate of ETR at high irradiance.

The simulations shown in Figure 7 were carried out with a low degree of non-photochemical quenching. As diatoms can show a high degree of quenching this was also investigated, as quenching might decrease the signal of the upper layers, therefore making the contribution of deep layer fluorescence more pronounced, but the effects were only noticable at irradiance exceeding 1000 μmol photons m^{-2} s^{-1} where the ETR-E curves showed an upward turn. Nevertheless, the effect was small and added only a 7% further increase in ETR at 2000 μmol photons m^{-2} s^{-1} (Forster and Kromkamp 2004). The question now is: can we use a variable fluorescence to measure total primary production if the true ETR is overestimated by the contribution of deep layer fluorescence to the fluorescence measured at the surface. As most of the chlorophyll profiles in estuarine sediments show either a homogeneous or exponential

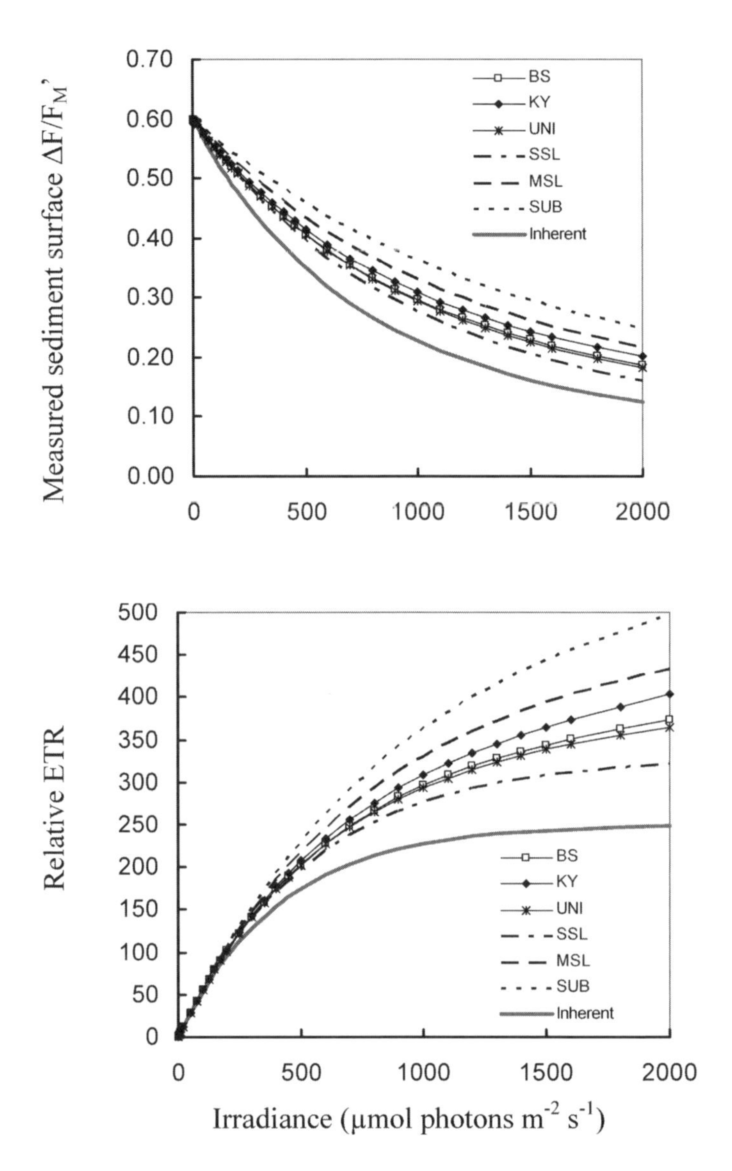

Figure 7. Change in $\Delta F/F_M$' (top panel) and rETR (bottom panel) with irradiance for different chlorophyll distributions with depth (see text). Data from Forster and kromkamp (2004)

decrease with depth (De Brouwer and Stal 2001a; Kelly *et al.* 2001; Perkins *et al.* 2003, own observations), the model analysis by Forster and Kromkamp (2004) suggest that the effect can be quantified and is not very dependent on the shape of the exponential decrease in chlorophyll with depth. However, this is not the case with a surface assemblage, so knowledge of the local system is required to reduce potential errors caused by deep layer fluorescence. The results of the SSL and MSL also demonstrates that it is impossible to measure the light climate within an algal surface assemblage accurately because the sensor can actually be bigger than the layer involved. But suppose we know that we have homogenous distribution with depth or an exponential decrease with depth in biomass, is it then possible to derive an accurate prediction of total primary production? This was investigated Forster and Kromkamp (2004) by calculation of depth integrated primary production (ΣPP) using the inherent photosynthetic properties and the sediment optical model against the depth integrated value of ($\Delta F/F_M$' x E_z x $F_{o,z}$). As can be seen in Figure 8, the nearly linear relationships between ΣPP and the composite term Σ($\Delta F/F_M$' x E_z x $F_{o,z}$) is very similar between the KY, BS and UNI profiles, corroborating earlier results by Serôdio (2003). However, the relationship between ΣPP and Σ($\Delta F/F_M$' x E_z x $F_{o,z}$) is not similar for all biomass depth distribution profiles, and as the results for the SUB profile indicate, are dependent on the attenuation coefficient of the sediment as well, again indicating that knowledge about the local biomass distributions is required for accurate depth integrated primary production estimates. However, with sufficient knowledge of the biomass distribution and sediment optics, surface PAM measurements can be used to estimate areal primary production.

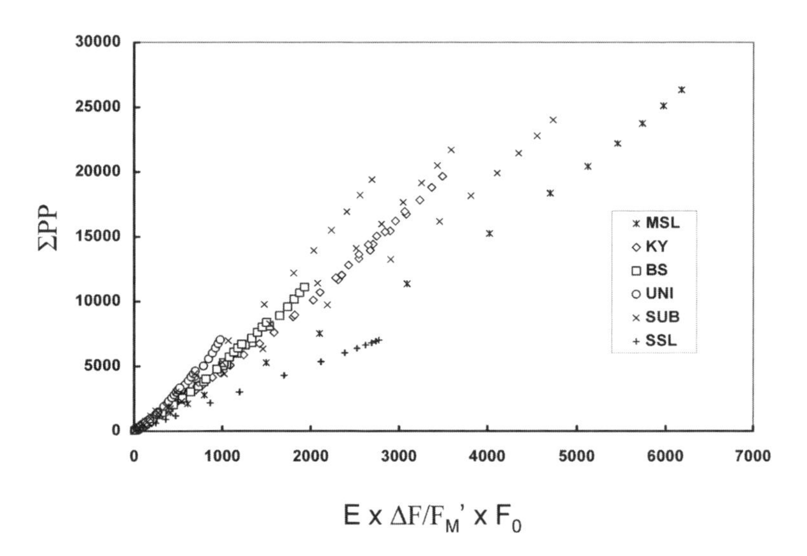

Figure 8. Comparison of calculated gross microphytobenthic primary production (ΣPP) and estimated primary production using the composite parameter E x $\Delta F/F_M$' x F_0 (Serodio 2003). From Foster and Kromkamp (2004)

Conclusions

Variable fluorescence measurements can be a useful way to estimate primary production of microphytobenthos, provided information about the depth distribution profiles and light attenuation coefficients is available. As in most cases MPB biomass in estuarine sediments will show an exponential decrease with depth during low tide, an a priori estimate of the overestimate of the inherent rates of electron transport can be assumed. It should be noted that for integration of radiocarbon measurements on suspensions the same knowledge is required. F_0 measurements can be taken as a proxy for algal biomass, but the relationship between F_0 and [chl] will depend very much on the biomass distribution in the sediment. Non-photochemical quenching will add to the uncertainty, as it will decrease the fluorescence yields, but not the [chl] concentration.

Acknowledgements

This study was supported by a grant from the EU-program HIMOM (EVK3-CT-2001-00052): A system of hierarchical monitoring methods for assessing changes in the biological and physical state of intertidal areas. This is NIOO-KNAW publication number 3543.

References

ALLAN, J.F., and T. PFANNSCHMIDT. 2000. Balancing the two photosystems: photosynthetic electron transfer governs transcriptionof reaction centre genes in chloroplasts. Phil. Trans. Royal Soc. London B ***355***: 1351-1360.

ASADA, A. 2000. The water-water cycle as alternative photon and electron sinks. Phil. Trans. Royal Soc. London B. ***355***: 1419-1431.

BAKER, N.R., K. OXBOROUGH, T. LAWSON, and J.I.L. MORISON. 2001. High resolution imaging of photosynthetic activities of tissues, cells and chloroplasts in leaves. J. Exp. Bot. ***52***: 615-621.

BARRANGUET, C., and J. KROMKAMP. 2000a. Estimating primary production rates from photosynthetic electron transport in estuarine microphytobenthos. Mar. Ecol- Progr. Ser. ***204***: 39-52.

—. 2000b. Estimating primary production rates from photosynthetic electron transport in estuarine microphytobenthos. Mar. Ecol. Prog. Ser. ***204***: 39-52.

BARRANGUET, C., J. KROMKAMP, and J. PEENE. 1998. Factors controlling primary production and photosynthetic characteristics of intertidal microphytobenthos. Mar. Ecol. Prog. Ser. ***173***: 117-126.

BASSI, R., and S. CAFFARRI. 2000. Lhc proteins and the regulation of photosynthetic light harvesting function by xanthophylls. Photosyn. Res. ***64***: 243-256.

BLANCHARD, G.F., and V. CARIOU-LE GALL. 1994. Photosynthetic characteristics of microphytobenthos in Marennes-Oléron Bay, France: preliminary results. J. Exp. Mar. Biol. Ecol. ***182***: 1-14.

BLANCHARD, G.F., J.M. GUARINI, P. RICHARD, P. GROS, and F. MORNET. 1996. Quantifying the short-term temperature effect on light-saturated photosynthesis of intertidal microphytobenthos. Mar. Ecol. Prog. Ser. ***134***: 309-313.

BLANCHARD, G.F., and P.A. MONTAGNA. 1992. Photosynthetic response of natural assemblages of marine benthic microalgae to sort- and long-term variations of incident irradiance in Baffin Bay, Texas. J. Phycol. ***28***: 7-14.

BLANCHARD, G.F. and others 2000. The effect of geomorphological structures on potential biostabilisation by microphytobenthos on intertidal mudflats. Cont. Shelf Res. ***20***: 1243-1256.

BOYNTON, W.R., W.M. KEMP, and C.W. KEEFE. 1982. A comparative analysis of nutrients and other factors influencing estuarine phytoplankton production, p. 69-90, Estuarine comparisons. Academic press Inc.

CAMPBELL, D., V. HURRY, A.D. CLARKE, P. GUSTAFSSON, and G. ÖQUIST. 1998. Chlorophyll fluorescence analysis of cyanobacterial photosynthesis and acclimation. Microbiol. Mol. Biol. Rev. ***62***: 667-683.

COLIJN, F., and V.N. DE JONGE. 1984. Primary production of microphytobenthos in the Ems-Dollard Estuary. Mar. Ecol. Prog. Ser. ***14***: 185-196.
DAU, H. 1994. Short-term adaptation of plants to changing light intensities and its relation to Photosystem II photochemistry and fluorescence. J. Photochem. Photobiol. B. ***26***: 3-27.
DAVISON, I.R. 1991. Environmental effects on algal photosynthesis: temperature. J. Phycol. ***27***: 2-8.
DE BROUWER, J., and L.J. STAL. 2001a. Short-term dynamics in microphytobenthos distribution and associated extracellular carbohydrates in surface sediments of an intertidal mudflat. Mar. Ecol. Prog. Ser. ***218***: 33-44.
DE BROUWER, J.F.C., S. BJELJIC, E.M.G.T. DE DECKERE, and L. J. STAL. 2000. Interplay between biology and sedimentology in a mudflat (Biezelingse Ham, Westerschelde, The Netherlands). Cont. Shelf Res. ***20***: 1159-1177.
DE BROUWER, J.F.C., and L.J. STAL. 2001b. Short-term dynamics in microphytobenthos distribution and associated extracellular carbohydrates in surface sediments of an intertidal mudflat. Mar. Ecol. Prog. Ser. ***218***: 33-44.
EPPING, E.H.G., and B.B. JORGENSEN. 1996. Light-enhanced oxygen respiration in benthic phototrophic communities. Mar. Ecol. Prog. Ser. ***139***: 193-203.
FALKOWSKI, P.G., K. WYMAN, A.C. LEY, and D.C. MAUZERALL. 1986. Relationship of steady-state photosynthesis to fluorescence in eucaryotic algae. Biochim. Biophys. Acta ***849***: 183-192.
FORSTER, R.M., and J.C. KROMKAMP. 2004. Effects of fluorescence from subsruface layers on the measurement of the quantum efficiench of photosystem II at the surface of an intact microphytobenthos assemblage. Mar. Ecol. Prog. Ser. ***284***: 9-22
FORSTER, R.M., and J.C. KROMKAMP. Estimating benthic primary production: scaling up from point measurements to the whole estuary. This book
GENTY, B., J.M. BRIANTAIS, and N.R. BAKER. 1989. The relationship between quantum yield of photosynthetic electron transport and quenching of chlorophyll fluorescence. Biochim. Biophys. Acta ***990***: 87-92.
GLUD, R.N. and others 1999a. Adaptation, test and in situ measurements with O_2 microopt(r)odes on benthic landers. Deep-Sea Res. I. ***46***: 171-183.
GLUD, R.N., M. KUHL, O. KOHLS, and N.B. RAMSING. 1999b. Heterogeneity of oxygen production and consumption in a photosynthetic microbial mat as studied by planar optodes. J. Phycol. ***35***: 270-279.
GLUD, R.N., N.B. RAMSING, and N.P. REVSBECH. 1992. Photosynthesis and photosynthesis-coupled respiration in natural biofilms quantified with oxygen microsensors. J. Phycol. ***28***: 51-60.
GLUD, R.N., C.M. SANTEGOEDS, D. DE BEER, O. KOHLS, and N.B. RAMSING. 1998. Oxygen dynamics at the base of a biofilm studied with planar optodes. Aquat. Microb. Ecol. ***14***: 223-233.
GORBUNOV, M.Y., Z.S. KOLBER, and P.G. FALKOWSKI. 1999. Measuring photosynthetic parameters in individual algal cells by Fast Repetition Rate fluorometry. Photosyn. Res. ***62***: 141-153.
GREEN, R.M., Z.S. KOLBER, D.G. SWIFT, N.W. TINDALE, and P.G. FALKOWSKI. 1994. Physiological limitation of phytoplankton photosynthesis in the eastern equatorial Pacific determined from variability in the quantum yield of fluorescence. Limnol. Oceanogr. ***39***: 1061-1074.
HARRISON, S. 1985. Heat exchanges in muddy intertidal sediments: Chicester harbour, west Sussex, England. Estuar. Coast. Shelf Sci. ***20***: 447-490.
HEIP, C.H.R., N.K. GOOSEN, P.M.J. HERMAN, J. KROMKAMP, J.J. MIDDELBURG, and K. SOETAERT. 1995. Production and consumption of biological particles in temperate tidal estuaries. Oceanog.. Mar. Biol. Ann. Rev. ***33***: 1-149.
HOLST, G., R.N. GLUD, M. KUHL, and I. KLIMANT. 1997. A microoptode array for fine-scale measurement of oxygen distribution. Sensors and Actuators B-Chemical ***38***: 122-129.
HONEYWILL, C., D.N. PATERSON, and S.E. HAGERTHEY. 2002. Determination of microphytobenthic biomass using pulse- amplitude modulated minimum fluorescence. Eur. J. Phycol. ***37***: 485-492.
HORTON, P., A.V. RUBAN, and M. WENTWORTH. 2000. Allosteric regulation of the light-harvesting system of photosystem II. Phil. Trans. Royal Soc. London B. ***355***: 361-370.
JÖNSSON, B. 1991. A C-14 Incubation Technique for Measuring Microphytobenthic Primary Productivity in Intact Sediment Cores. Limnol. Oceanogr. ***36***: 1485-1492.
KELLY, J.A., C. HONEYWILL, and D.M. PATERSON. 2001. Microscale analysis of chlorophyll-*a* in cohesive, intertidal sediments: the implications of microphytobenthos distribution. J. Mar. Biol. Ass. U.K. ***81***: 151-162.
KOLBER, Z., and P.G. FALKOWSKI. 1993. Use of active fluorescence to estimate phytoplankton photosynthesis in situ. Limnol. Oceanogr. ***38***: 1646-1665.
KOLBER, Z., J. ZEHR, and P.G. FALKOWSKI. 1988. Effects of growth irradiance and nitrogen limitation on photosynthetic energy conversion in photosystem II. Plant Physiol. ***88***: 923-929.

KOLBER, Z.S., O. PRÁSIL, and P.G. FALKOWSKI. 1998. Measurements of variable chlorophyll fluorescence using fast repetition rate techniques: defining methodology and experimental protocols. Biochim. Biophys. Acta ***1367***: 88-106.

KROMKAMP, J., C. BARRANGUET, and J. PEENE. 1998. Determination of microphytobenthos PSII quantum efficiency and photosynthetic activity by means of variable chlorophyll fluorescence. Mar. Ecol. Prog. Ser. ***162***: 45-55.

KROMKAMP, J., J. BOTTERWEG, and L.R. MUR. 1988. Buoyancy regulation in *Microcystis aeruginosa* grown at different temperatures. Fems Microbiol. Ecol. ***53***: 231-237.

KROMKAMP, J., and J. PEENE. 1999. Estimation of phytoplankton photosynthesis and nutrient limitation in the Eastern Scheldt estuary using variable fluorescence. Aquat. Ecol. ***33***: 101-104.

KROMKAMP, J., J. PEENE, P. VAN RIJSWIJK, A. SANDEE, and N. GOOSEN. 1995. Nutrients, llight and primary production by phytoplankton and microphytobenthos in the eutrophic turbid ,Westerschelde estuary (The Netherlands). Hydrobiologia ***311***: 9-19.

KROMKAMP, J.C., and R.M. FORSTER. 2003. The use of variable fluorescence measurements in aquatic ecosystems: differences between multiple and single turnover measuring protocols and suggested terminology. Eur. J. Phycol. ***38***: 103-112.

KÜHL, M., C. LASSEN, and B.B. JORGENSEN. 1994. Optical properties of microbial mats: light measurments with fiber-optic microprobes., p. 149-166. *In* L. J. Stal and P. Caumette [eds.], Microbial mats. NATO ASI Series. Springer Verlag.

LAZAR, D. 1999. Chlorophyll a fluorescence induction. Biochim. . Biophys. Acta B. ***1412***: 1-28.

LAZAR, D., P. TOMEK, P. ILIK, and J. NAUS. 2001. Determination of the antenna heterogeneity of Photosystem II by direct simultaneous fitting of several fluorescence rise curves measured with DCMU at different light intensities. Photosyn. Res. ***68***: 247-257.

LINDEBOOM, H.J., and B.H.H. DE BREE. 1982. Daily production and consumption in an eelgrass (*Zostera marina*) community in saline Lake Grevelingen: discrepancies between the O_2 and ^{14}C method. Neth. J..Sea Res. ***16***: 362-379.

LIPPEMEIER, S., R. HINTZE, K.H. VANSELOW, P. HARTIG, and F. COLIJN. 2001. In-line recording of PAM fluorescence of phytoplankton cultures as a new tool for studying effects of fluctuating nutrient supply on photosynthesis. Eur. J. Phycol. ***36***: 89-100.

MACINTYRE, H. L., and J. J. CULLEN. 1998. Fine-scale vertical resolution of chlorophyll and photosynthetic parameters in shallow-water benthos. Mar. Ecol. Prog. Ser. ***122***: 227-237.

MACINTYRE, H.L., R.J. GEIDER, and D.C. MILLER. 1996. Microphytobenthos: The ecological role of the "secret garden" of unvegetated, shallow-water marine habitats. I. Distribution, abundance and primary production. Estuaries ***19***: 186-201.

MAUZERALL, D. 1986. The optical cross section and absolute size of a photosynthetic unit. Photosynthesis.Research. ***10***: 163-170.

MELIS, A. 1999. Photosystem-II damage and repair cycle in chloroplasts: what modulates the rate of photodamage in vivo? Trends in Plant Science ***4***: 130-135.

MORRIS, E.P., and J.C. KROMKAMP. 2003. Influence of temperature on the relationship between oxygen- and fluorescence-based estimates of photosynthetic parameters in a marine benthic diatom (*Cylindrotheca closterium*). Eur. J. Phycol. ***38***: 133-142.

NEUBAUER, C., and U. SCHREIBER. 1987. The polyphasic rise of chlorophyll fluorescence upon onset of strong continuous illumination: I. Saturation characteristics and partial control by the photosystem II acceptor side. Verlag de Zeitschrift für Naturforschung ***42c***: 1246-1254.

NEUBAUER, C., and H.Y. YAMAMOTO. 1994. Membrane barriers and Mehler-peroxidase reaction limit the ascorbate available for violaxanthin de-epoxidase activity in intact chloroplasts. Photosyn.Res. ***39***: 137-147.

OHAD, I., N. KEREN, H. ZER, H. GONG, T.S. MOR, A. GAL, S. STAL, and Y. DOMOVICH 1994. Light-induced degradation of the photosystem II reaction centre D1 protein in vivo: an integrative aporach, p. 161-178. *In* N. R. Baker [ed.], Photoinhibition of photosynthesis: from molecular mechansims to the field. BIOS Scientifc Publishers.

OXBOROUGH, K., A.R.M. HANLON, G.J.C. UNDERWOOD, and N.R. BAKER. 2000. In vivo estimation of the photosystem II photochemical efficiency of individual microphytobenthic cells using high-resolution imaging of chlorophyll a fluorescence. Limnol. Oceanogr. ***45***: 1420-1425.

PARK, I.L., J.M. ANDERSON, and W.S. CHOW. 1996. Photoinactivation of functional photosystem II and D1-protein synthesis in vivo are independent of the modulation of the photosynthetic aparatus by growth irradiance. Planta ***198***: 300-309.

PARKHILL, J.P., G. MAILLET, and J.J. CULLEN. 2001. Fluorescence-based maximal quantum yield for PSII as a diagnostic of nutrient stress. J. Phycol. ***37***: 517-529.

PATERSON, D.M. 1986. The migratory behaviour of diatom assemblages in a laboratory tidal micoecosystem examined by low temperature scanning electron microscopy. Diatom Res. ***1***: 227-239.

—. 1989. Short-term changes in the erodibility of intertidal cohesive sediments related to the migratory behaviour of epipelic diatoms. Limnol. Oceanogr. ***34***: 223-234.
PATERSON, D.M., and K.S. BLACK. 1999. Water flow, sediment dynamics and benthic biology. Ad. Ecol. Res. ***29***: 155-193.
PATERSON, D.M., K.H. WILTSHIRE, M.J. BLACKBURN, and I.R. DAVIDSON. 1998. Microbiological mediation of spectral reflectance from intertidal chesive sediments. Limnol. Oceanogr. ***43***: 1207-1221.
PERKINS, R.G., C. HONEYWILL, M. CONSALVEY, H.A. AUSTIN, T.J. TOLHURST, and D.M. PATERSON. 2003. Changes in microphytobenthic chlorophyll a and EPS resulting from sediment compaction due to de-watering: opposing patterns in concentration and content. Cont. Shelf Res. ***23***: 575-586.
PERKINS, R.G., K. OXBOROUGH, A.R.M. HANLON, G.J.C. UNDERWOOD, and N.R. BAKER. 2002. Can chlorophyll fluorescence be used to estimate the rate of photosynthetic electron transport within microphytobenthic biofilms? Mar. Ecol. Prog. Ser. ***228***: 47-56.
PINCKNEY, J., and R.G. ZINGMARK. 1991. Effects of tidal stage and sun angles on intertidal benthic microalgal producivity. Marine Ecology Progress Series.
—. 1993. Biomass and production of benthic microalgal communities in estuarine habitats. Estuaries. ***16***: 887-897.
POMEROY, L.R. 1959. Algal productivity in salt marshes of Georgia. Limnol. Oceanogr. ***4***: 386-397.
POST, A., R. DE WIT, and L. MUR. 1985. Interaction between temperature and light intensity on growth and photosynthesis of the cyanobacterium *Oscillatoria agardhii*. J. Plankton Res. *7*: 487-495.
RASMUSSEN, M.B., K. HENRIKSEN, and A. JENSEN. 1983. Possible causes of temporal fluctuation in primary production of the microphytobenthos in the Danish Wadden Sea. Mar. Biol. ***73***: 109-114.
REVSBECH, N. P., and B. B. JØRGENSEN. 1983. Photosynthesis of benthic microflora measured with high spatial resolution by the oxygen microprofile method: capabilities and limitations of the method. Limnol. Oceanogr. ***28***: 749-756.
SAMSON, G., O. PRÁSIL, and B. YAAKOUBD. 1999. Photochemical and thermal phases of chlorophyll *a* fluorescence. Photosynthetica *37*: 163-182.
SCHREIBER, U., H. HORMANN, C. NEUBAUER, and C. KLUGHAMMER. 1995. Assessment of photosystem II photochemical quantum yield by chlorophyll fluorescence quenching analysis. Austr. J. Plant Physiol. ***22***: 209-220.
SCHREIBER, U., and C. NEUBAUER. 1990. O2-dependent electron flow, membrane energization and the mechanism of non-photochemical quenching of chlorophyll fluorescence. Photosyn. Res. ***25***: 279-293.
SCHREIBER, U., U. SCHLIWA, and W. BILGER. 1986. Continuous recording of photochemical and non-photochemical chlorophyll fluorescence quenching with a new type of modulation fluorometer. Photosyn. Res. ***10***: 51-62.
SEATON, G.G.R., and D.A. WALKER. 1990. Chlorophyll fluorescence as a measure of photosynhetic carbon assimilation. Proc. Royal Soc. London B. ***242***: 29-35.
SERÔDIO, J. 2003. A chlorophyll fluorescence index to estimate short-term rates of photosynthesis by intertidal microphytobenthos. J. Phycol. ***39***: 33-46.
SERÔDIO, J. 2004. Analysis of variable chlorophyll fluorescence in microphytobenthos assemblages: implications of the use of depth-integrated measurements. Aq. Microbial Ecol. ***36***: 137-152
SERÔDIO, J., J.M. DA SILVA, and F. CATARINO. 1997. Nondestructive tracing of migratory rhythms of intertidal benthic microaglae using *in vivo* chlorophyll fluorescence. J. Phycol. ***33***: 542-553.
—. 2001. Use of *in vivo* chlorophyll a fluorescence to quantify short-term variations in the productive biomass of intertidal microphytobenthos. Mar. Ecol. Prog. Ser. ***218***: 45-61.
SEURONT, L. and C. LETERME. Microscale patchiness in microphytobenthos distributions: evidence for a critical state. This book.
SMITH, D.J., and G.J.C. UNDERWOOD. 1998. Exopolymer production by intertidal epipelic diatoms. Limnology and Oceanography ***43***: 1578-1591.
SNEL, J.F.H., and H.H.A. DASSEN. 2000. Measurement of light and pH dependence of single-cell photosynthesis by fluorescence microscopy. Journal of Fluorescence ***10***: 269-273.
SUNDBACK, K., L. CARLSON, C. NILSSON, B. JONSSON, A. WULFF, and S. ODMARK. 1996. Response of benthic microbial mats to drifting green algal mats. Aquat. Microb. Ecol. ***10***: 195-208.
UNDERWOOD, G.J.C., and J. KROMKAMP. 1999. Primary production by phytoplankton and microphytobenthos in estuaries. Ad. Ecol. Res. ***29***: 93-153.
WILLIAMS, P.J.L., and J.E. ROBERTSON. 1991. Overall planktonic oxygen and carbon dioxide metabolisms: the proble of reconciling observations and calculations of photosynthetic quotients. J. Plankton Res. ***13***: 153-169.
WILLIAMS, P.J.L.B. 1993. Chemical and tracer methods of measuring plankton production, p. 21-36. *In* W.K.W. Li and S.Y. Maestrini [eds.], Measurment of primary production from the molecular to the global scale. Ices Marine Science Symposia. ICES.

WILTSHIRE, K.H., F. SCHROEDER, H.D. KNAUTH, and H. KAUSCH. 1996. Oxygen consumption and production rates and associated fluxes in sediment-water systems: A combination of microelctrode, incubation and modelling techniques. Archiv Hydrobiol. ***137***: 457-486.

WOLFSTEIN, K., F. COLIJN, and R. DOERFFER. 2000. Seasonal dynamics of Microphytobenthos biomass and photsynthetic characteristics in the Northern German Wadden Sea, obtained by the photosynthetic light dispensation system. Estuar. Coastal Shelf Sci. ***51***: 561-662.

YALLOP, M.L., B. DE WINDER, D.M. PATERSON, and L. J. STAL. 1994. Comparative structure, Primary production and bniogenic stabilization of cohesive and non-cohesive marine sediments inhabited by microphytobenthos. Estuar. Coastal Shelf Sci. ***39***: 565-582.

YOUNG, A.J., and H.A. FRANK. 1996. Energy transfer reaction involving carotenoids: quenching of chlorophyll fluorescence. J. Photochem. Photobiol. B ***36***: 3-15.

YOUNG, E.B., and J. BEARDALL. 2003. Rapid ammonium- and nitrate-induced perturbations to chl a fluorescence in nitrogen-stressed *Dunaliella tertiolecta* (Chlorophyta). J. Phycol. ***39***: 332-342.

Ronnie N. Glud

Microscale techniques to measure photosynthesis: A mini-review

Abstract

The benthic primary production of subtidal coastal sediments is grossly under sampled – especially in relation to its potential importance. This is partly due to lack of appropriate measuring techniques. Microsensor approaches represent a strong tool for describing and quantifying microscale benthic photosynthesis and related processes, however, spatial and temporal extrapolation from single point measurements in heterogeneous subtidal sediments is complicated. Complementary use of chambers and ^{14}C incubations can be helpful but these approaches have their own caveats and limitations. New approaches such as in situ PAM measurements and in situ PAM imagine may turn out to be very helpful in extrapolating microsensor data to larger scales – and robust intercalibrations between the techniques have been presented. Planar O_2 optodes and O_2 eddy-correlation measurements represent other very promising in situ approaches that overcome the limitations of microscale techniques for accessing benthic primary production. The present manuscript gives a brief review on available microscale techniques for quantifying subtidal benthic primary production

Introduction

The potential importance of subtidal benthic microphytes for local, regional and global carbon cycling is becoming increasingly recognized. In 2003, there existed a little more than 100 studies quantifying the benthic primary production in marine areas but most of these most have been performed in the intertidal zone (Cahoon 1999). Subtidal studies amount to less than 40 and only around 10 studies have been performed at water depth >5 m. Only 4 studies have been conducted in the Arctic area, hosting the relatively largest area of shallow-water, shelf sediment. Our current database on benthic microphytic activity in subtidal areas is very limited.

Apart from the relative few studies on benthic microphytic activity, the use of various techniques; ^{14}C-incubations, chamber incubations and microsensor approaches hamper the assessment of benthic microphytes activity. The ^{14}C incubation in principle measures the gross photosynthesis, however, inaccurate determination of the specific activity or slurry procedures at the best allow only a crude estimate of the in situ activity. Chamber or core incubations resolve a net-activity but typically average the activity of microphythic patches with that of bare sediment and the faunal respiration. Further, the approach is essentially a 'black box' and gives a very poor insight in the interstitial activity distribution and the photosynthesis related respiration. In contrast microsensor measurements can both resolve the net and the gross activity at a very high spatial and temporal resolution and thereby provide detailed insight in the microphytic activity. However, due to significant heterogeneity in natural systems extrapolation to larger areas might be very difficult, and detailed mapping of larger heterogeneous areas is sometimes an impossible task. Below the traditional microsensor technique is described in more detail and potential area extrapolation of subtidal microsensor data by the aid of other newly developed techniques is discussed.

Measuring benthic activity by microelectrodes.

Since the introduction of the Clark type oxygen microelectrode to aquatic biology our knowledge on the microscale oxygen dynamics at benthic interphases as increased tremendously (e.g. Revsbech *et al.* 1980; Revsbech and Jørgensen 1986; Jørgensen and Des Marais 1990). Due to their minute size (tip diameter < 2 µm) fast response time ($<$ 1s), and low stirring sensitivity ($<$1 %), the oxygen concentration at a given point and at a give time can be accurately measured (Gundersen *et al.* 1998, Glud *et al.* 2000). Vertical concentration profiles can easily be obtained and provide detailed insight in the microbial activity (Figure 1)

For dark incubated sediments the community dark respiration ($J_{dark,up}$) can be calculated from the linear concentration gradient obtained within the DBL as $J_{dark,up} = - D_0\ dC/dz$ (Sten-Knudsen 2002), where D_0 is the temperature and salinity corrected molecular diffusion coefficient for O_2 and C is the O_2 concentration at the depth z. Knowing the O_2 penetration depth (O_{pen}) it is a simple matter to calculate the average volume specific respiration rate. The net respiration at the respective depths can also be modelled from the curvature of the concentration profile within the sediment (Nielsen *et al.* 1990; Berg *et al.* 1998). This, however require a tortuosity corrected diffusion coefficient for the sediment, which typically is estimated from semi empirical relations between porosity and tortuosity (e.g. Ullman and Aller 1982; Iversen and Jørgensen 1993). However, in cases of high meiofaunal activity or high intercellular water content (e.g. microbial mats) such simplified approaches may underestimate the in situ transport coefficients and other alternative approaches must be applied (Aller and Aller 1992; Glud and Fenchel 1999; Berg *et al.* 2001; Wieland *et al* 2001).

In order to estimate the net-photosynthesis from microprofiles obtained in light the oxygen flux out of the photic zone can be calculated as: $J_{light,up} + J_{light,down} = D_0\ dC_{DBL}/dz_{DBL} + D_e\ dC_{SED}/dz_{SED}$, where D_e is the tortuosity corrected diffusion coefficient and dC/dZ refer to the gradients within the DBL and at the lower boundary of the photic zone, respectively (Fig 1). The lower boundary of the photic zone is typically indicated by the turning tangent of the concentration profile, but can also be defined from

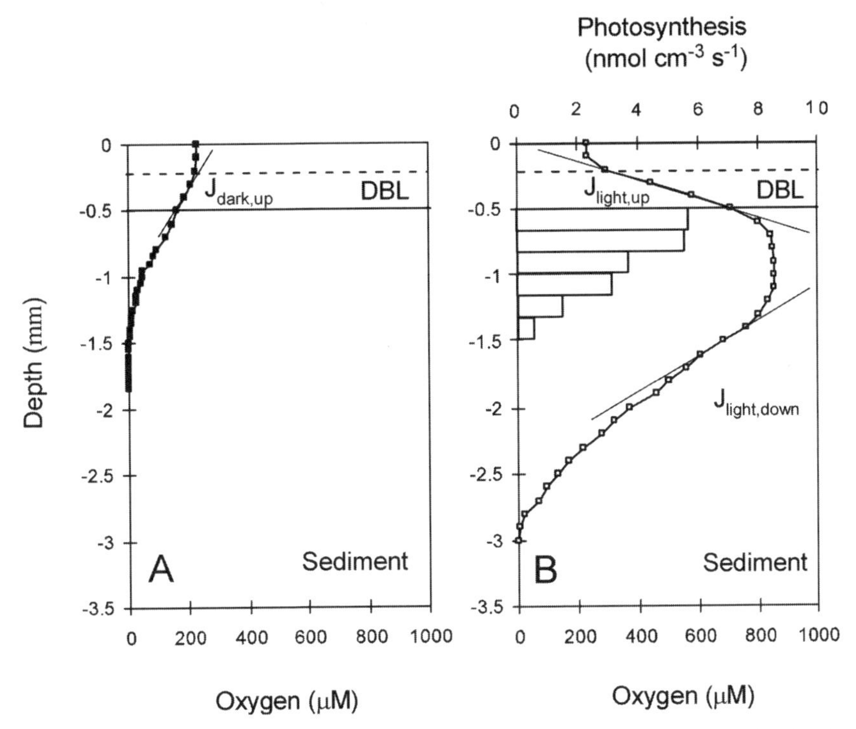

Figure 1. Two microprofiles measured in sediment cores recovered from a water depth of 1.2 m in Helsingør Habor, Denmark. The profiles were measured at the exact same spot in darkness and by a down-welling irradiance of 600 μE m^{-2} s^{-1}, respectively. The white blocks indicate the gross photosynthetic rates measured by the light-dark-shift approach at each depth. The temperature was 15 °C and the salinity 12 ‰

microprofiles of light or the gross photosynthetic rates (see below). In the present example the oxic zone in darkness is identical to the photic zone at the applied irradiance allowing a direct comparison between the activity in the two situations. In case the two zones cover different depth-spans this has to be corrected for (e.g. Kühl *et al.* 1996). Using the profiles above (and a measured porosity of 0.8) the community dark respiration amounted to 39.5 mmol m^{-2} d^{-1} and the net photosynthesis at the applied irradiance was 252.8 mmol m^{-2} d^{-1}. The downwards flux accounted for 25% of the net photosynthesis. This is on the high end of typical encountered values which usually are in the range of 10-20% (e.g. Kühl *et al* 1996; Epping and Jørgensen 1999; Wenzhöfer *et al* 2000; Fenchel and Glud 2000; Glud *et al.* 2002; Christensen *et al.* 2003). The relative fraction of the produced O_2 that diffuse downward primarily depends on the O_2 penetration depths and the thickness of the diffusive boundary layer (DBL) (Glud *et al.* 1992; Kühl *et al.* 1996).

The gross photosynthetic rates of microphytic communities can be measured at each depth by the so-called light-dark-shift technique (Revsbech and Jørgensen 1983). The approach is based on the fact that at steady state the O_2 production at a give point equals the respiration and the O_2 export away from that point. Assuming that the respiration and the O_2 export remain constant just after light eclipse it follows that the O_2 decrease in darkness equals the O_2 production rate in light (Revsbech and Jørgensen 1983). The two assumptions behind the technique have been carefully evaluated and it has been documented that the community respiration remains constant within the first 5 s after light eclipse (Glud *et al.* 1992). However, due to changes in the concentration gradient at the onset of darkness the export rate from a given point change within that period it is therefore essential to obtain the absolutely initial rate of decrease at each depth to resolve the correct activity distribution (Glud *et al.* 1992; Lassen *et al.* 1998). If a dark incubation longer than 1-2 s is applied the activity distribution will be smeared and high rates are underestimated while low rates are overestimated, the depth integrated activity does, however, remain constant (Glud *et al.* 1992). To obtain the initial O_2 decrease is usually now problem applying high quality microelectrodes and fast measuring equipment, but simple diffusion based modeling allow for correction if longer dark incubations have been used (Glud *et al.* 1992; Lassen *et al.* 1998). In Figure 1 the gross photosynthetic rate decreased almost exponentially with depth reflecting the biomass distribution and the light extinction. The depth integrated photosynthesis amounted to 338.8 mmol O_2 m^{-2} d^{-1} and by subtracting the net photosynthesis it followed that the respiration within the photic zone was 86 mmol m^{-2} d^{-1} in light or more than twice the community dark respiration. It has generally been found that the light induced enhancement of the aerobic 'respiration' of the photic zone is in the order of 35-80% (e.g. Canfield and Des Marais 1993; Glud *et al* 1999; Kühl *et al* 1996). Reasons for this is ascribed to a combination of photorespiration, stimulation of heterotrophic respiration following leakage of photsynthate to the carbon limited bacterial community and reoxidation of anaerobic metabolites (e.g. H_2S, FeS) accumulating during the dark period. To what extent the various O_2 consuming processes matter depends on the microbial community and a number of microenvironmental controls (e.g. irradiance, temperature, organic loading, salinity, pH). Simplified approaches comparing light and dark incubated communities to access the 'dark respiration' in light thus grossly underestimate the actual activity.

A detailed study performed at the site where the cores for Figure 1 were recovered showed that during a 12 h light period about 60% of the O_2 produced within the photic zone was consumed by respiration (or photorespiration). 12% was used for 'repaying' an oxygen-debt in the form of anaerobic metabolites (presumably FeS) accumulating during the previous dark period and the remaining 25% was released from the sediment (Fenchel and Glud 2000).

As demonstrated above O_2 microsensor studies can provide a very detailed insight in the benthic photosynthesis, respiration and the associated processes. However, to extrapolate findings to a lager sediment area is difficult when dealing with natural, heterogeneous sediments. To make an O_2 concentration microprofile and measuring the associated gross photosynthetic activity takes in the order of 20-40 min, consequently there is a limit to how many measurements that can be performed at given location, especially in situ where environmental controls might change during the measurment. Automated profiling systems may to some extent overcome this problem and have been successfully deployed; the approach does, however, require relatively sophisticated measuring equipment (e.g. Gundersen and Jørgensen 1990; Wenzhöfer *et al.* 2000; Glud *et al.* 2002). Intercalibration between microsensor data and other techniques being faster and/or covering larger areas represent an approach for microsensor based investigations to be extracted to the ecosystem level.

Newly developed pulse amplitude modulated (PAM) fluorometers – especially the so-called diving PAM – represent a very fast and simple way of measuring proxies for biomass and photosynthesis for benthic communities. Several studies have documented strong correlations between chl *a* content and the minimal fluorescence yield (F_0) of sediments and between the relative electron transport rate (ETR) between phosystem I and II and photosynthetic activity measured as ^{14}C incubation approaches, chamber incubations and microprofile data (e.g. Barranguet and Kromkamp 2000; Glud *et al* 2002;). However, at the physiological level the relations are poorly defined and far from universal and studies have documented that the relations are confounded by migration patterns, light acclimation or environmental stress (e.g. Perkins *et al.* 2001). Nevertheless extrapolation of microprofile data to ecosystem level has been successfully performed by the aid of PAM-measurements (Glud *et al.* 2002). Other contributions in the present volume provide a more detailed insight and discussion on the potential of PAM fluorometer measurements both on microscale and for imaging (Kromkamp and Forster, this volume; Oxborough *et al.*2000).

Measuring benthic activity by optodes

Another recent development applicable for accessing the activity of benthic communities are planar optodes (Glud *et al.* 1996). The technique is based on dynamic fluorescence quenching of oxygen acting on an immobilized fluorophore (Kautsky 1939). In the absence of O_2 the fluorophore absorbs light and emits the absorbed energy as fluorescence of a defined intensity and life time. In the presence of O_2, quenching decreases both the intensity and the lifetime of the fluorescent signal (Klimant *et al.* 1995; Hartmann *et al.* 1997).

The principle has been used to construct so-called microoptodes for traditional microprofile measurements and large scale sensors for hydrographic measurements (Klimant *et al.* 1995; Klimant *et al.* 1997; Tengberg *et al.* 2003). For planar optodes the oxygen quenchable fluorophore is immobilzed in plasticized PVC (or another immobilization agent like sol-gels) at the surface of a transparent support foil. For most planar optodes applied so far a ruthenium based complex has been applied, but other alternatives exists (Klimant *et al.* 1997; Precht *et al.* 2004, Oguri *et al.* 2006). In order to avoid scattering effects and saturation the planar optode is typically covered by an optical insulation provided by a 20 μm layer of black silicone. As the sensing layer has a similar thickness and the support foil is about 175 μm thick, the total thickness of planar optodes are around 200 μm. For laboratory work the sensors can be fixed at the inside of a flume channel and excitation light can be provided from the outside (Figure 2). Irrespective of using fluorescent intensity or fluorescent lifetime as the O_2 sensitive parameter the signal is collected by a 12 bit digital camera after passing an emission filter (Glud *et al.* 1996; Holst *et al.* 1998).

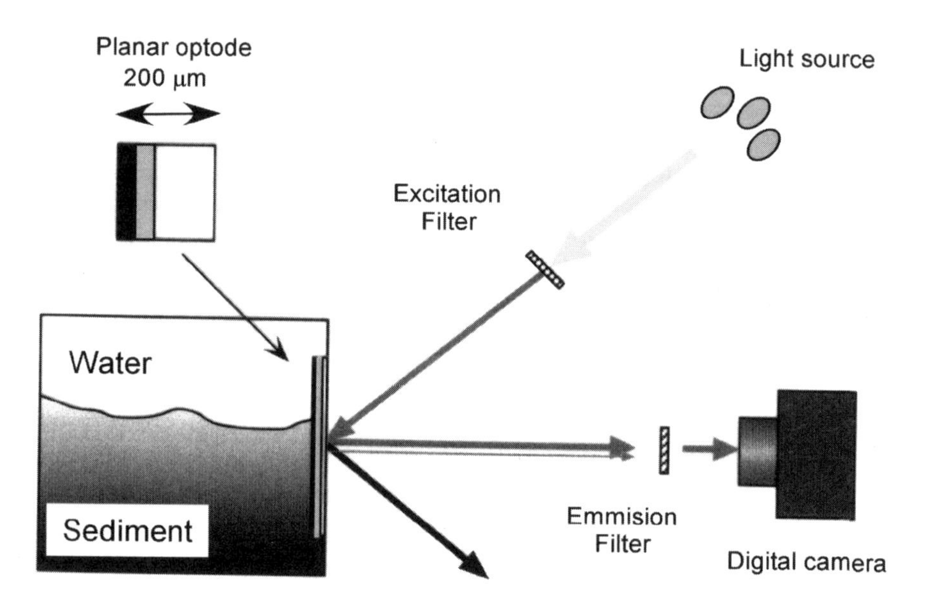

Figure 2. The basic measuring set-up for laboratory based planar optode measurements. As light source newly developed blue or bluegreen Light emitting diodes (LED) are now applied. More details are given in Glud *et al.* 1996, Holst *et al.* 1998, Glud *et al.* 2000)

In contrast to the Clark type microelectrodes the relationship between the O_2 concentration and the sensing signal is nonlinear, but can be expressed by a modified Stern-Volmer equation.

$$\frac{I}{I_0} = \frac{\tau}{\tau_0} = \alpha + \frac{1}{(1 = K_{SV}\ C)}\ (1 - \alpha)$$

where I_0 and I are the fluorescent intensities at anoxia and in the presence of O_2 at a concentration C, respectively, and τ and τ_0 are the equivalent fluorescent lifetimes. The α value represent the fraction of nonquenchable signal, while K_{sv} is the quenching coefficient (Klimant *et al.* 1995; Holst *et al.* 1998). In case of a well defined α (typically around 0.15) a two point calibration routine following the outline in Figure 3 can be applied, otherwise a three-point calibration routine is required. Calibrations can either be performed pixel by pixel or by applying universal constants for the image. Lifetime based sensing allow the use of transperent optodes facillitating alignment between O_2 distribution and structures in the sample (Holst and Grünwald 2001, Frederiksen and Glud 2006). A more detailed discussion on various approaches for quantifying fluorescent lifetime is given elsewhere (Holst *et al.* 1998; Glud *et al.* 2000).

The spatial resolution of obtained O_2 images depends on the optics in front of the camera, but a lower limit is given by the thickness of the layers of optical insulation layer and sensing chemistry allowing a horizontal diffusion within the planar sensor.

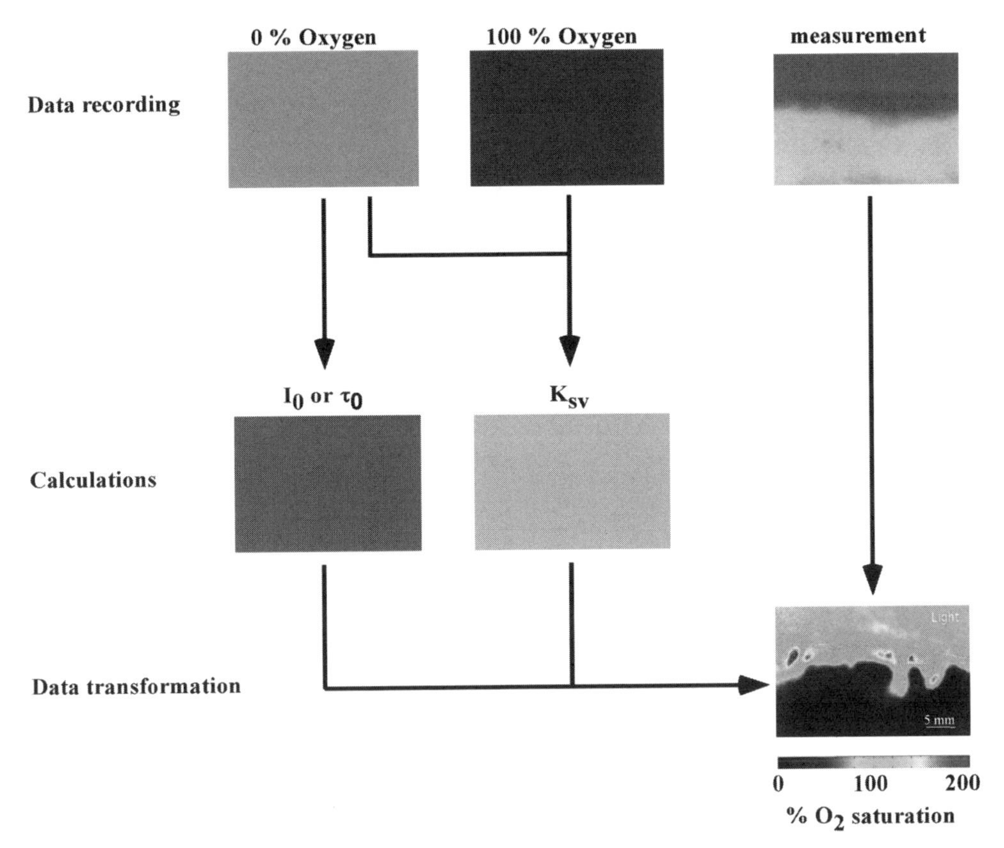

Figure 3. Knowing the α-value, images of the two constants required by the modified Stern-Volmer equation (I_0 (or τ_0) and K_{sv}) can be derived by a two-point calibration and measuring data can be transformed into O_2 images. The presented image was obtained at a diatom covered sediment and the horizontal heterogeneity is apparent, scale bars and sediment surface is included. For further discussion on the image please refer to Fenchel and Glud 2000.

This might be further confounded by light scattering. In our standard configuration images cover an area of 5 x 7 cm and as the camera chip has 640 x 480 pixels the spatial resolution is in the order of 100 x 100 μm. The temporal resolution is limited by the sensor response time but is usually in the order of 5-20 seconds. A single O_2 image by our system in principle contains 640 vertical microprofiles and images obtained in laminated photosynthetic active microbial mats have documented a relatively homogenous horizontal structure (Figure 4). Despite of this, sites only separated by a few mm still express different light responses (Figure 4). In principle all the simple calculations performed on traditionally obtained microprofiles (Figure 1) can be performed on such O_2 images. Thereby very detailed insights in the benthic O_2 dynamics i.e. net photosynthesis, respiration, distribution of hotspots and autotrophic-heterotrophic coupling can be provided (Glud *et al.* 1999). Applying the light-dark shift

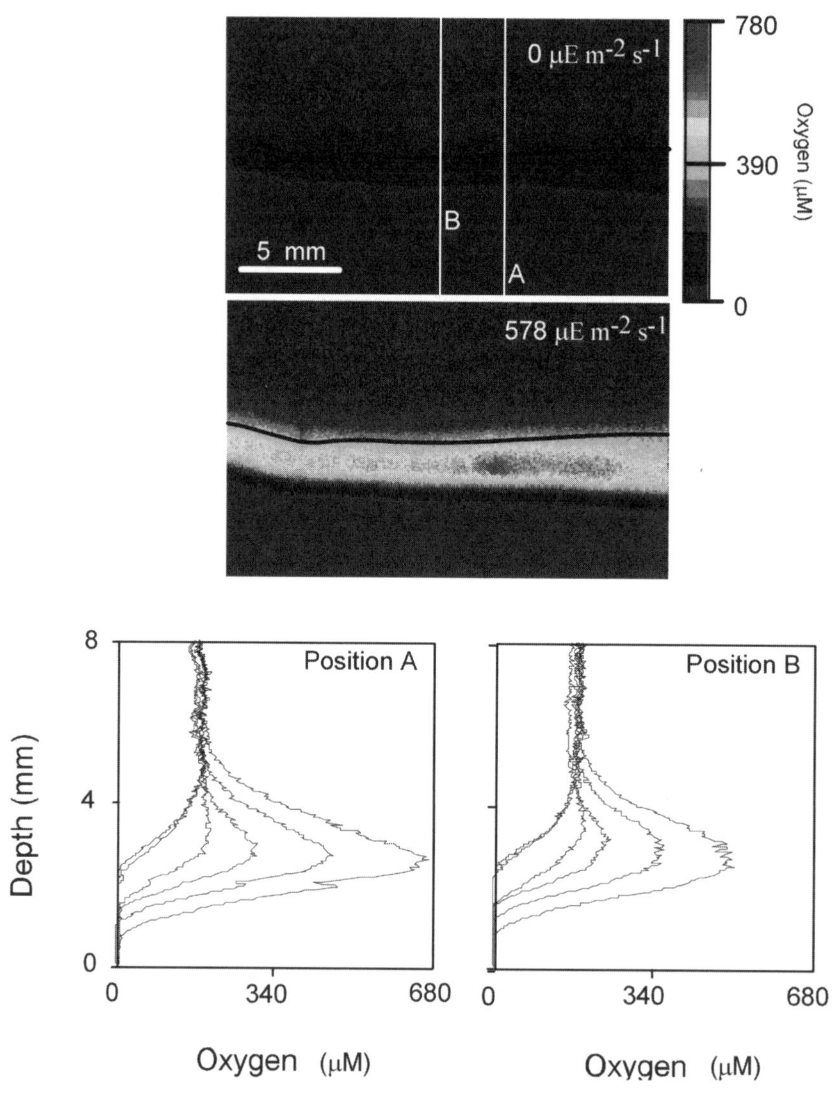

Figure 4. Two O_2 images obtained in a cyanobacteria dominated mat in darkness and at a down welling irradiance of 578 μE m^{-2} s^{-1}. Profiles extracted along the two lines indicated in the upper panel are shown in the lowest panel as they appear when the mat was exposed to 6 different light levels. (Redrawn from Glud *et al.* 1999)

technique also allow images of the gross photosynthetic activity to be resolved, but due to the relatively long response time of planar optodes the activity images are smeared – see above (Glud *et al.* 1999).

In contrast, as also outlined in Figure 3, measurements in subtidal diatom covered sediments have resolved a very high degree of horizontal heterogeneity (Figure 5). Hotspots of activity can be identified and by obtaining frequent images the temporal dynamic of the sediment-water interface can be documented (Wenzhöfer and Glud 2004). However, in extreme cases as shown in Figure 5 it becomes difficult to perform simple vertical calculations as the exchange rather has to be approached as radial diffusion between hotspots and the surroundings. To complicate matters even further this sediment was affected by intensive irrigation by *hediste diversicolor*. Software handling entire images and extracting exchange rates (or net activities) in such extreme cases still have to be developed. Simple 1D diffusion based calculations can of course still be performed on selected spots of the image. Another complication is that planar optodes operates along a wall. Hydrodynamically seen that can be a problem and extracting information on e.g. DBL dynamics should be done with caution.

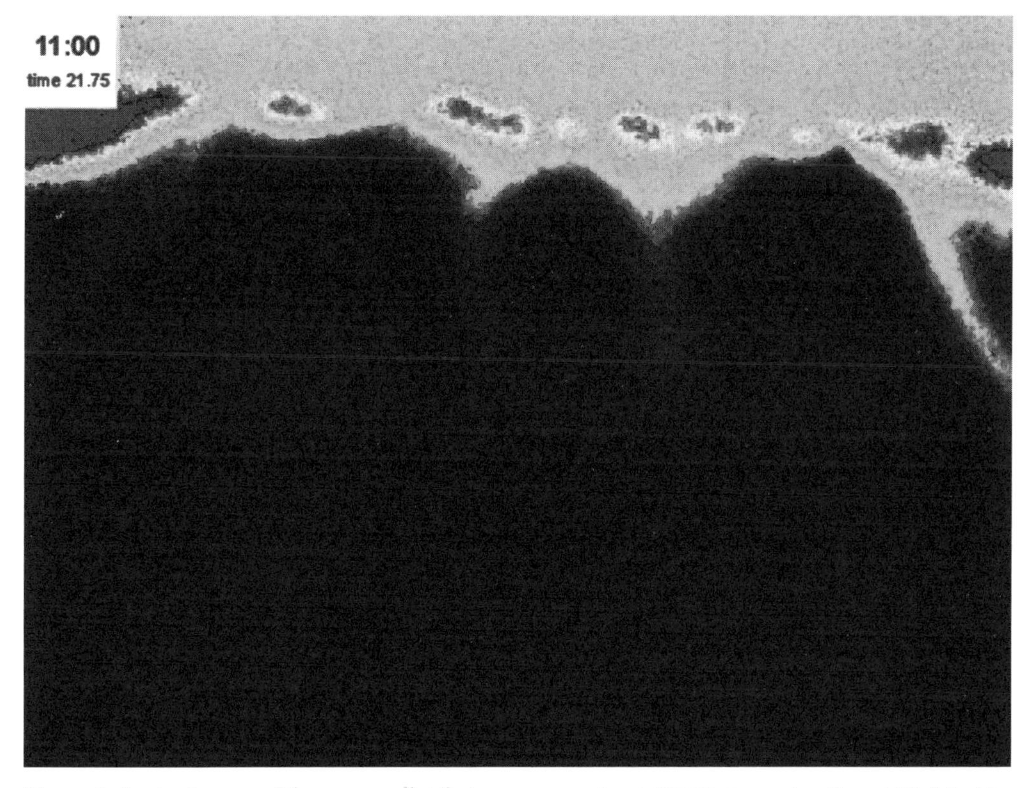

Figure 5. In situ images of the oxygen distribution measured in a diatom covered sediment inhabited by *hediste diversicolor*. The extensive horizontal variation is apparent along actively ventilated infauna burrows. (Data from Wenzhöfer and Glud 2004)

Despite the problems, which in many cases can be dealt with, planar O_2 optodes offers a possibility to obtain a more realistic representation of O_2 distribution in benthic microphytic communities. Images at a given location can be obtained within minutes and large areas can be scanned relatively fast. Thereby planar optodes complement traditional microelectrodes approaches and combined measurements can provide a very detailed insight in the benthic microphytic activity. Planar optodes can be adopted to inverted periscopes for on line in situ measurements (Glud *et al.* 2001). A new in situ approach based on eddy-correlation-techniques may also provide a very power full future tool to quantify benthic O_2 exchange rates (i.e. net photosynthesis) over large sediment areas non-invasively (Berg *et al.* 2003).

References

ALLER, R.C., and J.Y. AILER 1992. Meiofauna and solute transport in marine muds. Limnol. Oceanogr. **37**: 1018-1033.

BARRANGUET, C. and J. KROMKAMP 2000. Estimating primary production rates from photosynthetic electron transport in estuarine microphytobenthos. Mar. Ecol. Progr. Ser. **204**: 39-52.

BERG, P., N. RISGAARD-PETERSEN, and S. RYSGAARD. 1998. Interpretations of measured concentrations profiles in sediment porewater. Limnol. Oceanogr. **81**: 289-303.

BERG, P., H. RØY, F. JANSSEN, V. MAYER, B.B. JØRGENSEN, M. HUETTEL, D. DE BEER. 2003. Oxygen uptake by aquatic sediments measured with a novel non-invasive eddy-correlation technique. Mar. Ecol. Prog. Ser. **261**: 75-83.

BERG, P., S. RYSGAARD, P. FUNCH, and M.K. SEJR. 2001. Effects of bioturbation on solutes and solids in marine sediments. Aquat. Microb Ecol **26**: 81-94.

CAHOON, L.B. 1999. The role of benthic microalgae in neritic ecosystems. In: Oceanography and Marine Biology: An Annual Review (Eds Ansell, A.D., Gibson R.N., Barnes M.). **37**: 47-86

CANFIELD, D.E., and D.J. DES MARAIS. 1993. Biogeochemical cycles of carbon, sulfur and free oxygen in a microbial mat. Geochim. Cosmochim. Acta **57**: 3971-3984.

CHRISTENSEN, P.B., R.N. GLUD, T. DALSGAARD, and P. GILLESPIE. 2003. Implications of long line mussel farming on benthos, oxygen, and nitrogen dynamics in coastal sediments, Aquaculture **218**: 567-588.

EPPING, E.H.G., and B.B. JØRGENSEN. 1996 Light-enhanced oxygen respiration in benthic photrphic communities. Mar. Ecol. Progr. Ser. **139**: 193-203

FENCHEL, T., and R.N. GLUD. 2000. Benthic Primary production and O_2-CO_2 dynamics in a shallow water sediment: Spatial and temporal heterogeneity. Ophelia **53**: 159-171.

FREDERIKSEN, M.S. and R.N. GLUD. 2006 Oxygen dynamics in the rhizosphere of zostera marix: A two-dimensional planar optode study Limnol. Oceanogra **51**: 1072-1083.

GLUD R.N., N.B. RAMSING, and N.P. REVSBECH. 1992. Photosynthesis and photosynthesis coupled respiration in natural biofilms quantified with oxygen microelectrodes. J. Phycol. **28**: 51-60

GLUD, R.N., N.B. RAMSING, J.K. GUNDERSEN, and I. KLIMANT. 1996. Planar optrodes, a new tool for fine scale measurements of two-dimensional O_2 distribution in benthic communities. Mar. Ecol. Prog. Ser.**140**: 217-226.

GLUD, R.N., M. KÜHL, O. KOHLS, and N.B. RAMSING. 1999. Heterogeneity of oxygen production and consumption in a photosynthetic microbial mat as studied by planar optodes. J. Phycol. **35**: 270-279.

GLUD, R.N., and T. FENCHEL. 1999. The importance of ciliates for interstitial solute transport in benthic communities. Mar. Ecol. Prog. Ser. **186**: 87-93.

GLUD, R.N., J.K. GUNDERSEN, and N.B. RAMSING. 2000. Electrochemical and optical oxygen microsensors for in situ measurements. In: *In situ* Analytical Techniques for Water and Sediment. (eds.. J. Buffle & G Horvai). John Wiley & Sons, pp. 19-73.

GLUD, R.N., M. KÜHL, F. WENZHÖFER, and S. RYSGAARD. 2002. Benthic diatoms of a high Arctic fjord (Young Sound, NE Greenland): Importance for ecosystem primary production. Mar. Ecol. Prog. Ser. **238**: 15-29

GLUD, R.N., A. TENGBERG, M. KÜHL, P.O.J. HALL, I. KLIMANT, and G. HOST 2001. An in situ instrument for planar O_2 optode measurements at benthic interfaces. Limnol. Oceanogr. **46**: 2073-2080.

GUNDERSEN, J.K., AND B.B. JØRGENSEN. 1990. Microstructure of diffusive boundary layers and the oxygen uptake of the sea floor. Nature **345**: 604-607.

GUNDERSEN, J.K., N.B. RAMSING, and R.N. GLUD, 1998. Predicting the signal of oxygen microsensors from physical dimensions, temperature, salinity, and oxygen concentrations. Linmol. Oceanogr. **43**: 1932-1937

HARTMANN, P., W. ZIEGLER, G. HOLST, and D.W. LUBBERS. 1997. Oxygen flux fluorescence lifetime imaging. Sensors and Actuators B **38-39**: 110-115.

HOLST, G., O. KOHLS, I. KLIMANT, B. KÖNIG, M. KÜHL, and T. RICHTER. 1998. A modular luminescence lifetime imaging system for mapping oxygen distribution in biological samples. Sensors and Actuators B **51**: 163-170.

HOLST, G., and B. GRUMWALD. 2001 Luminescence lifetime imaging with transparent oxygen optodes. Sensors and Actuators B **74**: 78-90.

IVERSEN, N., and B.B. JØRGENSEN. 1993. Diffusion coefficients of sulfate and methane in marine sediments: Influence of porosity. Geochim. Cosmochim. Acta **57**: 571-578.

JØRGENSEN, B.B., and D.J. DES MARAIS. 1990. The diffusive boundary layer of sediments: Oxygen microgradients over a microbial mat. Limnol. Oceanogr. **35**: 1343-1355.

KAUTSKY, H. 1939. Quenching of luminescence by oxygen. Trans. Faraday Soc. **35**: 216-219.

KLIMANT, I., V. MEYER, and M. KÜHL. 1995. Fiber-optic oxygen microsensors, a new tool in aquatic biology. Limnol. Oceanogr. **40**: 1159-165.

KLIMANT, I., M. KÜHL, R.N. GLUD, and G. HOLST. 1997. Optical measurements of oxygen and other environmental parameters in microscale: Strategies and biological applications. Sensors and Actuators B **38-39**: 29-37.

KÜHL, M., R.N. GLUD, H. PLOUG, and N.B. RAMSING. 1996. Microenvironmental control of photosynthesis and photosynthesis-coupled respiration in an epilithic cyanobacterial biofilm. J. Phycol. **32**: 799-812.

LASSEN C., R.N. GLUD, N.B. RAMSING, and N.P. REVSBECH. 1998. A method to improve the spatial resolution of photosynthetic rates obtained by oxygen microsensors. J. Phycol. **34**: 89-93.

NIELSEN, L.P., P.C. CHRISTENSEN, N.P. REVSBECH, and J. SØRENSEN. 1990. Denitrification and oxygen respiration in biofilms studied with a microsensor for nitrous oxide and oxygen. Microb. Ecol. **19**: 63-72.

OGURI, K., H. KITAZATO, and R.N. GLUD. (2006) Planar oxygen optode used with platinum Octaetylporphyrin UV-LED excitation light source for high-resolution imaging in low oxygen environments. Mar. Chem. **100**: 95-107.

OXBOROUGH, K., A.R.M. HANLON, G.J.C. UNDERWOOD, and N.R. BAKER. 2000. In vivo estimation of the photosystem II photochemical efficiency of individual microphytobenthic cells using high-resolution imaging of chlorophyll a fluorescence. Limnol. Oceanogr. **45**: 1420-1425

PERKINS. R.G., G.J.C. UNDERWOOD, V. BROTAS, G.C. SNOW, B. JESUS, and L. RIBEIRO. 2001. Responses of microphytobenthos to light: Primary production and carbonhydrate allocation over an emersion period. Mar. Ecol. Prog. Ser. **223**: 101-112.

PRECHT, E., U. FRANKE, L POLERECKY, and M. HUETTEL. 2003. Oxygen dynamics in permeable sediments with wave driven pore water exchange. Limnol. Oceanogr. **49**: 693-705.

REVSBECH, N.P., J. SØRENSEN, T.H. BLACKBURN, and J.P. LOMHOLT. 1980. distribution of oxygen in marine sediments measured with microelectrodes. Limnol. Oceanogr. **25**: 403-411.

REVSBECH, N.P., and B.B. JØRGENSEN. 1983. Photosynthesis of benthic microflora measured with high spatial resolution by the oxygen microprofile method: Capabilities and limitation of the method. Limnol. Oceanogr. **28**: 749-756.

REVSBECH, N.P., and B.B. JØRGENSEN. 1986. Microelectrodes and their use in microbial ecology. Advances in Microbial Ecology, 9, Marshall K.C., editor, Plenum, New York, pp. 293-352.

STEN-KNUDSEN. 2002. Biological membranes: Theory of transport, potentials and electrical impulses, pp 671. Cambridge University Press, Cambridge.

TENGBERG A., J. HOVDENES, D. BARRANGER, O. BROCANDEL, R DIAZ, J. SARKKULA, C. HUBER, and A. SHANGELMAYER. 2003. Optodes to measure oxygen in the aquatic environment. Sea Tech. **44**: 2-8

ULLMAN, W.J., and R.C. ALLER. 1982. Diffusion coefficients in nearshore marine sediments. Limnol. Oceanogr. **27**: 552-556

WENZHÖFER, F., O. HOLBY, R.N. GLUD, H.K. NIELSEN, and J.K. GUNDERSEN. 2000. In situ microsensor studies of a hydrothermal vent at Milos (Greece). Mar. Chem. **69**: 43-54.

WENZHÖFER, F. and R.N. GLUD 2004. Small-scale spatial and temporal variability in coastal benthic O_2 dynamics: Effects of fauna activity. Limnol. Oceanogra. **49**: 1471-1481.

WIELAND, A., D. DE BEER, L.R. DAMGAARD, M KUHL, D. VAN DUSSCHOTEN, and H. VAN AS. 2001. Fine scale measurement of diffusivity in a microbial mat with nuclear magnetic resonance imaging. Limnol. Oceanogr. **46**: 248-259.

III. Extracellular polymeric substances: function and role in mudflats

J.F.C. de Brouwer, T.R. Neu and L.J. Stal

On the function of secretion of extracellular polymeric substances by benthic diatoms and their role in intertidal mudflats: A review of recent insights and views

Abstract

Benthic diatoms produce large amounts of extracellular polymeric substances (EPS). These EPS's are believed to play important roles in physiological as wel as ecological processes in intertidal environments. Nevertheless, the function of these EPS's and the role they play in mediating physical and biological processes are currently poorly understood. This paper attempts to review current insight and views on this topic. The focus is on the complexity of EPS, which reveals itself at different levels. It has become clear that diatoms secrete different types of EPS. Moreover, the composition of EPS may vary among different species. These variations lead to the formation of different but distinct and ordered tertiary EPS structures as was shown by confocal laser scanning microscopy of diatom aggregates. These structural aspects of EPS secreted by benthic diatoms should be taken into account in any research that aims at explaining its role in the ecology of intertidal systems.

Introduction

Intertidal mudflats are highly dynamic systems that are characterized by rapid changes in environmental variables. Both diurnal light- and tidal cycles impact mudflats, inducing rapid fluctuations in light, salinity, temperature, water content, oxygen, and other environmental parameters. Under submerged conditions, sediment particles are transported (eroded from the surface or deposited on top of the mudflat) caused by physical processes such as wave action and tidal currents. Hence, the sediment-water interface is a highly dynamic environment.

Notwithstanding these dynamic conditions, intertidal mudflats are highly productive systems in which epipelic diatoms are the most important primary producers (Underwood and Kromkamp 1999). They supply the organic matter for the benthic (Middelburg *et al.* 2000) and planktonic foodwebs (De Jonge and Van Beusekom 1992), and could provide up to 50% of total primary production in estuaries (Underwood and Kromkamp 1999). Epipelic diatoms secrete a considerable part of the photosynthetically fixed carbon as Extracellular Polymeric Substances (EPS) that consist for a large part of carbohydrates (Hoagland *et al.* 1993; Staats *et al.* 1999). Because of the copious secretion of EPS, diatoms become embedded in a matrix of these polymers which are attached to the sediment. These structures are known as diatom mats or biofilms and are common in intertidal mudflats. By living in a biofilm, epipelic diatoms create their own stable microenvironment that allows them to cope with the rapidly changing conditions in intertidal mudflats (Decho 1994). Moreover, diatom biofilms enhance the stability of the sediment surface by increasing the erosion threshold (Paterson 1997, and references therein, Kornman and De Deckere 1998). Hence, diatoms have profound effects on the morphodynamics of mudflats (Dyer 1998) and influence sediment transport, potentially over the scale of whole estuaries (Frostick and Mccave 1979).

In spite of the importance of EPS production by benthic diatoms, the factors that control this process as well as its function are poorly understood. The complex interactions that take place between the diatoms and the biological and physical environment they live in make its study difficult. Therefore an important part of the research has been carried out on laboratory cultures of isolated strains of benthic diatoms. While it is evidently much easier to study the ecophysiology of EPS-production under controlled conditions in pure cultures, it is a strong drawback that the organisms are removed from their natural environment. However, the combination of these two approaches may give clues that lead to the elucidation of the role of EPS both for the diatoms and the biofilm intrinsically as well as for intertidal mudflats as a whole and even at the scale of the estuarine ecosystem. This paper attempts to review current insights and views on the possible functions of EPS secreted by benthic diatoms in intertidal areas.

Methodological considerations of the study of EPS

The most common method to investigate EPS production by benthic diatoms is based on its isolation from a sample matrix. However, a universal method for the extraction of EPS does not exist and different investigators have developed different protocols for the isolation of EPS from cultures (Smith and Underwood 2000; Staats *et al.* 1999; Wustman *et al.* 1998) or from sediment samples (Chiovitti *et al.* 2003b; De Brouwer *et al.* 2000; Underwood and Paterson 2003; Underwood *et al.* 1995). Analysis of these fractions has mainly focused on the carbohydrate part of the EPS. Although carbohydrates often represent a major part of the EPS secreted by diatoms (Bhosle *et al.* 1995; Staats *et al.* 1999), it should be kept in mind that other components such as proteins, lipids and nucleic acids may form part of the EPS matrix (Lawrence *et al.* 2003). EPS-fractions that are recovered by any extraction are principally operationally defined and give *a priori* no information about their characteristics, their mechanisms of secretion or their possible functions. Moreover, it is largely unknown to what extend these operationally defined extracellular fractions contain contaminations originating from intracellular compounds. In diatom cultures, isolation of EPS from the culture supernatant has become an accepted method yielding a fraction that is sometimes referred to as 'colloidal EPS' (Smith and Underwood 2000) or as 'soluble EPS' (De Brouwer *et al.* 2002b). Centrifugation as a procedure to separate the cells and their extracellular polymeric compounds would likely avoid contamination by intracellular material (Staats *et al.* 1999), but it also results in incomplete recovery of EPS (De Brouwer *et al.* 2002b; Wustman *et al.* 1997). In order to also collect this more tightly bound EPS, sequential extraction protocols have been developed. However, more severe extraction procedures may damage the cells leading to contamination of the EPS by intracellular compounds. Hence, any protocol must be accompanied by rigorous controls and can not be carelessly employed for other organisms. Only few studies have discussed these problems in detail (Nielsen and Jahn 1999). Wustman *et al.* (1997) extracted EPS from different stalk forming diatoms and followed the extraction procedure by using differential interference contrast microscopy (DIC) while visualizing the EPS with alcian blue. They showed that extraction of the diatom *Achnanthes longipes* using water at 90 °C removed the bulk of the intracellular material while the extracellular stalks remained intact. These stalks were subsequently removed by hot bicarbonate extraction at 95 °C. These authors also showed that stalks of the freshwater diatom *Cymbella cistula* were removed by a treatment at 23 °C in 0.2 M EDTA and *trans*-1,2-diamino-cyclohexane-N,N,N,',N'-tetraacetic acid but this procedure did not remove the stalks of *A. longipes*. This indicates that the characteristics of the EPS vary among diatom species emphasizing that extraction protocols must be evaluated for each specific case. There are various methods that give clues about possible contamination by intracellular compounds. For instance, Staats *et al.* (1999) monitored the amount of protein in fractions that were extracted by warm (30 °C) water, reasoning that a compromised cell membrane would leak and proteins are readily water soluble. De Brouwer *et al.* (2002b) used the fluorescent marker DIBAC (bis-(1,3-dibarbituric acid) which enters cells with a compromised cell membrane, in order to monitor the cell integrity of *C. closterium* under different extraction protocols. Chiovitti *et al.* (2004) showed that the glucose rich material that was extracted using warm (30 °C) water consisted of a β-1,3-glucan, which lead them to conclude that it represented the internal storage glucan chrysolaminaran rather than EPS. This view is subject of debate (Chiovitti *et al.* 2004; De Brouwer and Stal 2004), but it shows the importance to identify the nature of these operationally defined fractions.

Extraction of EPS from sediment samples is even more complicated compared to the extraction of cultured cells. One of the main reasons for this is that sediments represent ill-defined biological matrices that have a wide range of interactions with the sedimentary environment. Sediment particles are coated with organic matter. The amount of organic matter adsorbed to sediment particles is inversely related to grain size (Bergamaschi *et al.* 1997). This is because finer sediment particles constitute a larger adsorptive area compared to coarser grains. Subsets of extracted organic matter, commonly analyzed as carbohydrate (Underwood *et al.* 1995), show a similar inverse relationship with grain size (De Brouwer *et al.* 2000; De Brouwer *et al.* 2003; Paterson *et al.* 2000). Therefore, the presence of carbohydrate does not only depend on phototrophic biomass (Underwood and Smith 1998) but also on sediment characteristics. This is especially the case when EPS production is low because of a low abundance of microphytobenthos. Additionally, intertidal sediments are inhabited by other organisms than microphytobenthos and these may also produce EPS. Such organisms include macrofauna (Meadows *et al.* 1990), meiofauna (Riemann and Schrage 1978) and bacteria (Dade *et al.* 1990). Although, current knowledge suggests that benthic diatoms are the major producers of EPS in intertidal mudflats (Underwood and Paterson 2003), the involvement of other organisms in the production, modification and degradation of EPS is currently largely unknown.

Another complicating factor in the extraction of EPS from sediments is related to sample handling. Sediment samples are often lyophilized prior to the extraction of EPS. The advantage of lyophilized samples is that water is removed creating a uniform sample matrix. However, several studies have shown that lyophilization causes lyses of the diatoms through which intracellular carbohydrates may be co-extracted (De Brouwer *et al.* 2000; Wigglesworth-Cooksey *et al.* 2001). Hence, the correlation between chlorophyll *a* and carbohydrate which is often reported for intertidal mudflats (Blanchard *et al.* 2000; Underwood and Smith 1998) is possibly in part attributed to intracellular compounds.

Although the extraction procedures for EPS from cultures of diatoms and field samples are easy to apply and may seem straightforward, the interpretation of the results remains difficult. There is an urgent need for a more comprehensive approach that allows tracing back the extracted material to its source. There is probably little to gain by optimizing the extraction procedures, hence our efforts should concentrate on the analytical side of the extracted material. A comprehensive analysis should give information about the origin (i.e. the organisms that produced it or diagenetically altered sedimentary organic matter) and allow the distinction between intra- and extracellular material. This may be done by using preparative techniques such as size fractionation (De Brouwer and Stal 2001; Wustman *et al.* 1998) or electrophoresis (Chiovitti *et al.* 2003a; Puskaric and Mortain-Bertrand 2003). In addition, microscope techniques can be used for exopolymer characterization or to monitor the extraction of EPS from diatoms.

By using Atomic Force Microscopy Higgins *et al.* (2002) identified two different types of EPS that differed in mechanical properties. The same technique was used to follow the extraction of EPS from *Pinnularia viridus* cells (Chiovitti *et al.* 2003b; Higgins *et al.* 2002). These authors showed that the differences in composition of the different carbohydrate fractions were associated with changes in the morphology and properties of the cell surface mucilage. Furthermore, fluorescence microscopy and confocal laser scanning microscopy have been used in combination with the application of fluorescently labeled lectins to localize EPS structures in isolated diatoms (De Brouwer *et al.* 2005; Neu 2000; Wustman *et al.* 1997) or natural phototrophic assemblages (Neu 2000; Norton *et al.* 1998). Although lectins recognize specific carbohydrate sequences in the EPS, it is not straightforward to distinguish between species specific EPS. By using antibodies raised against species specific EPS fractions it has been possible to analyze and localize the distribution of specific intracellular (Chiovitti *et al.* 2004) and extracellular glycoconjugates (Kawaguchi *et al.* 2003; Lind *et al.* 1997; Wustman *et al.* 1998). Furthermore, application of antibodies identified the role of extracellular glycoproteins in gliding movement and substratum adhesion (Lind *et al.* 1997).

Composition of EPS

In addition to quantitative information about exopolymers secreted by the diatoms, knowledge about the chemical composition may provide additional clues to understand the mechanism(s) of EPS-excretion and their function(s) in intertidal sediments. Generally, diatom-derived EPS-fractions are dominated by carbohydrates, the remaining part being mainly proteins and sulfate-groups (Bhosle *et al.* 1995; Hoagland *et al.* 1993; Staats *et al.* 1999; Wustman *et al.* 1997). Therefore, the majority of studies investigating the structure of EPS have focused on the analysis of carbohydrates originating from cultures of benthic diatoms (Allan *et al.* 1972; Chiovitti *et al.* 2003b; De Brouwer and Stal 2002; Staats *et al.* 1999; Underwood *et al.* 2004; Wustman *et al.* 1997) or from intertidal sediment samples (Cowie and Hedges 1984; De Brouwer *et al.* 2003; De Brouwer and Stal 2001; De Winder *et al.* 1999; Taylor *et al.* 1999). Meta-analyses of monosaccharide distributions of EPS-fractions originating from different species of benthic diatoms or from intertidal sediments indicated that considerable differences exist in the composition of EPS (Underwood and Paterson 2003). Cluster analysis identified two major EPS-types that could further be distinguished into 6 groupings within this dataset. Both types of EPS were represented by samples from diatom cultures as well as sediments. Based on this analysis, a conceptual model of EPS secretion was presented. This model proposed secretion of different groups of extracellular glycoconjugates, which could mainly be differentiated in (i) a light-dependent production of labile low molecular weight sugars, (ii) light-dependent production of a polymeric glucose-rich fraction closely attached to cells and, (iii) light-independent production of a more heterogeneous (colloidal) EPS that varied to some degree among different species.

In addition, it has become increasingly clear that an extracted EPS fraction does not necessarily contain one separate functional type of EPS but may consist of a mixture of different polymers. By using pyrolysis-mass spectrometry, Smith & Underwood (2000) showed differences in the composition of colloidal EPS produced during the logarithmic and stationary phase in cultures of 5 epipelic diatoms. The presence of two types of EPS in the colloidal EPS extracts of the diatom *Cylindrotheca closterium* were identified by precipitation of the EPS through a series of increasing alcohol concentrations (Underwood *et al.* 2004). In that study, nutrient replete cells produced a complex EPS while nutrient limited cells produced an additional EPS that had a less complex monosaccharide composition. Furthermore, examination of EPS-secretion by the diatom *Cylindrotheca fusiformis* under N- and P-limitation recovered at least three exopolymers with different size distributions of which two were different in monosaccharide composition (Magaletti *et al.* 2004). Wustman *et al.* (1997) observed that stalks produced by diatoms that caused fouling were composed of different parts that were apparently formed by distinctly different types of EPS. A detailed characterization of the water-insoluble, bicarbonate-extractable EPS-fraction of *A. longipes* showed the presence of polymers in three size ranges that were different in composition. Subsequent use of enzyme-linked immunosorbent assays (ELISA) showed the different locations of these different polymers, and indicated that their concerted action results in the formation of highly structured stalks.

By using a similar approach it was shown that two extracellular glycoproteins of a size >200 kDa secreted by the diatom *Craspedostauros australis* were involved in adhesion of the cell to a substratum as well as in cell motility while two smaller glycoproteins (87 and 112 kDa) formed a non-adhesive cell surface mucilage. The above mentioned examples not only show that benthic diatoms are able to secrete compositionally distinct types of EPS that may serve different functions, they also indicate that self assembly of extracellular components lead to the formation of complex structural biocomposites. It has been emphasized that the formation of these extracellular complexes rather than the secretion of a single substance are important for diatom functionality such as motility and adhesion to a substratum (Wetherbee *et al.* 1998).

Another important aspect of EPS-secretion concerns the uniformity of EPS-composition among different species of benthic diatoms. Variations in monosaccharide composition have been observed among EPS obtained from different benthic diatoms. Unfortunately, experimental conditions under which the diatoms were grown as well as the extraction protocols were not uniform. This might have influenced the outcome of the analyses of EPS and it is therefore not opportune to draw conclusions based solely on the comparison of monosaccharide composition. Only few studies have been published that used linkage analysis to identify substitution patterns in the EPS of diatoms (Chiovitti *et al.* 2003a; Chiovitti *et al.* 2003b; Wustman *et al.* 1997). In general, these investigations showed the presence of complex heterogeneous glycoconjugates. Comparison of glycoconjugates from different diatoms suggests that their structure is highly species specific.

Another approach is to combine destructive structural analysis of carbohydrate fractions from different diatom species and non-destructive visualization of glycoconjugates using lectins. Lectins are proteins that lack enzymatic or immunogenic activity and that possess specific carbohydrate and protein binding sites (Neu and Lawrence 1999). Lectins have mostly been characterized by its specificity for certain monosaccharides, although it has become increasingly evident that complex glycoconjugates are probably the more competitive binding sites. Monosaccharide compositions in four different carbohydrate fractions extracted from three different species of benthic diatoms are shown in Table 1. The soluble EPS fraction from all three species showed a heterogeneous distribution of monosaccharides with galactose, glucose, mannose/xylose and glucuronic acid being the most abundant sugars, each of which contributed >10% to the carbohydrate composition. In contrast, bound EPS and internal sugar fractions were composed almost exclusively of glucose which represented 75-96% of the carbohydrate. Also, the residual sugars were dominated by glucose or by xylose/mannose but this material showed a considerable variation among different species. Considering that the extracted fractions may represent mixtures of glycoconjugates it is difficult to judge if the material present in the extractants provide structurally unique exopolymers. In order to solve this problem, a set of 15 lectins was applied to the cultures of the 3 diatom species and the intensity of labeling and the localization of the labels were qualitatively evaluated (Table 2). The responses of the various lectins differed greatly among the extracellular matrices of the three species of diatoms. In addition, the localization of the

Table 1. Monosaccharide distributions of carbohydrate fractions extracted from early stationary phase cultures the diatoms *Cylindroteca closterium*, *Navicula mutica* and *Nitzschia brevissima*. Extraction procedure and monosaccharide analysis was conducted according to De Brouwer and Stal (2002) extended with direct methanolyses of the residual cell pellet to obtain the residual sugar fraction. Values represent averages and standard deviations of 3 replicate measurements.

Monosaccharide	Soluble EPS	Bound EPS	Internal sugars	Residual sugars
Cylindrotheca closterium				
Fucose	3.4 (0.2)	1.0 (0.2)	1.1 (0.7)	1.0 (0.2)
Rhamnose	14.3 (1.4)	7.0 (0.8)	2.8 (0.4)	5.1 (0.2)
Arabinose	0.5 (0.06)	0.2 (0.03)	n.d.	n.d.
Glucoseamine	1.3 (0.1)	0.1 (0.03)	0.1 (0.03)	0.5 (0.03)
Galactose	19.9 (2.7)	4.3 (0.2)	3.1 (0.6)	16.3 (0.5)
Glucose	13.8 (0.14)	77.7 (2.7)	86.7 (2.1)	64.4 (3.5)
Mannose/xylose	14.5 (0.3)	3.3 (0.4)	3.3 (0.3)	3.4 (2.4)
Galacturonic acid	5.3 (4.1)	1.5 (0.4)	1.2 (0.1)	3.4 (0.3)
Glucuronic acid	27.0 (0.5)	4.9 (1.3)	1.6 (0.2)	5.9 (0.7)
Navicula mutica				
Fucose	0.6 (0.04)	0.03 (0.03)	0.7 (0.2)	1.2 (0.2)
Rhamnose	9.5 (0.3)	2.5 (1.3)	n.d.	4.1 (0.03)
Arabinose	0.3 (0.01)	0.1 (0.01)	n.d.	n.d.
Glucoseamine	1.2 (0.07)	0.2 (0.1)	0.02 (0.03)	0.7 (0.06)
Galactose	11.0 (0.15)	1.4 (0.7)	1.6 (0.6)	17.7 (1.5)
Glucose	49.4 (1.2)	93.3 (3.0)	94.2 (2.2)	28.6 (6.5)
Mannose/xylose	11.7 (0.5)	0.7 (0.6)	3.5 (1.4)	43 (5.3)
Galacturonic acid	1.3 (0.4)	n.d.	n.d.	0.6 (1.0)
Glucuronic acid	15.0 (1.1)	1.8 (0.5)	0.05 (0.08)	4.2 (0.1)
Nitzschia cf. *brevissima*				
Fucose	7.3 (0.8)	3.4 (0.4)	n.d.	0.6 (0.2)
Rhamnose	7.9 (0.3)	3.6 (1.1)	n.d.	0.5 (0.1)
Arabinose	1.5 (0.1)	1.6 (0.2)	n.d.	n.d.
Glucoseamine	0.7 (0.09)	n.d.	n.d.	n.d.
Galactose	17.5 (0.6)	9.8 (1.8)	3.6 (1.4)	4.0 (0.4)
Glucose	21.0 (3.5)	75.2 (2.4)	96.0 (1.6)	91.3 (1.6)
Mannose/xylose	27.9 (1.4)	6.4 (2.0)	0.4 (0.3)	2.2 (1.1)
Galacturonic acid	4.0 (0.4)	n.d.	n.d	n.d.
Glucuronic acid	12.0 (1.8)	n.d.	n.d	1.5 (0.03)

Table 2. Response and localization of 15 different lectins applied to axenic cultures of the benthic diatoms *C. closterium*, *N. mutica* and *N.* cf. *brevissima*. Labelling response was assessed by qualitatively assessing the visualizations; X: no signal, –: low signal intensity, ±: intermediate signal intensity, +: good signal intensity, ++: excellent signal intensity. Localization coding; CS: cell surface, M: matrix staining, S: staining of bright extracellular spots, (p): partial staining.

Lectin	*C. closterium*		*N. mutica*		*N.* cf. *brevissima*	
Aleuria aurantia	+	CS	++	CS+M(p)	X	
Amaryllis	–	CS	±	CS(p)+M	–	CS(p)
Concanavaline A	±	CS	X		++	M
Helix aspersa	+	CS	X		+	S
Iberis amara	+	CS(p)	+	M(p)	++	CS+M+S
Lens culinaris	–	CS(p)	+	CS(p)+M(p)	++	M
Limulis polyphenus	X		X		-	CS(p)
Lycopersicon esculentum	X		-	CS	+	M+CS(p)
Maackia amurensis	X		X		-	S
Phaseolus coccineus	X		X		++	S+M
Sambuca nigra	X		X		X	
Solanum tuberosum	X		X		X	
Urtica dioica	X		+	CS	-	CS
Vicia sativa	–	CS	++	CS	±	CS(p)+S
Wheat germ agglutinin	X		+	CS+M(p)	X	

labels was distinctively different (Table 2, Figure 1). These results strongly suggest that the extracellular material secreted by these benthic diatoms contained highly specific glyconjugates. This agrees with earlier work that showed that different types of EPS may have a specific localization within the extracellular matrix of diatoms (Lind *et al.* 1997; Wustman *et al.* 1998).

Function and mechanism of EPS-secretion

Diatoms secrete substantial amounts of photosynthetically fixed carbon as extracellular carbohydrates. Estimates reported for epipelic diatoms range between 30-73% of the fixed CO_2 being secreted as EPS (Goto *et al.* 1999; Middelburg *et al.* 2000; Smith & Underwood 2000; de Brouwer *et al* 2000; Wolfstein *et al.* 2003). In epipelic diatoms EPS is secreted from a long narrow slit (raphe) in the silica frustule. Edgar & Picket-Heaps (1984) suggested that the role of EPS in gliding movement was to attach the diatom to a substratum. They conceived that the EPS strands were displaced parallel to the raphe as a result of contractions of microfilaments of the cytoskeleton, which would provide the force for gliding. Subsequently, Webster *et al.* (1985) identified this

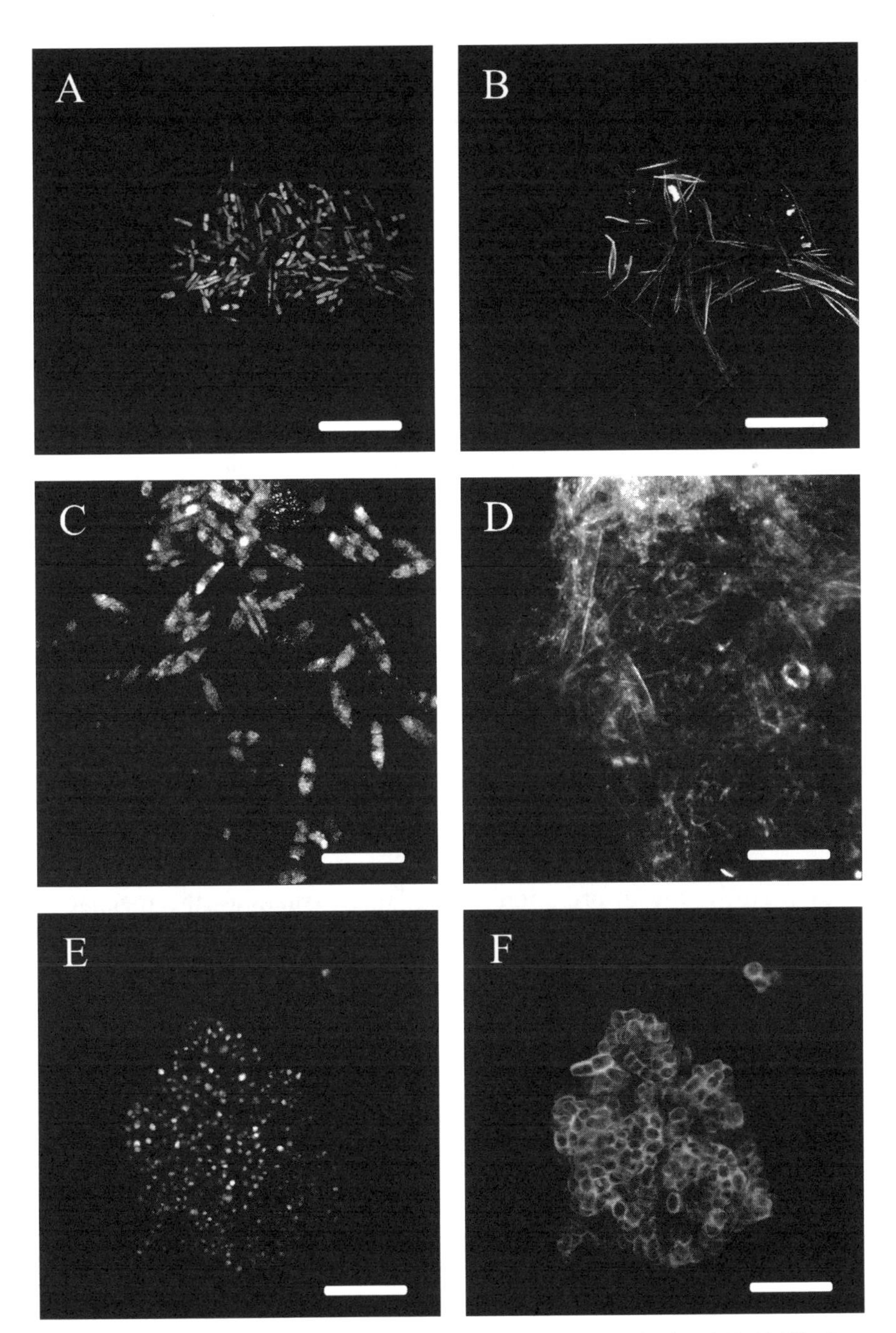

Figure 1. Maximum intensity images of chlorophyll *a* (A, C, E) and lectin signal (B, D, F,) of aggregates from axenic cultures of the diatoms *Cylindrotheca closterium* (A, B) stained with *Iberis amara*, *Nitzschia* cf. *brevissima* (C, D) stained with concanavaline A and *Navicula mutica* stained with the lectin *Iberis Amara*. Scale bars: A-D: 40 μm; E-F: 20 μm.

EPS as actin- and tubuline-based microfilaments that were involved in gliding motility of the diatom *Amphora coffeaeformis.* This role of EPS was experimentally confirmed by Lind *et al.* (1997) who used antibodies to inhibit substratum adhesion and gliding of the diatom *Stauroneis decipiens.* Some researchers have observed that diatoms in intertidal mudflats migrate in response to light and to the tide (Paterson 1986; Pinkney & Zingmark 1991; Serôdio *et al.* 1997; Smith & Underwood 1998; Underwood & Smith 1998a), although others did not observe such migration so that it does not seem to be a general mechanism. Since diatoms also migrate in dark, EPS secreted for motility must occur independent of light. That this is indeed the case has been demonstrated in cultures (Smith & Underwood, 2000; de Brouwer & Stal 2003) as well as under natural conditions in intertidal mudflats (Smith & Underwood 1998).

Hitherto, the amount of EPS required for gliding has not been quantified but Edgar & Pickett-Heaps (1984) conceived that motility probably requires only small quantities of EPS, and hence, would represent a low metabolic expense. Since benthic diatoms secrete large amounts of EPS it is attractive to suppose that the production of this material serves other purposes. Because it was found that EPS production was light dependent and coupled to photosynthesis (de Winder *et al.* 1999; Staats *et al.* 2000a) it was conceived that it was the result of unbalanced growth caused by the depletion of an essential nutrient. This view is supported by the results from culture experiments (Lewin 1955; Myklestad & Haug 1972; Bhosle *et al.* 1995; Staats *et al.* 2000b) that demonstrated that nutrient depletion (particularly nitrogen and phosphorous) enhanced exopolymer production. Mass balance calculations agreed also with the occurrence of unbalanced growth due to nutrient shortage (Ruddy *et al.* 1998a). By coupling nitrogen limitation to carbohydrate production, the model of Ruddy *et al.* (1998b) accurately described short-term microphytobenthos and exopolymer dynamics in intertidal mudflat sediments.

Currently, hardly any information is available explaining the mechanisms by which EPS is secreted and describing the possible pathways of its production. By using radioactive labeled $^{14}CO_2$, it was shown that the transfer of photosynthetically fixed carbon into the extracellular pools occurred within 30 min. This was shown both for a culture of *C. closterium* as well as in a field sample of microphytobenthos (Wolfstein *et al.* 2002). This suggests a close coupling between photosynthesis and EPS secretion as was already emphasized by de Winder *et al.* (1999) and Staats *et al.* (2000a). However, this was not confirmed by Underwood *et al.* (2004), who reported for the same species that the flow of photosynthetically fixed carbon to EPS occurred with a delay of 3-4 h, suggesting an uncoupling of photosynthesis and EPS-production. By using inhibitors that block glucan synthesis, glucan catabolism and protein synthesis, Underwood *et al.* (2004) suggested at least two pathways for EPS production. One pathway was dependent on the synthesis and subsequent catabolism of intracellular glucan, while the other pathway was independent of glucan catabolism. This is strong evidence that diatoms may excrete different types of EPS as the product of different metabolic pathways and that therefore follow different dynamics and differ in composition (see previous section).

Other parameters may also affect the production of EPS. Cooksey (1981) for instance demonstrated attachment of diatoms to a surface was inhibited in the absence of Ca^{2+}. This was accompanied by a large decrease in EPS-secretion (K.E. Cooksey, personal communication). Moreover, it was observed that adjustment of the salinity of the medium had an effect on growth and EPS release in diatom cultures (Tokuda 1969; Allan *et al.* 1972) and also changed the composition of the EPS (Allan *et al.* 1972). Furthermore, effects of light levels and temperature may alter the production of extracellular material. Staats *et al.* (2000a) observed that short-term production of EPS occurred above a photon irradiance of 15 µmol $m^{-2}s^{-1}$ and they suggested that a certain minimum amount of light was required to allow accumulation of exopolysaccharides. However, radioactive ^{14}C-labelling experiments over a broad range of photon irradiances (7-1200 µmol $m^{-2}s^{-1}$) resulted in only labeled EPS, and Wolfstein *et al.* (2002) concluded that this light threshold for EPS-production must have been below 7 µmol $m^{-2}s^{-1}$. Temperature is another factor that affects EPS production. Wolfstein and Stal (2002) found that the rates of carbon fixation and EPS excretion were maximal at 25 °C. This value agrees with the optimum temperature for photosynthesis in microphytobenthos (Blanchard *et al.* 1996). Light and temperature affected the patterns and quantities of the different EPS-fractions (i.e. colloidal and attached EPS) in different ways, which once again suggest that these fractions are produced and secreted by different pathways (De Brouwer and Stal 2002).

The roles of EPS in intertidal coastal areas

Sediment stabilization

EPS produced by benthic assemblages in intertidal mudflats plays an important role in stabilization of surface sediments. Several studies have shown the relation between the presence of diatom biofilms and a decrease in sediment erosion (De Brouwer *et al.* 2000; Paterson 1989; Sutherland *et al.* 1998; Underwood and Paterson 1993). It is, however, not easy to experimentally verify the specific role of EPS in the mediation of biogenic sediment stabilization, and experiments often lead to contradictory results. For example, Underwood and Paterson (1993) found that colloidal carbohydrate was the best biochemical predictor of sediment stability, but other studies indicated that no such relation existed (De Brouwer *et al.* 2000; Defew *et al.* 2002; Paterson *et al.* 2000). (Yallop *et al.* 2000) found that sediment stability correlated with several variables including chlorophyll *a*, extracellular carbohydrate and EPS fractions, water content and bacterial biomass. By using multiple regression analysis these authors calculated that sediment stability was best predicted by using a combination of chlorophyll *a*, colloidal EPS and water content. This indicates that multiple processes and their interactions are involved in the process of biogenic stabilization in intertidal sediments. Biogenic stabilization is the result of the secretion of EPS that leads to the formation of a cohesive organic matrix in which diatoms and sediment particles are embedded (Taylor and Paterson 1998). Besides the physical effect of EPS as a 'glue' by which sediment particles stick together, a decrease in bottom roughness resulting from the formation of the biofilm probably has an additional effect on sediment stabilization (Paterson and Black 1999).

An approach that directly assesses the role of EPS in sediment stabilization is to add it to sediment and subsequently measure changes in sediment behavior. Several studies have demonstrated that bacterial EPS modified sediment characteristics (Dade *et al.* 1990; Tolhurst *et al.* 2002). De Brouwer *et al.* (2002a) and Perkins *et al.* (2004) used EPS that was previously extracted from intertidal sediments and subsequently added to sediments in known amounts. De Brouwer *et al.* (2005) used EPS from pure cultures of various benthic diatoms and added it to muddy sediments. Surprisingly, EPS derived from benthic diatoms or from natural diatom biofilms did not affect sediment characteristics in the way it was shown for bacterial EPS. Diatom-derived EPS partly adsorbed to sediment particles but did not change the sediment properties (De Brouwer *et al.* 2002a). Calcium-ions enhanced the adsorption of EPS, which suggests that cation-bridging is an important mechanism to bind EPS to sediment particles. Furthermore, it was shown that uronic acids play a role in sediment-EPS interactions (Dade *et al.* 1990; De Brouwer *et al.* 2005), which further emphasizes the importance of cation bridging in adsorption of EPS to the sediment. Perkins *et al.* (2004) found that the addition of EPS affected sediment properties when it was subjected to desiccation, its overall effect being an increase in the threshold of sediment erosion, probably because of increased electrostatic binding between sediment particles (Chenu 1993). The addition of EPS resulted in an increased retention of water while erosion thresholds were comparable to controls. The interpretation of these results was difficult because it was observed that the presence of salts alone (naturally present in the EPS extracts) also led to increased sediment stability.

Although it was not possible to show convincingly that EPS alone is able to stabilize mud, this does not necessarily exclude its role in sediment morphogenesis. It should be taken into account that EPS in the biofilms forms structured entities (Decho 1994) and it is likely that this structure is lost upon extraction (De Brouwer *et al.* 2002a; Underwood and Paterson 2003). Visualization of diatom EPS using carbohydrate-specific lectins (Figure 1) showed that exopolymers are present as ordered structures. For the diatom *C. closterium* only cell surface glycoconjugates were visible, while for the other two benthic diatom species the set of lectins also visualized matrix-structures. The morphological characteristics of the matrix glycoconjugates varied between species. *N. mutica* formed a honeycomb type of structure where the algae where associated mainly with EPS that was located closely to the organisms. In contrast, *N.* cf. *brevissima* formed an extensive matrix existing of channels as well as amorphous structures. Because the composition and structure of EPS varies among diatoms, it is likely that biogenic sediment stabilization is also species dependent. Culture experiments have confirmed this (De Brouwer *et al.* 2005; Holland *et al.* 1974). Furthermore, the relations between extracellular carbohydrate fractions and critical shear stress for axenic cultures of the diatoms *Cylindrotheca closterium* and *Nitzschia* cf. *brevissima* showed that, although these species secreted similar amounts of EPS, the effect on the rheological properties (i.e. critical shear stress) of the sediments were notably different. Hence, it was concluded that the addition of isolated EPS from these diatoms did not affect sediment stability. It was conceived that in addition to the quantity and chemical composition of the EPS, the assembly of exopolymers in ordered three dimensional matrix structures is essential to increase the erosion threshold of muddy sediments.

Currently, it is unknown which fraction of EPS is actually responsible for sediment stabilization. It has been suggested that EPS related to motility is important for the stabilization of intertidal sediment (Paterson 1989). On the one hand, extracellular polymers that putatively serve for motility and adhesion to a substratum are highly cohesive (Higgins *et al.* 2002), and this suggests that it could glue sediment particles. On the other hand, it is known that motility trails of cultured diatoms usually detach shortly after secretion and are not likely to form such extended structures as those depicted in Figure 1. This view was confirmed by Wigglesworth-Cooksey and Cooksey (2004) who noted that the footpath and motility extracellular glycoconjugates produced by *Amphora coffeaeformis* and a *Navicula* sp. appeared to have an effect only over short distance in order to establish a physical contact between the diatom and a substratum (Wetherbee *et al.* 1998). However, in addition polymers with a different composition were secreted forming an extensive matrix. This EPS was more likely to mediate sediment stabilization because it exerted its effect over much greater distances and was potentially able to embed sediments into the EPS-matrix. By using a model sediment system, Wigglesworth-Cooksey and Cooksey (2004) also observed that sediment stability (measured as hydraulic conductivity) induced by *A. coffeaeformis* was closely correlated with the accumulation extracellular matrix material that was not soluble in 0.5 M $NaHCO_3$ at 90 °C (Wigglesworth-Cooksey *et al.* 2001). This polymer was mainly produced under PO_4^{3-}-limiting conditions, suggesting that nutrient limitation could play a role in secretion of matrix EPS and thus sediment stabilization. Similarly, the erosion rate of a sediment inoculated with a culture of the diatom *Nitzschia curvilineata* was highly correlated with the bulk carbohydrate to chlorophyll *a* ratio, which was considered as an indicator of the physiological state of the diatoms (Sutherland *et al.* 1998). Also in this study stationary phase conditions appeared to trigger production of bulk EPS as well as the decrease in erosion rate.

EPS as a food source

Exopolymers are a potential food source for other organisms inhabiting intertidal environments. In general, utilization of organic carbon by bacteria occurs rapidly. When extracellular fractions originating from a community of benthic diatoms were added to sediment slurries, it was found that 50% was utilized within 24 h (Goto *et al.* 2001). For extracellular polymers isolated from the pelagic diatom *Thalassiosira* a decrease of 50% in EPS concentration was obtained after 11-25 days of incubation (Aluwihare and Repeta 1999; Giroldo *et al.* 2003). The difference in this utilization rate between the pelagic and benthic systems is perhaps the results of a much higher density of bacteria in the sediment. The degradability of extracellular organic matter is dependent on its composition and structure. Nevertheless, Goto *et al* (2001) found little variation in the decomposition of various extracellular carbohydrate fractions that were isolated from different diatom species as well as a natural microphytobenthos assemblage. This indicates that compositional variations in the extracellular material were not limiting the initial degradation of the extracellular carbon pool, suggesting that a wide variety of carbohydrate hydrolyzing enzymes may be present in sediments to enable a rapid degradation of bioavailable glycoconjugates.

Also *in situ* studies in intertidal mudflats indicate that transfer of algal derived extracellular material occurs rapidly. Middelburg *et al.* (2000) showed that the transfer of photosynthetically fixed carbon into bacterial phospholipid fatty acids occurred within 4 h. Although these authors did not specifically investigate the role of EPS they concluded that transfer of organic carbon to bacteria occurred via extracellular material, which was estimated to represent 40 % of photosynthetically fixed carbon. This confirms the observation that a short-term coupling exists between bacterial production and extracellular compounds released by diatoms in phototrophic biofilms (Van Duyl *et al.* 1999). Water-extractable carbohydrates appeared to play an important role in bacterial dynamics. Its rapid utilization by diatoms suggests that this water-extractable carbohydrate represents a highly labile pool of carbon. Indeed it was found that the EPS produced in the surface layer of the sediment during tidal emersion consisted of polymers that consisted predominantly of glucose (De Brouwer and Stal 2001; Taylor *et al.* 1999). Glucose is preferentially degraded by bacteria (Giroldo *et al.* 2003; King 1986) indicating that this photosynthetically produced EPS fraction is important for transfer of carbon within the microbial foodweb.

Utilization of EPS is not necessarily restricted to bacteria. Heterotrophy has been observed among diatoms and some observations indicate that diatoms decompose their own EPS, utilizing it as an energy storage when deprived of light (Smith and Underwood 2000; Staats *et al.* 2000). Furthermore, meio- and macrofauna may also utilize EPS, however not much information on this subject is available. Current knowledge (Decho and Lopez 1993; Hoskins *et al.* 2003) suggest that certain animals were able to efficiently utilize algal as well as bacterial EPS. However, it should be emphasized that EPS is a poor food source with respect to its nutritional value. It is rather a valuable energy source in combination with other, more nutritional, compounds. Further research is necessary to identify the different consumers that utilize this labile carbon source in intertidal environments.

Acknowledgements

The authors would like to Ute Kuhlicke and Ute Wollenzien for technical assistance. This work was funded by the Schure-Beijerinck-Popping fund. (SBP/JK/2002-17) and IOP Milieutechnology/Zware metalen project number IZW99121. This is publication 3545 of the Netherlands Institute of Ecology, Yerseke, the Netherlands.

References

Allan, G.G., J. Lewin, and P. Johnson. 1972. Marine polymers. IV Diatom polysaccharides. Bot. Marina **15**: 102-108.

Aluwihare, L.I., and D.J. Repeta. 1999. A comparison of the chemical characteristics of oceanic DOM and extracellular DOM produced by marine algae. Mar. Ecol. Prog. Ser. **185**: 105-117.

BERGAMASCHI, B.A., E. TSAMAKIS, R.G. KEIL, T.I. EGLINTON, D.B. MONTLUÇON, and J.I. HEDGES. 1997. The effect of grain size and surface area on organic matter, lignin and carbohydrate concentration, and molecular composition in Peru Margin sediments. Geochimica et Cosmochimica Acta **61**: 1247-1260.

BHOSLE, N.B., S.S. SAWANT, and A.B. WAGH. 1995. Isolation and partial chemical analysis of exopolysaccharides from the marine fouling diatom *Navicula subinflata*. Bot. Marina **38**: 103-110.

BLANCHARD, G.F., J.M. GUARINI, P. RICHARD, P. GROS, and F. MORNET. 1996. Quantifying the short-term temperature effect on light-saturated photosynthesis of intertidal microphytobenthos. Marine Ecology-Progress Series **134**: 309-313.

BLANCHARD, G.F. and others 2000. The effect of geomorphological structures on potential biostabilisation by microphytobenthos on intertidal mudflats. Cont. Shelf Res. **20**: 1243-1256.

CHENU, C. 1993. Clay- or sand-polysaccharide associations as models for the interface between micro-organisms and soil: water related properties and microstructure. Geoderma **56**: 143-156.

CHIOVITTI, A., A. BACIC, J. BURKE, and R. WETHERBEE. 2003a. Heterogeneous xylose-rich glycans are associated with extracellular glycoproteins from the biofouling diatom Craspedostauros australis (Bacillariophyceae). Eur. J. Phycol. **38**: 351-360.

CHIOVITTI, A., M.J. HIGGINS, R.E. HARPER, R. WETHERBEE, and A. BACIC. 2003b. The complex polysaccharides of the raphid diatom Pinnularia viridis (Bacillariophyceae). J. Phycol. **39**: 543-554.

CHIOVITTI, A., P. MOLINO, S.A. CRAWFORD, R. TENG, T. SPURCK, and R. WETHERBEE. 2004. The glucans extracted with warm water from diatoms are mainly derived from intracellular chrysolaminaran and not exttracellular polysaccharides. Eur. J. Phycol. **39**: 000-000.

COOKSEY, K.E. 1981. Requirement for calcium in adhesion of a fouling diatom to glass. Appl. Environ. Microbiol. **41**: 1378-1382.

COWIE, G.L., and J.I. HEDGES. 1984. Carbohydrate sources in a coastal marine environment. Geochimica et Cosmochimica Acta **48**: 2075-2087.

DADE, W. B. and others 1990. Effects of bacterial exopolymer adhesion on the entrainment of sand. Geomicrobiol. J. **8**: 1-16.

DE BROUWER, J.F.C., S. BJELIC, E.M.G.T. DE DECKERE, and L.J. STAL. 2000. Interplay between biology and sedimentology in an intertidal mudflat (Biezelingse Ham, Westerschelde, The Netherlands. Cont. Shelf Res. **20**: 1159-1177.

DE BROUWER, J.F.C., E. DE DECKERE, and L.J. STAL. 2003. Distribution of extracellular carbohydrates in three intertidal mudflats in Western Europe. Estuar. Coastal Shelf Sci. **56**: 313-324.

DE BROUWER, J.F.C., G.K. RUDDY, T.E.R. JONES, and L.J. STAL. 2002a. Sorption of EPS to sediment particles and the effect on the rheology of sediment slurries. Biogeochem. **61**: 57-71.

DE BROUWER, J.F.C., and L.J. STAL. 2001. Short term dynamics in microphytobenthos distribution and associated extracellular carbohydrates in surface sediments of an intertidal mudflat. Mar. Ecol. Prog. Ser. **218**: 33-44.

—. 2002. Daily fluctuations of exopolymers in cultures of the benthic diatoms *Cylindrotheca closterium* and *Nitzschia* sp. (Bacillariophyceae). J. Phycol. **38**: 464-472.

—. 2004. Does warm-water extraction of benthic diatoms yield extracellular polymeric substances or does it extract intracellular chrysolaminaran? Eur. J. Phycol. **39**: 129-131.

DE BROUWER, J.F.C., K. WOLFSTEIN, G.K. RUDDY, T.E.R. JONES, and L.J. STAL. 2005. Biogenic stabilization of intertidal sediments; the importance of extracellular polymeric substances produced by benthic diatoms. Microbial ecology **49**: 501-512.

DE BROUWER, J.F.C., K. WOLFSTEIN, and L.J. STAL. 2002b. Physical characterization and diel dynamics of different fractions of extracellular polysaccharides in an axenic culture of a benthic diatom. Eur. J. Phycol. **37**: 37-44.

DE JONGE, V.N., and J.E.E. VAN BEUSEKOM. 1992. Contribution of resuspended microphytobenthos biomass to total phytoplankton in the Ems estuary and its possible role for grazers. Neth. J. Sea Res. **30**: 91-105.

DE WINDER, B., N. STAATS, L.J. STAL, and D.M. PATERSON. 1999. Carbohydrate secretion by phototrophic communities in tidal sediments. J. Sea Res. **42**: 131-146.

DECHO, A.W. 1994. Molecular scale events influencing the macro-scale cohesiveness of exopolymers, p. 135-148. *In* W.E. Krumbein, D.M. Paterson and L.J. Stal [eds.], Biostabilization of sediment. BIS Verlag.

DECHO, A.W., and G.R. LOPEZ. 1993. Exopolymer microenvironments of microbial flora: Multiple and interactive effects on trophic relationships. Limnol. Oceanogr. **38**: 1633-1645.

DEFEW, E.C., T.J. TOLHURST, and D.M. PATERSON. 2002. Site-specific features influence sediment stability of intertidal flats. Hydrol. Earth System Sci. **6**: 971-981.

FROSTICK, L.E., and I.N. MCCAVE. 1979. Seasonal shifts of sediment within an estuary mediated by algal growth. Estuar. Coastal Mar. Sci. **9**: 569-576.

GIROLDO, D., A.A.H. VIEIRA, and B.S. PAULSEN. 2003. Relative increase of deoxy sugars during microbial degradation of an extracellular polysaccharide released by a tropical freshwater Thalassiosira sp (Bacillariophyceae). J. Phycol. **39**: 1109-1115.
GOTO, N., O. MITAMURA, and H. TERAI. 2001. Biodegradation of photosynthetically produced extracellular organic carbon from intertidal benthic algae. J. Exp. Mar. Biol. Ecol. **257**: 73-86.
HIGGINS, M.J., S.A. CRAWFORD, P. MULVANEY, and R. WETHERBEE. 2002. Characterization of the adhesive mucilages secreted by live diatom cells using Atomic Force Microscopy. Protist **153**: 25-38.
HOAGLAND, K.D., J.R. ROSOWSKI, M.R. GRETZ, and S.C. ROEMER. 1993. Diatom extracellular polymeric substances: function, fine structure, chemistry, and physiology. J. Phycol. **29**: 537-566.
HOLLAND, A.F., R.G. ZINGMARK, and J.M. DEAN. 1974. Quantitative evidence concerning the stabilization of sediments by marine benthic diatoms. Mar. Biol. **27**: 191-196.
HOSKINS, D.L., S.E. STANCYK, and A.W. DECHO. 2003. Utilization of algal and bacterial extracellular polymeric secretions (EPS) by the deposit-feeding brittlestar Amphipholis gracillima (Echinodermata). Mar. Ecol. Prog. Ser. **247**: 93-101.
KAWAGUCHI, T., H.A. SAYEGH, and A.W. DECHO. 2003. Development of an indirect competitive enzyme-linked immunosorbent assay to detect extracellular polymeric substances (EPS) secreted by the marine stromatolite-forming cyanobacteria *Schizothrix sp*. J. immunoassay immunochem. **24**: 29-39.
KING, G.M. 1986. characterization of β-glucosidase activity in intertidal marine sediments. Appl. Environ. Microbiol. **51**: 373-380.
KORNMAN, B., and E.M.G.T. DE DECKERE. 1998. Temporal variation in sediment erodibility and suspended sediment dynamics in the Dollard estuary, p. 231-241. *In* K. S. Black, D. M. Paterson and A. Cramp [eds.], Sedimentary processes in the intertidal zone. Special publications. The Geological Society.
LAWRENCE, J.R. and others 2003. Scanning transmission X-ray, laser scanning, and transmission electron microscopy mapping of the exopolymeric matrix of microbial biofilms. Appl. Environ. Microbiol. **69**: 5543-5554.
LIND, J.L., K. HEIMANN, E.A. MILLER, C. VAN VLIET, N.J. HOOGENRAAD, and R. WETHERBEE. 1997. Substratum adhesion and gliding in a diatom are mediated by extracellular proteoglycans. Planta **203**: 213-221.
MAGALETTI, E., R. URBANI, P. SIST, C.R. FERRARI, and A.M. CICERO. 2004. Abundance and chemical characterization of extracellular carbohydrates released by the marine diatom Cylindrotheca fusiformis under N- and P-limitation. Eur. J. Phycol. **39**: 133-142.
MEADOWS, P.S., J. TAIT, and S.A. HUSSAIN. 1990. Effects of estuarine infauna on sediment stability and particle sedimentation. Hydrobiologia **190**: 263-266.
MIDDELBURG, J.J., C. BARRANGUET, H.T.S. BOSCHKER, P.M.J. HERMAN, T. MOENS, and C.H.R. HEIP. 2000. The fate of intertidal microphytobenthos carbon: An in situ ^{13}C-labeling study. Limnol. Oceanogr. **45**: 1224-1234.
NEU, T.R. 2000. *In situ* cell and glycoconjugate distribution in river snow studied by confocal laser scanning microscopy. Aquat. Microb. Ecol. **21**: 85-95.
NEU, T.R., and J.R. LAWRENCE. 1999. Lectin binding analysis in biofilm systems, p. 145-151. *In* R.J. Doyle [ed.], Biofilms. Methods in enzymology. Academic press.
NIELSEN, P.H., and A. JAHN. 1999. Extraction of EPS, p. 49-72. *In* J. Wingender, T.R. Neu and H.C. Flemming [eds.], Microbial extracellular polymeric substances. Springer.
NORTON, T.A., J. POPE, C.J. VELTKAMP, B. BANKS, C.V. HOWARD, and S.J. HAWKINS. 1998. Using confocal laser scanning microscopy, scanning electron microscopy and phase contrast light microscopy to examine marine biofilms. Aquat. Microb. Ecol. **16**: 199-204.
PATERSON, D.M. 1989. Short-term changes in the erodibility of intertidal cohesive sediments related to the migratory behavior of epipelic diatoms. Limnol. Oceanogr. **34**: 223-234.
—. 1997. Biological mediation of sediment erodibility, p. 215-229. *In* N. Burt, R. Parker and J. Watts [eds.], Cohesive Sediments. Wiley.
PATERSON, D.M., and K.S. BLACK. 1999. Water flow, sediment dynamics and benthic biology, p. 155-193, Advances in Ecological Research, Vol 29. Advances in Ecological Research.
PATERSON, D.M., T.J. TOLHURST, J.A. KELLY, C. HONEYWILL, E.M.G.T. DE DECKERE, V. HUET, S.A. SHAYLER, K.S. BLACK, J. DE BROUWER, and I. DAVIDSON. Variations in sediment stability and sediment properties across the Skeffling mudflat, Humber estuary, UK. Cont. Shelf Res. **20**: 1373-1396.
PERKINS, R.G., D.M. PATERSON, H. SUN, J. WATSON, and M.A. PLAYER. 2004. Extracellular polymeric substances: quantification and use in erosion experiments. Cont. Shelf Res.**24**: 1623-1635.
PUSKARIC, S., and A. MORTAIN-BERTRAND. 2003. Physiology of diatom Skeletonema costatum (Grev.) Cleve photosynthetic extracellular release: evidence for a novel coupling between marine bacteria and phytoplankton. J. Plankton Res. **25**: 1227-1235.

RIEMANN, F., and M. SCHRAGE. 1978. The mucus-trap hypothesis on feeding of aquatic nematodes and implications for biodegradation and sediment texture. Oecologia **34**: 75-88.
SMITH, D.J., and G.J.C. UNDERWOOD. 2000. The production of extracellular carbohydrates by estuarine benthic diatoms: the effects of growth phase and light and dark treatment. J. Phycol. **36**: 321-333.
STAATS, N., B. DE WINDER, L.J. STAL, and L.R. MUR. 1999. Isolation and characterization of extracellular polysaccharides from the epipelic diatoms *Cylindrotheca closterium* and *Navicula salinarum*. Eur. J. Phycol. **34**: 161-169.
STAATS, N., L.J. STAL, B. DE WINDER, and L.R. MUR. 2000. Oxygenic photosynthesis as driving process in exopolysaccharide production of benthic diatoms. Mar. Ecol. Prog. Ser. **193**: 261-269.
SUTHERLAND, T.F., J. GRANT, and C.L. AMOS. 1998. the effect of carbohydrate production by the diatom *Nitzschia curvilineata* on the erodibility of sediment. Limnol. Oceanogr. **43**: 65-72.
TAYLOR, I.S., and D.M. PATERSON. 1998. Microspatial variation in carbohydrate concentrations with depth in the upper oncentrati of intertidal cohesive sediments. Estuar. Coastal Shelf Sci. **46**: 359-370.
TAYLOR, I.S., D.M. PATERSON, and A. MEHLERT. 1999. The quantitative variability and monosaccharide composition of sediment carbohydrates associated with intertidal diatom assemblages. Biogeochem. **45**: 303-327.
TOLHURST, T.J., T. GUST, and D.M. PATERSON. 2002. The influence of an extracellular polymeric substance (EPS) on cohesive sediment stability, p. 409-425. *In* J. C. Winterwerp and C. Kranenburg [eds.], Fine sediment dynamics in the marine environment. Proceedings in marine science. Elsevier Science.
UNDERWOOD, G.J.C., M. BOULCOTT, C.A. RAINES, and K. WALDRON. 2004. Environmental effects on exopolymer production by marine benthic diatoms: Dynamics, changes in composition, and pathways of production. J. Phycol. **40**: 293-304.
UNDERWOOD, G.J.C., and J. KROMKAMP. 1999. Primary production by phytoplankton and microphytobenthos in estuaries. Adv. Ecol. Res. **29**: 93-153.
UNDERWOOD, G.J.C., and D.M. PATERSON. 1993. Seasonal changes in diatom biomass, sediment stability and biogenic stabilization in the Severn estuary. J. Mar. Biol. Ass. UK **73**: 871-887.
—. 2003. The importance of extracellular carbohydrate production by marine epipelic diatoms. Advances in Botanical Research **40**: 183-240.
UNDERWOOD, G.J.C., D.M. PATERSON, and R.J. PARKES. 1995. The measurement of microbial carbohydrate exopolymers from intertidal sediments. Limnol. Oceanogr. **40**: 1243-1253.
UNDERWOOD, G.J.C., and D.J. SMITH. 1998. Predicting epipelic diatom exopolymer concentrations in intertidal sediments from sediment chlorophyll *a*. Microbial Ecol. **35**: 116-125.
VAN DUYL, F.C., B. DE WINDER, A.J. KOP, and U. WOLLENZIEN. 1999. Tidal coupling between carbohydrate concentrations and bacterial activities in diatom-inhabited intertidal mudflats. Mar. Ecol. Prog. Ser. **191**: 19-32.
WETHERBEE, R., J.L. LIND, and J. BURKE. 1998. The first kiss: establishment and control of initial adhesion by raphic diatoms. J. Phycol. **34**: 9-15.
WIGGLESWORTH-COOKSEY, B., D. BERGLUND, and K.E. COOKSEY. 2001. Cell-cell and cell-surface interactions in an illuminated biofilm: implications for marine sediment stabilization. Geochem. T. **10**: 75-82.
WIGGLESWORTH-COOKSEY, B., and K.E. COOKSEY. 2005. Use of fluorescently-conjugated lectins to study cell-cell interactions in model marine biofilms. Appl. Environ. Microbiol. **71**: 428-435
WOLFSTEIN, K., J.F.C. DE BROUWER, and L.J. STAL. 2002. Biochemical partitioning of photosynthetically fixed carbon by benthic diatoms during short-term incubations at different irradiances. Mar. Ecol. Prog. Ser. **245**: 21-31.
WOLFSTEIN, K., and L.J. STAL. 2002. Production of extracellular polymeric substances (EPS) by benthic diatoms: effect of irradiance and temperature Mar. Ecol. Prog. Ser. **236**: 13-22.
WUSTMAN, B.A., M.R. GRETZ, and K.D. HOAGLAND. 1997. Extracellular matrix assembly in diatoms (Bacillariophyceae) I. A model of adhesives based on chemical characterization and localization of polysaccharides from the marine diatom *Achnanthes longipes* and other diatoms. Plant Physiol. **113**: 1059-1069.
WUSTMAN, B.A., J. LIND, R. WETHERBEE, and M.R. GRETZ. 1998. Extracellular matrix assembly in diatoms (Bacillariophyceae) III. Organization of fucoglucuronogalactans within the adhesive stalks of *Achnanthes longipes*. Plant Physiol. **116**: 1431-1441.
YALLOP, M.L., D.M. PATERSON, and P. WELLSBURY. 2000. Interrelationships between rates of microbial production, exopolymer production, microbial biomass, and sediment stability in biofilms of intertidal sediments. Microbial Ecol. **39**: 116-127.

IV. Use of stable isotopes in foodweb research

Padma Maddi, Kevin R. Carman, Brian Fry and Bjoern Wissel

Use of primary production by harpacticoid copepods in a Louisiana salt-marsh food web

Abstract

We used stable isotopes ($\delta^{13}C$ and $\delta^{15}N$) to examine temporal (quarterly sampling throughout a year) and interspecific variation in the use of primary producers by three harpacticoid copepod taxa (*Coullana* sp, *Pseudostenhelia wellsi*, and laophontids) in a low-salinity Louisiana salt marsh. Microphytobenthos (MPB), phytoplankton, and *Spartina alterniflora* are the major primary producers in this system. The natural $\delta^{13}C$ values of harpacticoids in summer suggested strong dependence on phytoplankton (*Coullana* sp.) or a mixture of phytoplankton and MPB (*P. wellsi* and laophontids). During fall, winter, and spring, however, $\delta^{13}C$ values of harpacticoids were significantly enriched relative to summer; the copepod isotope values were generally slightly more enriched than MPB, but intermediate between $\delta^{13}C$ values of *Spartina* and phytoplankton. Such intermediate values could be an indication of a mixed diet of *Spartina* and phytoplankton, a diet comprised primarily of MPB, or some combination of all three food sources. The dual-isotope approach ($\delta^{13}C$ and $\delta^{15}N$) did little to resolve this uncertainty because $\delta^{15}N$ values of primary producers were similar. Isotope-addition experiments were conducted at each sampling period in which MPB were labeled with additions of $NaH^{13}CO_3$. Uptake of added ^{13}C by harpacticoids was analyzed using a 3-source mixing model, which was based on ^{13}C uptake by MPB in combination with natural $\delta^{13}C$ values of *Spartina* and phytoplankton. Mixing-model results verified the importance of phytoplankton (61-71%) and MPB (15-37%) in summer; *Spartina* contributed only1-14% to copepod diets. In winter and summer, phytoplankton contributed substantially, but to a lesser extent to copepod diets (36-59%). However, *Spartina* constituted > 50% of the diet of *P. wellsi*, and 22-48% of the diets of other copepods. During winter and summer, MPB contributed minimally (6-13%) to the diets of *P. wellsi* and *Coullana* sp. Collectively our data indicate strong temporal and interspecific variation in copepod diets; diets are dominated by phytoplankton in summer, but *Spartina* (presumably in the form of detritus) is an important component of the diet in other seasons. Our observations are in contrast with the prevailing dogma that MPB is the primary source of nutrition for salt-marsh invertebrates, and that *Spartina* contributes minimally to salt-marsh food webs.

Introduction

Relatively little is known about the food resources exploited by estuarine meiofaunal invertebrates (< 1 mm in size). ^{14}C-grazing studies suggest that meiofaunal grazing can have a significant impact on microphytobenthos (MPB) biomass and production in estuarine environments (Blanchard 1991; Montagna 1995; Carman *et al.* 1997; Pinckney *et al.* 2003), implying that MPB contribute significantly to meiofaunal diets. However, in addition to MPB, various studies indicate that some meiofauna may also consume phytoplankton (Decho 1986; Pace and Carman 1996; Buffan-Dubau and Carman 2000a) or vascular-plant (e.g., *Spartina alterniflora*) detritus (Couch 1989; Carman and Fry 2002). As currently employed, ^{14}C-grazing studies can be used to asses the consumption rate of a particular food source, such as MPB, but provide no information on the relative contributions of other potential food sources. Although poorly understood, it is clear that feeding strategies and dietary preferences vary substantially among meiofaunal taxa and individual species (*e.g.*, Carman and Thistle 1985; Pace and Carman 1996; Buffan-Dubau and Carman 2000a; Moens *et al.* 2002). Almost nothing is known about seasonal variation in the diets of meiofauna.

Stable isotopes (*e.g.*, ^{13}C and ^{15}N) can, in theory, be used to simultaneously determine the relative contributions of multiple food sources to consumer diets (Peterson and Fry 1987). Stable-isotope studies of estuarine salt-marsh food webs have led to the general conclusion that MPB primary production supports much of the secondary production by fish and macrofaunal invertebrates, and that production by vascular plants (*e.g.*, *Spartina alterniflora*) is a relatively minor source of food (*e.g.*, Sullivan and Moncreiff 1990; Currin *et al.* 1995; Currin *et al.* 2003). Such conclusions are consistent with observations that MPB are more labile and nutritious than detrital material derived from vascular plants (Miller *et al.* 1999). Similarly detailed information is lacking for meiofaunal-sized animals. Because of their small size, large numbers (100's to 1000's) of meiofaunal individuals were required to provide the biomass needed for a single determination of stable-isotope content (^{13}C and/or ^{15}N; Couch 1989; Riera *et al.* 1996; Middelburg *et al.* 2000), and thus only a few studies have attempted to use stable isotopes in the study of meiofaunal food webs. Couch (1989) concluded that *S. alterniflora* detritus was the primary source of nutrition for meiofaunal harpacticoid copepods and nematodes, but the few other stable-isotope studies of meiofauna have concluded that they rely heavily on MPB (Riera *et al.* 1996; Middelburg *et al.* 2002; Moens *et al.* 2002). Recent methodological developments allow for the accurate measurement of ^{13}C and ^{15}N on relatively small numbers (5-60) of meiofaunal animals (Carman and Fry 2002), which presents the opportunity to use stable isotopes for species-level analyses of meiofaunal diets.

However, in estuarine salt marshes, ^{13}C values of both MPB and many consumers (meiofaunal and macrofaunal) are typically intermediate between those of *Spartina* (which is more enriched with ^{13}C) and phytoplankton (which is more depleted). When three or more food sources are available, the diets of consumers with intermediate ^{13}C values cannot be unambiguously determined because of uncertainty as to whether consumers are using primarily MPB, a combination of *Spartina* and phytoplankton, or possibly all three food sources (Figure 1).

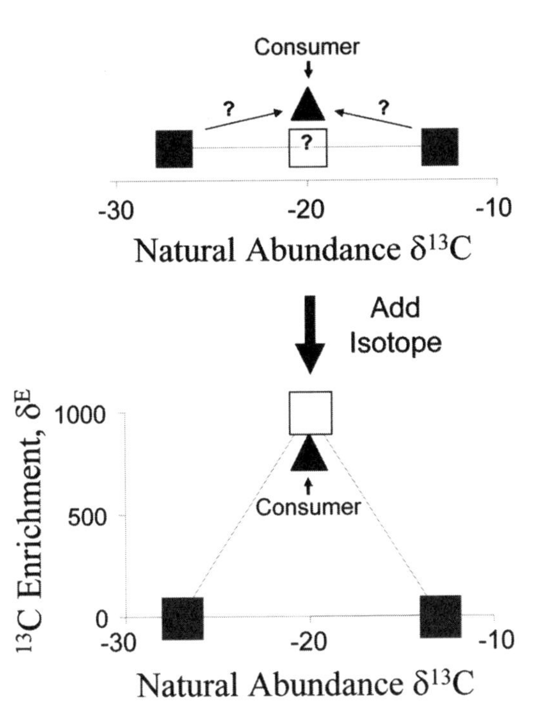

Figure 1. Stable-isotope mixing model. Determination of consumer food source(s) is uncertain when three or more food sources are available and consumer isotope values are intermediate between the most ^{13}C-enriched (*e.g.*, *Spartina*) and most ^{13}C-depleted (*e.g.*, phytoplankton) food sources. This uncertainty can be partially resolved by labeling the food source with the intermediate isotope value (*e.g.*, MPB). Uptake of label by consumers is proportional to the degree to which they feed on the intermediate food source.

In situ labeling of MPB with added ^{13}C has been used to study the dynamics of MPB consumption by meiofauna (Herman *et al.* 2000; Middelburg *et al.* 2000; Carman and Fry 2002; Moens *et al.* 2002). *In situ* labeling of MPB can, in principle, also be used to resolve uncertainty in food sources described above because it can strongly differentiate animals that consume MPB as food from those that do not (Figure 1).

In the present study, we measured natural stable-isotope (^{13}C and ^{15}N) values of three harpacticoid copepods and conducted simultaneous isotope-addition experiments ($NaH^{13}CO_3$) throughout the course of a year. These observations were used to address three questions: (1) Does meiofaunal consumption of MPB (and other sources of primary production) vary temporally and (2) among species, and (3) can ^{13}C labeling of the natural MPB assemblage be used to resolve uncertainty regarding meiofaunal consumption of MPB?

Methods and materials

The study site was an intertidal mud-flat surrounded by *Spartina alterniflora* cord grass located in Terrebonne Bay estuary (29° 15' N, 91° 21' W) near Cocodrie, Louisiana, USA. Tidal amplitudes are low (approximately 0.3 m), salinity ranges from 3-15 psu, and dissolved oxygen ranges from 3-8 mg L^{-1}. Samples to determine abundance and isotopic composition of phytoplankton, MPB, and meiofauna were collected at low tide during summer (07.02.01), fall (10.19.01), winter (02.11.02) and spring (04.24.02). Two poles were placed on either end of a 10-m transect, approximately 2 m away from and parallel to the marsh edge. Four replicate cores were collected from the mudflat at randomly determined intervals along the transect. On each sampling date, a parallel $NaH^{13}CO_3$ tracer-addition experiment was performed to evaluate the role of MPB as food for meiofauna (described below).

Major food sources

Phytoplankton: Suspended particulate material (SPM) (consisting of a mixture of microalgae, zooplankton, microorganisms, and nonliving organic material) was used as a proxy for phytoplankton. Four replicate samples of surface water were collected using 1-L plastic bottles. Water (50-100 mL) was filtered using GF/F Whatman filters for pigment analysis. Phytoplankton abundance was determined from HPLC (High Performance Liquid Chromatography) analysis of Chl *a* (described below).

For pigment analysis, filters were extracted in 5 mL 100% HPLC-grade acetone (Fisher Scientific). Acetone extracts were filtered using syringe filters (Sun International; diameter: 13 mm; pore size: 0.2 μm) twice to remove particulates. Extracts were diluted (66 μL sample + 44 μL water) with HPLC quality water (Fisher Scientific) to improve the sharpness of peaks, and photopigments were analyzed using HPLC (Wright *et al.* 1991).

For stable-isotope analysis, GF/F filters were pre-combusted (450 °C, 2 h) to remove excess carbon and SPM was concentrated until the filters clogged. Filters were dried and 4.5-mm circles were removed from each filter using a handheld punch. Six circles from each replicate were used for analysis of stable isotopes.

MPB: The top 1 cm of sediment from 3.2 cm i.d. butyrate cores was collected for analysis of meiofauna and MPB. Sediment was homogenized with a spatula, and a sub-sample (approximately 300 mg wet wt.) was collected for HPLC analysis of Chl *a* (Buffan-Dubau and Carman, 2000b). The remaining sample was fixed in 4% formaldehyde for analysis of meiofauna.

$\delta^{13}C$ values for MPB were determined from acetone extracts of sediment that were dried on precombusted 4.5 mm GF/F filters. Acetone extracts were used because they include an enrichment of photosynthetic pigments; however, they also include photosynthetic pigments from both living and dead (detrital) algal material, as well as other moderately polar organic material. Therefore, while the $\delta^{13}C$ values of acetone extracts are considered as a proxy for MPB values, they do not represent a pure MPB sample. To determine bias associated with analysis of $\delta^{13}C$ in acetone extracts, a separate experiment was carried out for two seasons (winter and spring) using water collected for phytoplankton. Surface water was filtered on pre-combusted GF/F filters, which were dried and cut in half. One half of the filter was used to measure $\delta^{13}C$ in non-acetone-extracted material, and the second half was used to measure $\delta^{13}C$ in acetone extracts. The average difference between the extracted and un extracted $\delta^{13}C$ values (4‰) was added to $\delta^{13}C$ values of the acetone extract to obtain an estimated $\delta^{13}C$ value of MPB.

Spartina alterniflora: Fresh leaf blades of *S. alterniflora* were collected along the marsh edge. The leaf blades were washed and oven dried and finely powdered using a Wig-L-Bug grinder before analyzing for $\delta^{13}C$ and $\delta^{15}N$.

Meiofauna

Sediment fixed in 4% formaldehyde was used for extraction of meiofauna used in the isotope analyses. In the laboratory sediment samples were washed through a 63-µm sieve and stained with Rose Bengal. Nematodes, ostracods, *Streblospio benedicti*, *Tanypus clavatus* (a chironomid larva), and three harpacticoid taxa (*Coullana* sp., *Pseudostenhelia wellsi*, and laophontid copepods) were separated for stable-isotope analysis. Laophontids at this site consist of two species, *Onychocamptus mohammed*, (approximately 80%) and *Paronchocamptus huntsmani* (approximately 20%) (J. Fleeger, personal communication), the identities of which can be determine only with detailed microscopic analysis. Here, we focus on stable-isotope analyses of the three harpacticoid taxa. Based on the observations of Carman and Fry (2002), the necessary number of animals (10-20 individuals) were handpicked picked using a tungsten-wire probe and cleared of attached debris. Copepods were transferred to small tin cups (3 x 5 mm), dried, and analyzed as described below.

Tracer-addition experiment

A parallel tracer-addition experiment was performed in summer and winter to determine the short-term uptake of added $NaH^{13}CO_3$. Eight cores (7.5 cm i.d.) were collected at random locations on the 10-m transect. Overlying water was removed without disturbing the surface sediment. Twenty milliliters of GF/F-filtered marsh water containing 72 mg of $NaH^{13}CO_3$ (Middelburg *et al.* 2000) was added to each core. Four of the eight cores were covered with aluminum foil and used as dark controls to measure uptake of ^{13}C through non-photosynthetic processes. All the cores were incubated for 4 h in ambient sunlight. After 4 h, the top 1 cm of sediment was harvested and homogenized with a spatula; a sub-sample (approximately 5 mg) was collected to measure MPB ^{13}C incorporation and pigment analysis (HPLC). The remaining sediment was fixed with 4% formaldehyde and processed as described to determine ^{13}C incorporation by meiofauna.

Stable-isotopic analysis

Samples were analyzed for $\delta^{13}C$ and $\delta^{15}N$ using a Carlo Erba NA 1500 elemental analyzer linked to a Finnigan Delta Plus ratio mass spectrometer. The elemental analyzer was modified for small-biomass samples as described by Carman and Fry (2002). Glycine and bovine liver were used as reference standards in combustion analysis. These standards as well as procedural blanks were used to correct for background values of C and N in the samples (Fry *et al.* 1992).

Isotope ratios were expressed as δ values (‰):

$$\delta^{13}C, \delta^{15}N = [(R_{sample} - R_{standard})/R_{standard}] \times 1000$$

where R = $^{13}C/^{12}C$ or $^{15}N/^{14}N$. Peedee Belemnite and atmospheric nitrogen were used as carbon and nitrogen isotope standards, respectively. For mixing-model calculations, consumer isotopic values were adjusted for fractionation by subtracting 0.5‰ for carbon, and 2.2‰ for nitrogen (McCutchan *et al.* 2003).

Values from tracer-addition experiments were expressed as ^{13}C enrichment (δ^E) by correcting $\delta^{13}C$ values from samples incubated in the light ($\delta^{13}C_{light}$) with the average values of samples incubated in the dark ($\delta^{13}C_{dark}$) to account for non-photosynthetic uptake of ^{13}C by copepods:

$$\delta^E = (((\delta^{13}C_{light} + 1000)/\ (\delta^{13}C_{dark} + 1000))-1) * 1000$$

Mixing-model

The percent contributions of *Spartina*, MPB, and phytoplankton were calculated by using a 3-source mixing model that included natural $\delta^{13}C$ and $\delta^{E13}C$ from tracer-addition experiments. Interpretations of the 3-source mixing model were based on the following assumptions:

1. In tracer-addition experiments, uptake of ^{13}C by meiofauna was only through consumption of MPB and ^{13}C-uptake by MPB was $\geq$ uptake by consumers.
2. Uptake of ^{13}C by *Spartina* and phytoplankton was zero in tracer-addition experiments.
3. *Spartina*, MPB, and phytoplankton were the major food sources available to meiofauna. Thus, the $\delta^{13}C$ and δ^E values of copepods were within the range of these three food sources.

Based on assumption 1, the δ^E values of MPB were adjusted according to the isotopically most-enriched consumer in each season. Source contributions were calculated from simultaneous solution of three mass-balance equations:

(1) $f_1 + f_2 + f_3 = 1$
(2) $f_1N_1 + f_2N_2 + f_3N_3 = N$ of consumer
(3) $f_1E_1 + f_2E_2 + f_3E_3 = E$ of consumer

Where,

f = Fractional contribution of a food source
N = Natural carbon isotope value ($\delta^{13}C$)
E = Enriched carbon isotope value (δ^E)
Subscripts 1-3 refer to food sources 1-3

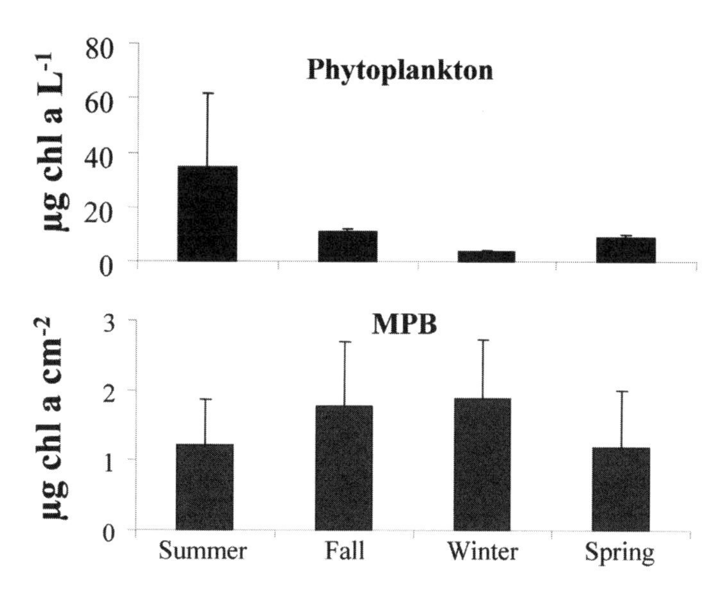

Figure 2. Temporal variation in the biomass of phytoplankton and MPB. Bars are means + 1 SD (N=4).

Results

Microalgal abundances

Although temporal variation was not significant (ANOVA, $p = 0.49$), MPB biomass as estimated from Chl *a* was highest in fall and winter when mud-flats are exposed during the day (Figure 2). MPB Chl *a* concentrations ranged from a minimum of 1.2 ± 0.8 (spring) to a maximum of 1.9 ± 0.8 µg cm^{-2} (winter). Phytoplankton abundance varied significantly among seasons (ANOVA, $p = 0.03$) and was approximately 3x higher in summer (34.6 ± 27.1 µg Chl *a* L^{-1}) than in other seasons (Figure 2).

Isotopic compositions of potential food sources

Natural $\delta^{13}C$ values varied among food sources and seasons (Figure 3, Table 1). Throughout the year, *Spartina* was consistently most enriched in ^{13}C, and phytoplankton was most depleted in ^{13}C. Isotopic values of MPB were more ^{13}C-enriched than phytoplankton in all seasons except fall, and the $\delta^{13}C$ of *Spartina* values were relatively constant throughout the year (~-13‰). Both phytoplankton and MPB showed similar temporal variation in $\delta^{13}C$, with most enriched values in winter and most depleted values in spring.

$\delta^{15}N$ values were available only for phytoplankton and *Spartina*. No significant differences were found between $\delta^{15}N$ values of *Spartina* and phytoplankton at any time of the year, and seasonal variation was small (Table 1).

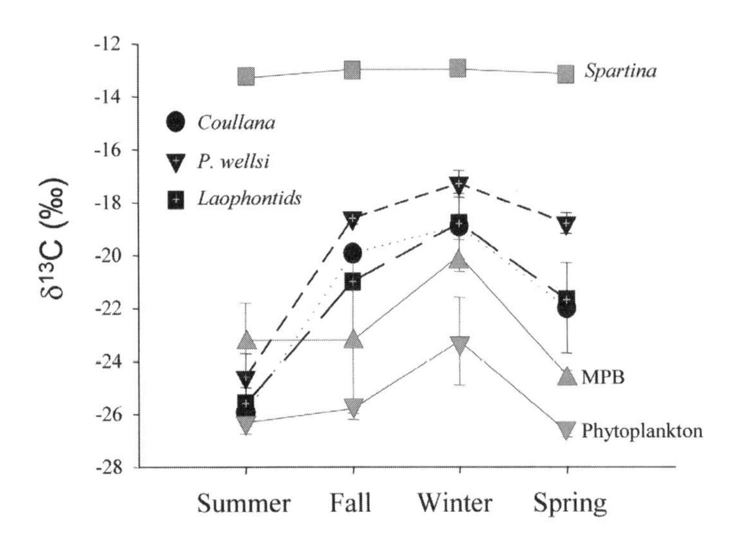

Figure 3. Temporal variation in the $\delta^{13}C$ of primary producers (*Spartina*, MPB, and phytoplankton) and three harpacticoid copepod taxa (*Coullana* sp., *Pseudostenhelia wellsi*, and laophontids). Values are means ± 1 SD (N=4).

Table 1. Natural $\delta^{13}C$ and $\delta^{15}N$ values of harpacticoid copepods (*Coullana* sp., *Pseudostenhelia wellsi*, and laophontids) and primary producers (MPB, phytoplankton, and *Spartina*) in four seasons. Values are means (SD), N=4.

	Summer 2001		Fall 2001		Winter 2002		Spring 2002	
Consumers	$\delta^{13}C$	$\delta^{15}N$	$\delta^{13}C$	$\delta^{15}N$	$\delta^{13}C$	$\delta^{15}N$	$\delta^{13}C$	$\delta^{15}N$
Coullana sp.	-25.4 (0.5)	8.8 (1.1)	-19.4 (0.3)	7.4 (0.2)	-18.4 (1.2)	8.3 (0.8)	-21.5 (1.7)	7.6 (0.5)
P. wellsi	-24.1 (0.9)	9.1 (1.2)	-18.1 (0.2)	7.6 (0.6)	-16.8 (0.5)	7.6 (0.3)	-18.3 (0.4)	7.5 (0.5)
Laophontids	-25.1 (0.6)	11.4 (0.8)	-20.5 (0.2)	ND	-18.3 (0.3)	8.9 (0.8)	-21.2 (0.1)	ND
Food Sources								
MPB	-27.2 (1.4)	ND	-27.2 (3.0)	ND	-25.3 (0.6)	ND	-27.2 (0.2)	ND
Phytoplankton	-26.3(0.4)	6.7 (0.4)	-25.8 (0.2)	6.7 (0.8)	-23.3 (1.7)	5.2 (0.3)	-26.7 (0.2)	5.5 (0.9)
Spartina	-13.3 (0.0)	6.7 (0.2)	-13.0 (0.1)	5.8 (0.1)	-13.0 (0.1)	6.3 (0.3)	-13.2 (0.1)	6.0 (0.1)

Phytoplankton $\delta^{13}C$ values (-26.7 to -23.3‰) were consistent with values from a previous study at this site (Carman and Fry 2002). Our phytoplankton $\delta^{13}C$ values were lower than many estuarine literature values (*e.g.*, Currin *et al.* 1995; Riera *et al.* 1996), but similar to those reported in low-salinity systems (Deegan and Garritt 1997; Wainright *et al.* 2000). Our $\delta^{15}N$ phytoplankton values were consistent with published values (*e.g.*, Currin *et al.* 1995; Fogel *et al.* 1989). Our *Spartina* values $\delta^{13}C$ (-13.3 to -13.0‰) and $\delta^{15}N$ (5.8 - 6.7‰) values were consistent with literature values (*e.g.,* Couch 1989; Peterson and Howarth 1987; Sullivan and Moncreiff 1990; Currin *et al.* 1995). The proxy $\delta^{13}C$ values of MPB measured in this study were lighter (-24.6 to -20.0‰) than most existing literature values for salt marshes (e.g., -16 to -18‰: Haines 1976; Currin *et al.* 1995), but similar to those reported by Sullivan and Moncreiff (1990) in a Mississippi salt marsh (-20.6‰). $\delta^{13}C$ values of MPB in winter were enriched by ~4‰ relative to other seasons, and temporal variation of MPB $\delta^{13}C$ values resembled that observed for phytoplankton.

Isotopic compositions of consumers

The $\delta^{13}C$ values of copepods varied temporally, and all three copepod taxa showed similar temporal variation (Figure 3, Table 1). Copepod $\delta^{13}C$ values were most depleted in summer and most enriched in winter. Although *Coullana* sp., *P. wellsi,* and laophontids showed similar seasonal trends, *P. wellsi* values were slightly heavier than those of the other copepods. In summer, the $\delta^{13}C$ values of copepods were most similar to those of phytoplankton, while in other seasons they were intermediate among food sources.

$\delta^{15}N$ values are not discussed separately, but are presented in dual-isotope plots with $\delta^{13}C$ (Figure 4). $\delta^{15}N$ were available in all seasons only for *Coullana* sp. and *P. wellsi.* Because $\delta^{15}N$ values were not measured for MPB in this study, we used published values for this study site as an estimate (Carman and Fry 2002).

In summer, the dual-isotope composition of *Coullana* sp. was closely aligned with phytoplankton (Figure 4). Laophontid copepods and *P. wellsi* isotope compositions were distinctly different from each other, but generally intermediate between MPB and phytoplankton isotope compositions. Summer $\delta^{13}C$ values of copepods contrasted markedly with those from other seasons; in all other seasons, copepod $\delta^{13}C$ values were intermediate between the $\delta^{13}C$ values of *Spartina* and phytoplankton.

Tracer-addition experiments

Uptake of ^{13}C by copepods in tracer-addition experiments varied among seasons and among taxa (Table 2; data were unavailable for fall). Uptake of ^{13}C by *Coullana* sp. was highest in summer and lowest in winter; uptake by *P. wellsi* was highest in summer and spring, and lowest in winter; uptake by laophontids was highest in winter and spring and lowest in summer.

As described in Methods and Materials, data from uptake of ^{13}C by copepods in tracer-addition experiments were used in a 3-source mixing model to determine the relative contributions of phytoplankton, MPB, and *Spartina*. Results in Table 2 are shown incorporated in the 3-source mixing models of Figure 5. Results of the model calculations (Figure 6) support the inference from natural $\delta^{13}C$ (Figure 3) that copepods depended heavily on phytoplankton (61-71%) and secondarily on MPB (15-37%) in summer; *Spartina* contributed only1-14% to copepod diets in summer. In winter, phytoplankton contributed substantially, but to a lesser extent to copepod diets (36-54%). However, *Spartina* constituted > 50% of the diet of *P. wellsi* in winter, and comprised 26-40% of the diets of *Coullana* and laophontids. During winter, MPB contributed minimally (6%) to the diets of *P. wellsi* and *Coullana* sp., but comprised a greater fraction of laophontid diets (38%).

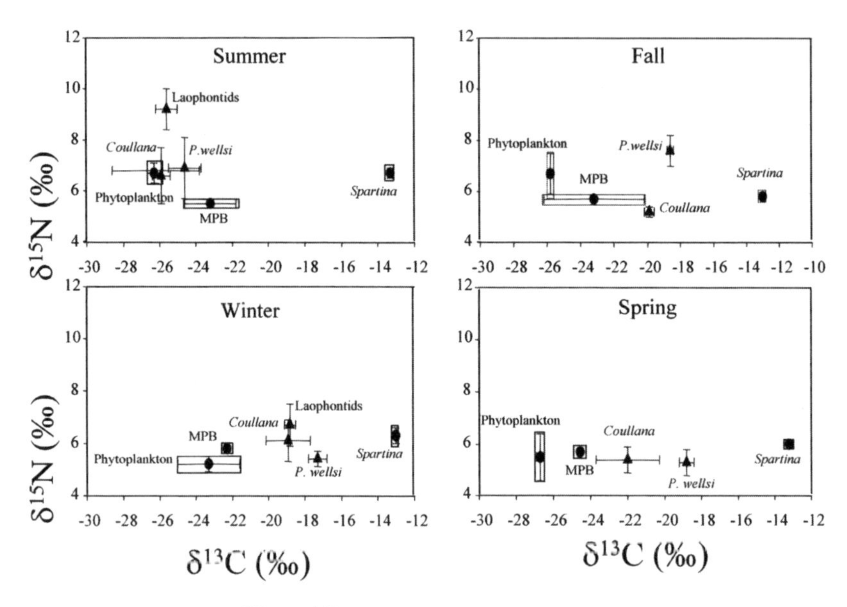

Figure 4. Dual-isotope plots of $\delta^{13}C$ v. $\delta^{15}N$ for primary producers (*Spartina*, MPB, and phytoplankton) and three copepod taxa (*Coullana* sp., *Pseudostenhelia wellsi*, and laophontids). Separate plots are shown for each season. $\delta^{15}N$ values of MPB were not measured in this study, but were estimated from a previous study (Carman and Fry 2002). Values are means ± 1 SD (N=4).

Table 2. Uptake of ^{13}C by consumers and MPB in isotope-addition experiments conducted in summer and winter. Values are mean (SD). Although only uptake by copepods is discussed in this paper, several other taxa were examined in the broader experiment. The taxon with the maximum uptake (highest δ^{E} values; shown in bold) in a season was used as the MPB value in the 3-source mixing model (see text). Maximum uptake in summer was observed in *Tanypus clavatus* (a chironomid larva), and maximum uptake in winter was observed in ostracods (mixed assemblage of species).

Consumers	Summer 2001 δ^{E}	Winter 2002 δ^{E}
Coullana sp.	247 (133)	65 (21)
P. wellsi	137 (21)	51 (10)
Laophontid	177 (37)	403 (132)
Ostracod	133 (175)	**897 (190)**
T. clavatus	**531 (19)**	336 (135)
MPB	184 (137)	414 (308)

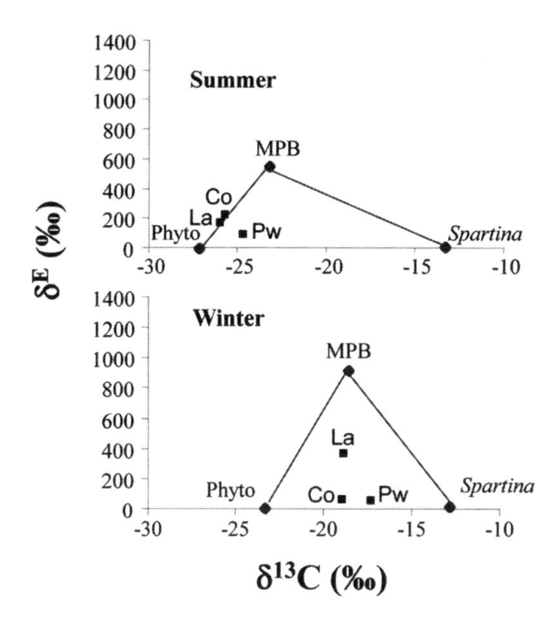

Figure 5. Three-source mixing models used to estimate contributions of MPB, phytoplankton (Phyto), and *Spartina* to the diets of three copepod taxa (*Coullana* sp., *Pseudostenhelia wellsi*, and laophontids). The Y-axis depicts ^{13}C values in copepods and food sources in isotope-enrichment experiments. 'MPB' enrichment values were taken as the maximum value observed in all grazers examined (see text for explanation; values shown in bold in Table 2). The X-axis shows the natural $\delta^{13}C$ values for phytoplankton (Phyto) and *Spartina*. 'Co' = *Coullana* sp., 'Pw' = *Pseudostenhelia wellsi*, and 'La' = laophontids.

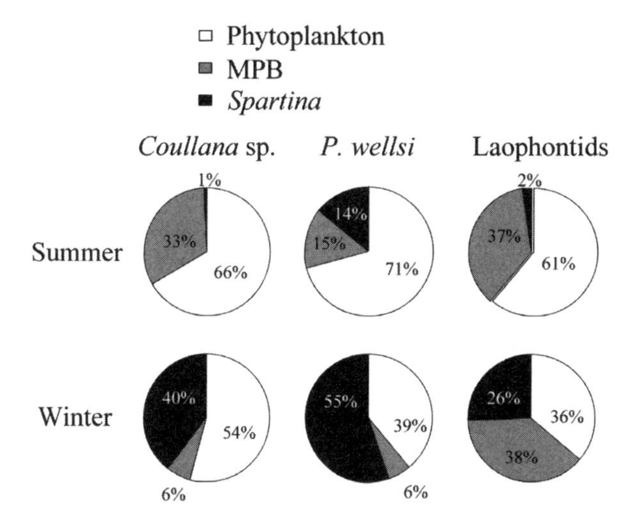

Figure 6. Results of 3-source mixing model showing calculated relative contributions of phytoplankton, MPB, and *Spartina* to the diets of three copepod taxa (*Coullana* sp., *Pseudostenhelia wellsi*, and laophontids) in three seasons. The mixing model is graphically illustrated in Figure 5, and calculations are described in the text.

Discussion

A major limitation in the use of stable isotopes in food-web studies lies in the interpretation of consumer-isotope values that are intermediate between three or more potential food sources. When this situation occurs, it is difficult to determine the contribution of each food source because the isotopic values of consumers may reflect consumption of a single food source or a mixture of two or more food sources. In the current study, the $\delta^{13}C$ isotopic values of potential food sources were generally well separated, with the exception of MPB and phytoplankton in fall. In summer, both single-isotope ($\delta^{13}C$) and dual-isotope ($\delta^{13}C$ v $\delta^{15}N$) plots strongly indicated that phytoplankton was the primary source of nutrition for *Coullana* sp. Natural-isotope values for *P. wellsi* and laophontids also indicated a significant dietary contribution from phytoplankton; however, natural-isotope compositions indicated that the diets of *P. wellsi* and laophontids also consisted of either a substantial amount of MPB or a relatively small amount of *Spartina*. During other seasons, natural $\delta^{13}C$ and $\delta^{15}N$ compositions indicated that the diets of all copepods differed substantially from summer; however, neither single- nor dual-isotope values provided conclusive evidence of copepod diets. Natural-isotope values did, however, indicate that *Spartina* biomass contributed significantly to the diets of all copepods during fall, winter, and spring.

Tracer-addition experiments were designed to assess meiofaunal consumption of MPB, and to help resolve ambiguities in the interpretations of natural-isotope compositions. ^{13}C uptake by copepods in tracer-addition experiments reflect feeding activity over a relatively short (4-h) time, whereas natural $\delta^{13}C$ (and $\delta^{15}N$) values reflect integrated dietary preferences over a longer period (days to weeks). Thus, it is possible that the use of data from tracer-addition experiments in the 3-source mixing model results in a bias toward short-term feeding activity that may not be representative of longer-term feeding habits. Although it is known that meiofaunal feeding activity may vary throughout a diel period (Buffan-Dubau and Carman 2000a), it is not known if meiofauna qualitatively change the type of food that they are consuming over short time scales. Adult *Coullana* females feed most actively at noon (Buffan-Dubau and Carman 2000a), but *Coullana* is negatively phototactic (Harris 1977). It is therefore possible that *Coullana* were not feeding at maximum rates at the time of tracer-addition experiments (midday) due to high light exposure to the cores. However, we note that low ^{13}C uptake in summer relative to winter correlated well with that of natural-isotope values, which indicated a strong dependence on phytoplankton in summer. Uptake of ^{13}C in tracer-addition experiments was assumed to be related to feeding on live, photosynthetically active MPB. The generally low uptake of ^{13}C observed for *P. wellsi* is consistent with the conclusion of Pace and Carman (1996) that *P. wellsi* feeds primarily on detrital algae (based on gut-pigment analysis).

Future studies would benefit from better characterization of $\delta^{13}C$ and δ^{E} values for MPB. As noted in Methods, acetone extracts of sediment include a variety of organic materials that are not from MPB, and thus $\delta^{13}C$ values obtained by this method are almost certainly not an accurate representation of MPB ^{13}C content. Better physical separation of MPB is needed.

Acetone extracts of sediment yielded δ^E values that were lower than the δ^E values of several grazers (Table 2). This observation is contrary to food-web tracer theory, which predicts that isotope concentration in algae that incorporate label must be greater than isotope concentration in consumers that ingest the labeled algae (Daro 1978). We compensated for this problem by using the maximum δ^E values observed in grazers as an estimate of δ^E values in MPB. However, determination of δ^E values for MPB, and thus the accuracy of the 3-source mixing model, would be improved by better separation of MPB.

Also, we generally observed high variability among replicates in the enrichment experiments. To gain statistical confidence in final estimates of food source contributions (e.g., Figure 6), the use of 8-10 replicates would seem advisable in future studies.

Nevertheless, our observations indicate that copepod diets do change substantially over the course of a year. In winter, laophontids showed greater ^{13}C uptake than did *Coullana* sp. and *P. wellsi*,, and the mixing model indicated that MPB contributed more to laophontid diets (38% of total diet) than to the diets of *P. wellsi* or *Coullana* sp. (6% of total diet). For all three copepods, our observations indicated a significant shift from predominant use of microalgal (phytoplankton and possibly MPB) food in summer, to diets that had significant contributions from *Spartina* in other seasons. Thus, we conclude that the diets of copepods do vary during the course of the year. This observation indicates that food-web studies conducted at a single time (typically summer) should be interpreted with caution.

Based on habitat preference and feeding mechanisms, four feeding types of harpacticoid copepods have been identified (Marcotte 1983). *Coullana* sp. belongs to a group that feeds by sorting 3-dimensional food particles (*e.g.*, clusters of diatoms) from the sediment; laophontids belong to a feeding type that feeds on the edges of food particles such as grass blades or sediment particles; and *P. wellsi* belongs to a group that gleans food particles from the surface of detritus and clay floccules. Although, the three harpacticoid copepods technically belong to different feeding groups, our observations do not indicate major differences in their feeding strategies.

However, natural stable-isotope analyses and results of the 3-source mixing model indicate subtle, but potentially important dietary differences among copepods. For example, natural $\delta^{13}C$ values of *P. wellsi* were consistently more enriched than those of other species, and mixing-model results indicated that *P. wellsi* relied more on *Spartina* detritus than did other species.

Conclusions

Copepods exhibited strong temporal variability in their utilization of food sources as well as inter-specific differences in diet. While our observations suggest that algae contribute significantly to the diets of all three copepod taxa, MPB is not necessarily the principal source of algal food. In summer in particular, phytoplankton appears to be a principal source of food. Preliminary mixing-model results suggested that MPB never contributed more than 38% to the diets of copepods. Further, our data suggest that *Spartina* (presumably in the form of detritus) contributes significantly to copepod diets during much of the year. The latter observation contrasts with various food-web studies of estuarine macrofauna, which conclude that primary production from MPB is the principal source of nutrition for primary consumers.

More generally, our observations highlight the difficulties associated with interpreting food webs using natural-isotope analyses when three or more sources of primary production are available. It is relatively straightforward to determine the diets of consumers that have natural-isotope values similar to those of food sources with most enriched or depleted isotope values (phytoplankton or *Spartina* in this study). However, when consumer isotopic values are intermediate, food-source contributions cannot be determined with confidence. We show that labeling of one food source can help resolve ambiguity in interpreting natural-isotope compositions, and suggest that future studies should expand upon this approach by labeling multiple food sources. For example, experimental manipulations with isotopically enriched *Spartina* detritus would be useful for determining more conclusively whether or not *Spartina* contributes significantly to copepod diets, and possibly the diets of other consumers. We note, however, that data obtained from enrichment studies can be highly variable, which in turn will yield mixing-model results with broad confidence intervals. Increased replication would help to remedy this problem.

Acknowledgements

This study was supported by grants from by NSF (OCE-9818453; KC) and Sigma-Xi (PM). J. Fleeger provided helpful comments on the manuscript. S. Silva, S. Mahon, and T. Marshall provided field assistance.

Appendix 1. Notation and calculations for use with enriched samples.

a. Expressing enrichment as δ^E. Most ecologists calculate isotopic enrichment as $\Delta\delta$, subtracting a control δ value for an unenriched sample from the measured δ value for the enriched sample. However, the correct way to express the isotopic contrast between two samples is actually via the atom % notation (explained in section b, below), or using isotopic ratios of the two samples and a variant of the δ definition. Here we express ‰ enrichment as δ^E rather than $\Delta\delta$, with

$$\delta^E = \{[(\delta1+1000)/(\delta2+1000)]-1\}1000.$$

To give an example that shows the difference between enrichments calculated via δ^E vs. $\Delta\delta$, consider an isotope-enriched algal sample has a $\delta^{13}C$ value of 600‰ and a control $\delta^{13}C$ sample of this same species has a value of -21‰. The value for δ^E is 634‰, close to, but not the same as the incorrect $\Delta\delta$ enrichment value of 621‰ obtained by simple subtraction (600-(-21)) = 621‰).

The above formula for δ^E can be derived from the measured δ values of the two samples:

$$\delta1 = [(R1/R) - 1]1000$$
$$\delta2 = [(R2/R) - 1]1000$$

where R is the $^{13}C/^{12}C$ ratio in the standard, and R1 and R2 are the isotopic ratios in the samples. We define δ^E as:

$$\delta^E = [(R1/R2) - 1]1000$$

We can solve these equations for δ^E, first rearranging the above definitions to solve for R1 and R2:

$$R1 = R(\delta 1 + 1000)/1000$$
$$R2 = R(\delta 2 + 1000)/1000$$

The next step is to divide R1 by R2 and cancel R and 1000 values,

$$R1/R2 = (\delta 1 + 1000)/(\delta 2 + 1000)$$

Finally, to obtain δ^E, subtract 1 and multiply the entire result by 1000,

$$\delta^E = \{[(\delta 1 + 1000)/(\delta 2 + 1000)] - 1\}1000.$$

b. Atom % and mixing calculations for highly enriched samples. For samples with added isotope, δ values become increasingly unreliable in simple mixing equations. So, it is better to switch to atom percent that is a direct measure of "% isotope", and use the atom percent values in the normal mixing equations. Spreadsheets make it easy to convert δ values to atom percent values, starting from three equations:

1. the δ definition, $\delta = [(R_{SAMPLE}/R_{STANDARD}) - 1]1000$, where $R = {}^{H}F/{}^{L}F$, F = fractional abundance of the heavy isotope ${}^{H}F$ or light isotope ${}^{L}F$,
2. the sum of the fractions of the light and heavy isotope add to one: ${}^{H}F + {}^{L}F = 1$.
3. atom % heavy isotope = $100 * {}^{H}F$.

These equations can be solved for atom % percent of heavy isotope,

$$\text{atom \% } {}^{13}C = 100*(\delta + 1000)/[(\delta + 1000 + (1000/R_{STANDARD})]$$

where $R_{STANDARD}$ is the known isotope ratio of the standard, e.g., 0.0112372 for carbon isotopes.

Example: an aquatic insect from an isotope enrichment experiment has a $\delta^{13}C$ value of 350‰, and two potential foods measure 80 and 700‰. What are the source contributions for these two foods? Solution: convert the values to atom % values, and substitute the atom percent values for δ value into the normal two source mixing equation used for samples that are not enriched:

$$f = \text{fractional contribution of source } 1 = (\delta_{SAMPLE} - \delta_{SOURCE2})/(\delta_{SOURCE1} - \delta_{SOURCE2}).$$

δ values for source 1, source 2 and the sample are 80, 700, and 350‰ respectively, and the corresponding atom % (%^{13}C) values are 1.19907%, 1.87451% and 1.49435%. The solution is f = (1.49435-1.87451)/(1.19907-1.87451) = 0.5628, the fraction contributed by the 80‰ source 1, and the fraction contributed by the 700‰ source 2 is 1-f, or 0.4372.

We can compare this f = 0.5628 answer to that obtained with the original δ values, f = (350-700)/(80-700) = 0.5645, finding an almost identical result. So, in this case, using the δ values in the mixing equation is acceptable. However, in experiments with much higher isotope enrichments, differences between the two methods start to exceed 0.01 in the final fractional result (e.g., for the case when ^{13}C-enriched values are 80, 3900 and 1711 for source 1, source 2 and the sample, f = 0.5628 when calculated with atom %, but f = 0.5730 when calculated with δ values; the difference in these f values = 0.0102). The results based on atom % are always the correct results, and especially as ^{13}C enrichments become larger than a few thousand ‰, calculations should be based on atom %. (Note: for nitrogen, the point at which differences in f start to exceed 0.01 is higher, about 11,500‰, with this higher limit due to a difference in the isotope value of the N vs. C standard used in the atom % equation above).

References

BLANCHARD, G.F. 1991. Measurements of meiofauna grazing rates on microphytobenthos: is primary production a limiting factor? J. Mar. Biol. Ecol. **147**: 37-46.

BUFFAN-DUBAU, E. and K.R. CARMAN. 2000a. Diel feeding behavior of meiofauna and their relationships with microalgal resources. Limnol. Oceanogr. **45**: 381-395.

— 2000b. Extraction of benthic microalgal pigments for HPLC analyses. Mar. Ecol. Prog. Ser. **204**: 293-297.
CARMAN, K.R., J.W. FLEEGER, and S.M. POMARICO. 1997. Response of a benthic food web to hydrocarbon contamination. Limnol. Oceanogr. **42**: 561-571.
CARMAN, K.R. and B. FRY. 2002. Small-sample methods for $\delta^{13}C$ and $\delta^{15}N$ analysis of the diets of marsh meiofaunal species using natural-abundance and tracer-addition isotope techniques. Mar. Ecol. Prog. Ser. **240**: 85-92.
CARMAN, K.R. and D. THISTLE. 1985. Microbial food partitioning by three species of benthic copepods. Mar. Biol. **88**: 143-148.
COUCH, C.A. 1989. Carbon and nitrogen stable isotopes of meiobenthos and their food sources. Estuar. Coast. Shelf Sci. **28**: 433-441.
CURRIN, C.A, S.C. WAINRIGHT, K.W. ABLE, M.P. WEISTEIN, and C.M. FULLER. 2003. Determination of food support and trophic position of the mummichog, Fundulus heteroclitus, in New Jersey smooth cordgrass (Spartina alterniflora), common reed (Phragmites australis) and restored salt marshes. Estuaries 26: 495-510.
CURRIN, C.A., S.Y. NEWELL, and H.W. PAERL. 1995. The role of standing dead *Spartina alterniflora* and benthic microalgae in salt-marsh food webs: considerations based on multiple stable isotope analysis. Mar. Ecol. Prog. Ser. **121**: 99-116.
DARO, M.H. 1978. A simplified ^{14}C method for grazing measurements on natural planktonic populations. Helg.Wiss. Meeresunters. **31**: 241-248.
DECHO, A.W. 1986. Water cover influences on diatom ingestion rates by meiobenthic copepods. Mar. Ecol. Prog. Ser. **33**: 139-146.
DEEGAN, L.A. and R.H. GARRITT. 1997. Evidence for spatial variability in estuarine food webs. Mar. Ecol. Prog. Ser. **147**: 31-47.
FOGEL, M.L., K.E. SPRAGUE, A.P. GIZE, and W.F. ROBERT. 1989. Diagenesis of organic matter in Georgia salt marshes. Estuar. Coast. Shelf Sci. **28**: 211-230.
FRY, B., W. BRAND, F.J. MERSCH, K. THOLKE, and R. GARRITT. 1992. Automated analysis system for coupled $\delta^{13}C$ and $\delta^{15}N$ measurements. Anal. Chem. **64**: 288-291.
HARRIS, R.P. 1977. Some aspects of the biology of the harpacticoid copepod *Scottalana Canadensis* (Willey) maintained in laboratories culture. Chesapeake Sci. **18**: 245-252.
HERMAN, P.M.J., J.J. MIDDELBURG, J. WIDDOWS, C.H. LUCAS, and C. HEIP. 2000 Stable isotopes as trophic tracers: combining field sampling and manipulative labelling of food resources for macrobenthos. Mar. Ecol. Prog. Ser. **204**: 79-92.
MARCOTTE, B.M. 1983. The imperatives of copepod diversity: perception, cognition, competition and predation. Ph.D. Dissertation, University of South Carolina.
MCCUTHCAN, J.H., W.M. LEWIS, C. KENDALL, and C.C. MCGRATH. 2003. Variation in trophic shift for stable isotope ratios of carbon, nitrogen and sulfur. Oikos **102**: 378 – 390.
MIDDELBURG, J.J., C. BARRANGUET, H.T.S. BOSCHKER, P.M.J. HERMAN, T. MOENS, and C.H.R. HEIP. 2000. The fate of intertidal microphytobenthos carbon: An in situ ^{13}C-labelling study. Limnol. Oceanogr. **45**: 1224-1234.
MILLER, D.C., GEIDER, R.J., and MACLNTYRE, H.L. 1996. Microphytobenthos: The ecological role of the 'secret garden' of unvegetated, shallow-water marine habitats. II. Role in sediment stability and shallow-water food-webs. Estuaries **19**: 202-212.
MOENS, T., C. LUYTEN, J.J. MIDDELBURG, P.M.J. HERMAN, and M. VINCX. 2002. Tracing organic matter sources of estuarine tidal flat nematodes with stable carbon isotopes. Mar. Ecol. Prog. Ser. **234**: 127-137.
MONTGANA, P. 1995. Rates of metazoan, meiofaunal microbivory: A review. Vie Milieu **45**: 1-9.
PACE, M.C. and K.R. CARMAN. 1996. Interspecific differences among meiobenthic copepods in the use of microalgal food resources. Mar. Ecol. Prog. Ser. **143**: 77-86.
PETERSON, B.J. and R.W. HOWARTH. 1987. Sulfur, carbon and nitrogen isotopes used to trace organic matter flow in the salt-marsh estuaries of Sapelo Island, Georgia. Limnol. Oceanogr. **32**: 1195-1213.
PETERSON, B.J. and B. FRY. 1987. Stable isotopes in ecosystem studies. Ann. Rev. Ecol. Sys. **18**: 293-320.
PINCKNEY, J.L., K.R. CARMAN, S.E. LUMSDEN, and S.N. HYMEL. 2003. Microalgal-meiofaunal trophic relationships in muddy intertidal estuarine sediments. Aquat. Microb. Ecol. **31**: 99-108.
RIERA, P., P. RICHARD, A. GREMARE, and G. BLANCHARD. 1996. Food source of inter-tidal nematodes in the Bay of Marennes Oleron (France), as determined by dual stable isotope analysis. Mar. Ecol. Prog. Ser. **143**: 303-309.
SULLIVAN, M.J. and C.A. MONCREIFF. 1990. Edaphic algae are an important component of salt marsh food-webs: evidence from multiple stable isotope analyses. Mar. Ecol. Prog. Ser. **62**: 149-159.

WAINRIGHT, S.C., M.P. WEISTEIN, K.W. ABLE, and C.A. CURRIN. 2000. Relative importance of benthic microalgae, phytoplankton and the detritus of smooth cord grass *Spartina alterniflora* and the common reed *Phragmites australis* to brackish marsh food webs. Mar. Ecol. Prog. Ser. **200**: 77-91.

WRIGHT, S.W., S.W. JEFFREY, R.F.C. MANTOURA, C.A. LLEWELLYN, T. BJORNLAND, D. REPETA, and N. WELSCHMEYER. 1991. Improved HPLC method for the analysis of chlorophylls and carotenoids from marine phytoplankton. Mar. Ecol. Prog. Ser. **77**: 183-196

V. Upscaling primary production

Gérard F. Blanchard, Tony Agion, Jean-Marc Guarini, Olivier Herlory
and Pierre Richard

Analysis of the short-term dynamics of microphytobenthos biomass on intertidal mudflats

Abstract

The short-term dynamics of microphytobenthos on intertidal mudflats was studied over a 3-month period in late Winter-early Spring. Biomass changes during diurnal emersions and between diurnal emersions were assessed and statistically analysed. It turned out that most of the time there were an increase of biomass during diurnal emersions and a decrease of biomass during immersions and nocturnal emersions, but there was always a high variability in the magnitude of the biomass changes. As a whole, the increase of biomass during diurnal emersions and the decrease of biomass between diurnal emersions were significantly higher during Spring tides than during Neap tides, thus pointing out the importance of the coupling between biological and physical processes in the functioning of the primary production system. In addition, biomass-dependent processes were evidenced: there was an inverse relationship between the net biomass accumulation during diurnal emersion and the amount of biomass present at the beginning of the diurnal emersion. This relationship clearly shows that microphytobenthic biomass tends to converge towards an equilibrium point where production equals grazing. Therefore, both the characteristics of the semi-diurnal tidal cycle and the biomass-dependent processes explain a significant part of the observed variability of the short-term biomass changes. Based on these results, a conceptual model of the functioning of the primary poduction system is proposed which strongly indicates that the overall productivity of the system is related to the loss rates (grazing and resuspension).

Introduction

Intertidal mudflats are highly productive ecosystems associated to estuaries and semi-enclosed bays. In western Europe, these geomorphological structures are mostly devoid of macrophytes, but exhibit nevertheless a high primary productivity due to the presence of microphytobenthos. This benthic microalgal compartment is mainly composed of diatoms, closely attached to sand grains (epipsammon) or free and motile in fine muds (epipelon). When the mud is very fine, the dominance of these epipelic diatoms at the surface of mudflats gives rise to a particular and efficient system of primary production. Guarini *et al.* (2000a) have first provided a conceptual representation of this system and have derived from it a mathematical model of the short-term dynamics of the biomass (Guarini *et al.* 2000b; Guarini *et al.* this book).

Schematically, this conceptual model is shown in Figure 1; the production system can thus be described by three different states:

- The first state represents the diurnal emersion phase when the mudflat is exposed to the atmosphere and when sunlight reaches the sediment surface. This state can be characterized by a 2-compartment model: a bottom compartment which is the top 1 cm of the sediment, and a surface compartment which is the epipelic diatom biofilm occurring only during diurnal emersion after the upward migration of microalgae at the surface of the sediment. This biofilm structure has been fully described by Paterson (1989). Primary production only occurs during this phase and is entirely due to the biofilm; the effect of vertical migration on the constitution of the biofilm and on the dynamics of primary production has been clearly demonstrated by Pinckney and Zingmark (1991) and by Serôdio *et al.* (1997). Therefore, a significant increase of microalgal biomass can be expected over the course of emersion if the primary production rate overcomes the grazing rate by dcposivores.
- The second state represents the nocturnal emersion phase when the mudflat is still exposed to the air, but there is no light. Microalgae are dispersed in the bottom compartment after downward migration and there is no biofilm at the surface of the sediment. There is no primary production but there might be grazing during this phase; as a result, microalgal biomass can be expected to decrease.
- The third state represents the immersion phase when the mudflat is covered with turbid water and light does not reach the sediment surface. There is no production, but there is grazing and also the possibility of resuspension of microalgae into the water column (see de Jonge and van Beusekom 1992; de Jonge and van Beusekom 1995, for the resuspension issue). Therefore, a decrease of microalgal biomass can be expected during this phase.

From this theoretical analysis, it is clear that high-frequency variations are likely to be a prominent feature of intertidal microphytobenthic biomass and, as such, reflect the close coupling between biological and physical processes which might contribute to explain the high level of productivity of this epipelic diatom community. Unfortunately, only very few investigations have addressed so far this short time scale (Blanchard *et al.*, 1998; Blanchard *et al.*, 2001; Blanchard *et al.*, 2002), because most previous studies were interested in monthly and seasonal variations (see Underwood and Kromkamp 1999; McIntyre *et al.* 1996; Miller *et al.* 1996; Colijn and de Jonge 1984, for reviews). The available information is nevertheless sufficiently relevant to point out a series of oscillations of the biomass at short-term scale – consistent with the conceptual framework synthesized in Figure 1, due to increases during diurnal emersions and decreases during immersions. As these datasets only concerned particular situations, there is still a need to extend our knowledge to the

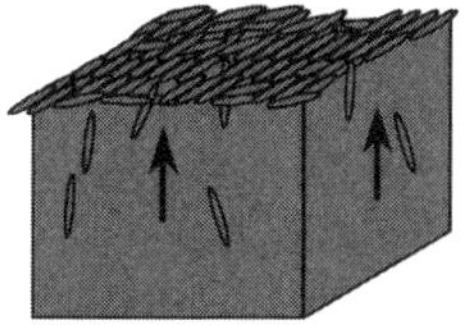

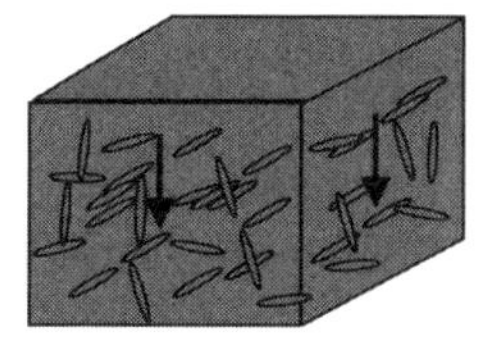

Figure 1. Simple schematic and conceptual representation of the primary production system on bare intertidal mudflats. Three states of the system can be distinguished: diurnal emersion, nocturnal emersion, immersion. For each state, the main processes responsible of the biomass changes are indicated: primary production, grazing, resuspension. Two compartments are taken into account: the surface compartment or biofilm when it is formed only during diurnal emersion; the bottom compartment which is the top 1 cm of the sediment.

full diversity of meteorological and tidal situations encountered on mudflats, and to fully characterize the emergent properties of this intertidal primary production system.

Therefore, this study aims at (i) describing and quantifying the high-frequency variability of intertidal microphytobenthos biomass over a period long enough to encompass different tidal and meteorological situations, (ii) identifying some of the underlying controlling factors, and (iii) bringing out some of the emergent properties of the system.

Material and methods

The study was undertaken in Aiguillon Bay which is located along the French Atlantic coast (47°00' N 1°05' W). It is dominated by bare intertidal mudflats composed of very fine muds. Sampling was carried out during three 16-day periods over 3 months in late Winter-early Spring: 17 February-4 March, 17 March-1 April, 14-29 April, respectively. Each period encompassed a Spring-Neap-Spring tidal cycle. For each of these 3 periods, biomass was measured at the beginning and at the end of every diurnal emersion periods and, for each sampling time (beginning or end of low tides), 6 replicate cores (10 cm inner diameter) were taken within 1 m^2. The same m^2 was sampled repeatedly for 4 days. Biomass was assessed as the mean chlorophyll *a* concentration in the top 1 cm of these 6 cores.

This sampling design allows to calculate, on the same sampling unit, the biomass difference during diurnal emersions and between 2 successive diurnal emersions (i.e. 2 immersions and 1 nocturnal emersion). In parallel, relevant environmental parameters were recorded: irradiance and rainfall during diurnal emersions, wind speed during immersions, temperature.

In Aiguillon Bay, low tide occurs at midday during Spring tides, so that there is only one long diurnal emersion period in the middle of the day with a maximum sunlight input at the surface of the sediment; on the contrary, during Neap tides, high tide occurs at midday so that there are 2 short diurnal emersion periods, one early in the morning and the other one late in the afternoon, with a total amount of light reaching the sediment surface lower than during Spring tides.

Results and discussion

Only 2 other studies dealt with the short-term variations of microphytobenthos biomass on intertidal mudflats (Blanchard *et al.* 1998; Blanchard *et al.* 2002). They clearly pointed out that the dynamics was characterized by series of oscillations corresponding to increases of biomass during diurnal emersions and decreases of biomass during the other phases of the tidal cycle. This is also what we found in the present study (Figure 2). However, as in the previous studies, oscillations are not strictly regular; sometimes, there can be decreases of biomass during diurnal emersions and the magnitude of the biomass changes both during and between diurnal emersions is generally highly variable. It is therefore necessary to analyse thoroughly this variability and to find out which processes and factors may be responsible to be able to elaborate a consistent conceptual model of the functioning and dynamics of the primary production system of intertidal mudflats. The present investigation is the first one to provide such a detailed analysis.

Description and quantification of short-term biomass variations

The average Chl *a* concentration for the 3 periods was 135, 160 and 138 mg m^{-2}, respectively. The difference between the beginning and the end of every diurnal emersion for the 3 periods was calculated and reported in Figure 3A. This histogram thus shows the distribution of net biomass changes (n=48). When there was a net accumulation of biomass during emersion, primary production was higher than grazing; the biomass increase was most of the time between 0 and 30 mg m^{-2}, but it could be up to 60 mg m^{-2}. When there was a net loss of biomass, grazing was higher than production; the net loss was mostly up to 10 mg m^{-2}. The cumulative distribution shows that one third of the observations corresponds to a net loss of biomass while two thirds correspond to a net increase. So, during diurnal emersions of the whole

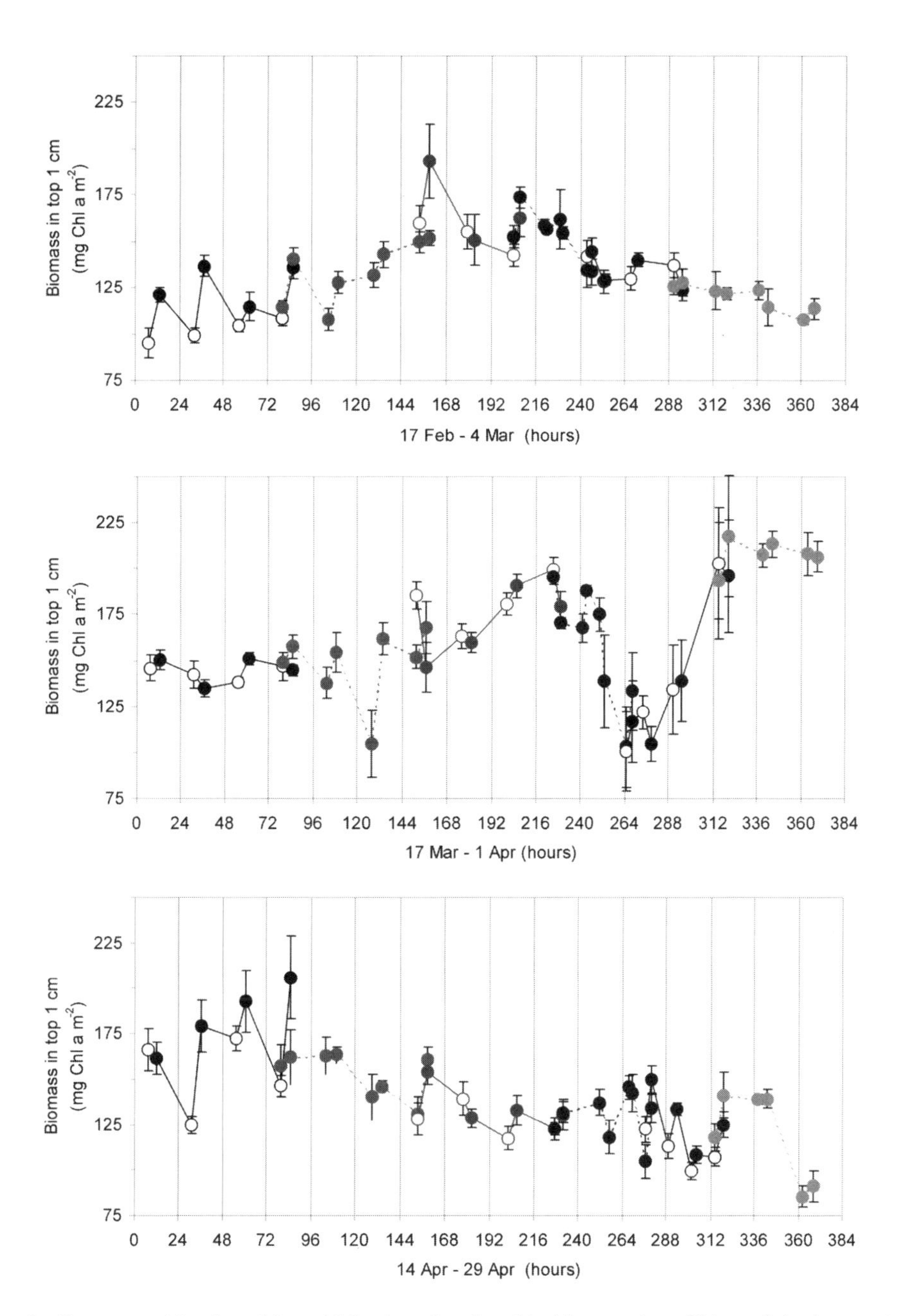

Figure 2. Short-term kinetics of intertidal microphytobenthic biomass (mg Chl *a* m^{-2} in the top 1 cm). Biomass was measured at the beginning and at the end of every diurnal emersions during three 16-day periods: 17 February-4 March, 17 March-1 April, 14-29 April (the X-axis is given in hours, the origin being the first sample taken). Empty circles represent biomass at the beginning of diurnal emersions; full circles represent biomass at the end of diurnal emersions. Solid and dotted lines connect samples taken within the same m^2.

3-month period, biomass increases were more frequent than biomass decreases, but the biomass loss or accumulation was always subject to a great deal of variability.

Likewise, the difference in biomass between successive diurnal emersions (during the 2 immersions and the nocturnal emersion separating 2 diurnal emersions) was calculated and reported in Figure 3B (n=48). The histogram shows that the change can be negative due to grazing and resuspension into the water column, or positive presumably due to sediment advection from other neighboring areas since there was no primary production. Two thirds of the available observations showed a loss of biomass and one third showed an increase. The decrease can be up to about 50 mg m^{-2}. So, most of the time there was a decrease of biomass between diurnal emersions.

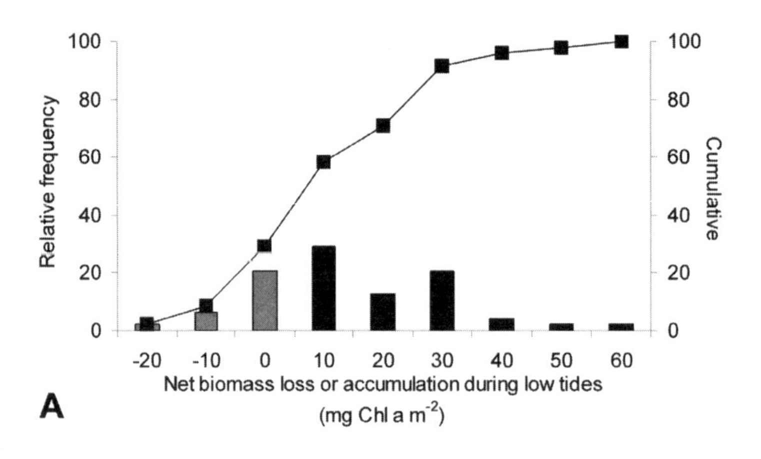

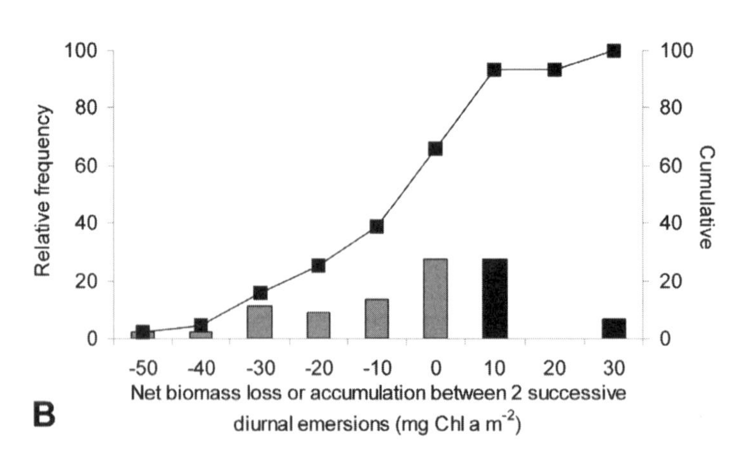

Figure 3. Variability of biomass changes. A. Frequency distribution of the biomass change (mg Chl *a* m^{-2}) during diurnal emersions over the 3-month period (n=48). B. Frequency distribution of the biomass changes (mg Chl *a* m^{-2}) between successive diurnal emersions over the 3-month period (n=48). The right-hand Y-axis gives the cumulative distribution.

Identification of the controlling factors

None of the physical parameters monitored during the sampling period, taken individually, can explain the observed variability of the biomass change during or between diurnal emersions. However, there appears to be an effect of the characteristics of the semi-diurnal tidal cycle – which reflects a combination of all these factors – on the amount of net biomass accumulated or lost during and between diurnal emersions, respectively.

When plotting the biomass changes as a function of the tidal range, it turns out that both the gain of biomass during emersions and the loss of biomass between emersions increase during Spring tides (Figure 4A). Calculations further show (Figure 5) that the average gain of biomass during emersions for large tidal coefficients (>75; 13.0±3.5 mg m^{-2}) was significantly higher (one side t-test, P=0.03) than for small tidal coefficients (<75; 4.6±2.6 mg m^{-2}), and that the average loss of biomass between emersions was significantly higher (one-side t-test, P=0.01) for large tidal coefficients (>75; 13.5±3.1 mg m^{-2}) than for small tidal coefficients (<75; 0.7±4.7 mg m^{-2}).

This may be explained by the fact that during Spring tides diurnal emersions are much longer and occur in the middle of the day, thus receiving a higher input of sunlight at the surface of the mud and increasing the probability of higher rates of primary production by the microphytobenthic biofilm; meanwhile, tidal currents are also higher during immersions, thus increasing the probability of biomass resuspension. As a result, over the course of the 14-day lunar cycle, the difference of the biomass level between emersions and immersions is expected to be higher during Spring tides.

As those biomass changes (during and between diurnal emersions) are functionally related through the tidal cycle, this means that the accumulation of biomass during diurnal emersions was directly proportional to the immediate previous loss during immersions (Figure 4B; P<0.001). This clearly points out the close coupling between biological (primary production) and physical (resuspension) processes.

Besides, apart the direct effects of physical factors, it has also been shown that the dynamics of microphytobenthos in intertidal mudflats can be controlled by biomass-dependent processes (Blanchard *et al.* 2001). This was clearly evidenced in the February and March datasets (Figure 6). When the biomass change during diurnal emersions is plotted against the level of biomass at the beginning of those emersions, we can observe an inverse relationship (P<0.05): the higher the biomass already present, the lower the net accumulation during emersion; there can even be a loss of biomass due to grazing when conditions for primary production are poor. In each case, biomass converges towards an equilibrium value where production is equal to grazing, and which was different from one season to the other (ca. 150 and 185 mg Chl *a* m^{-2} in February and March, respectively). This biomass-dependent process is therefore susceptible to explain a significant part of the observed variability in the level of biomass change during diurnal emersions (Figure 3).

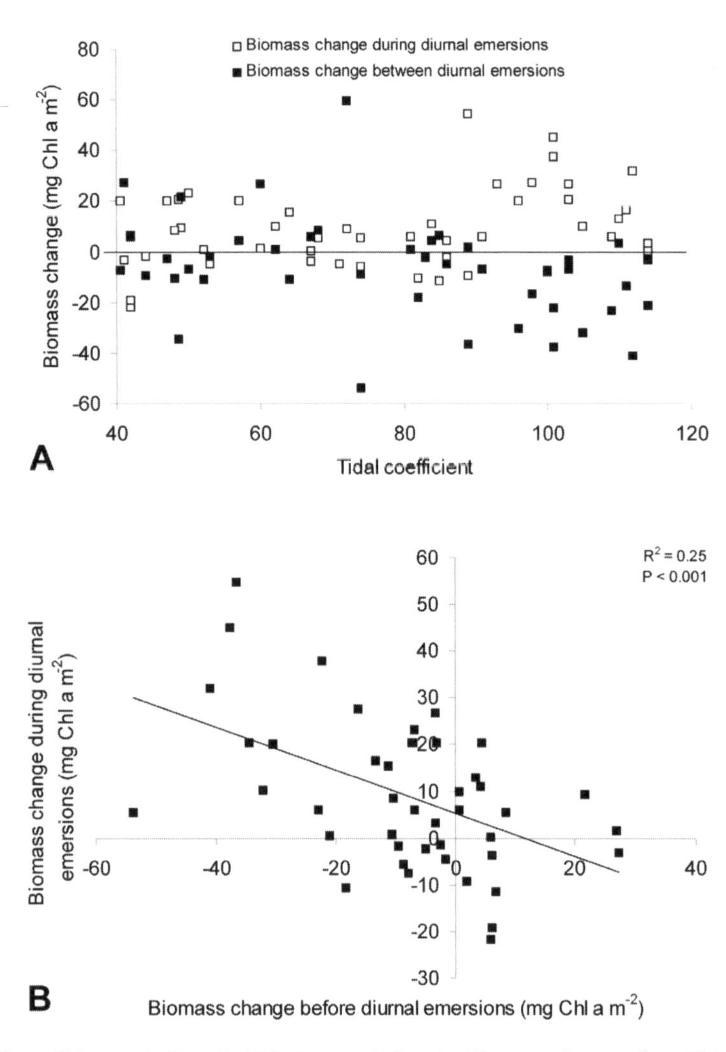

Figure 4. Effect of the semi-diurnal tidal characteristics. A. Biomass changes (mg Chl *a* m^{-2}) during diurnal emersions (empty squares) and between successive diurnal emersions (full squares) over the whole sampling period as a function of the tidal coefficient (index of tidal range: small coefficients for Neap tides and large coefficients for Spring tides). B. Inverse linear relationship ($P<0.001$; $R^2=0.25$) between the change of biomass during diurnal emersions and the change of biomass during the preceeding immersions and nocturnal emersions (mg Chl *a* m^{-2}).

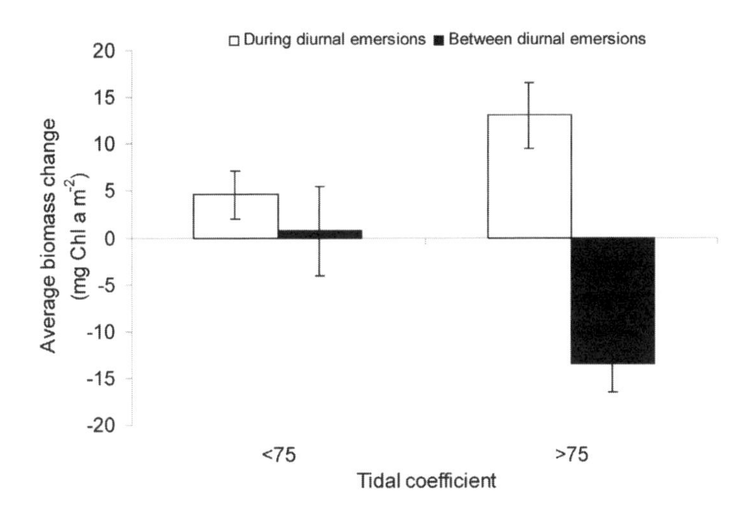

Figure 5. Difference between Neap and Spring tides. Average biomass change (mg Chl *a* m^{-2}) (±SD) during (empty bars) and between (full bars) diurnal emersions for tidal coefficients smaller (including Neap tides) and larger (including Spring tides) than 75, respectively.

Emergent properties of the intertidal primary production system

When looking at the whole data series and taking into account the effect of the semi-diurnal tidal characteristics as well as the occurrence of biomass-dependent processes, general properties of the primary production system on intertidal mudflats emerge.

It is obvious that microphytobenthos biomass converges towards an equilibrium point (Figure 6) with a slope which sets the maximum limit of net production (Figure 7A; theoretically, this slope depends on the value of the equilibrium point and the growth rate of microphytobenthos). At the equilibrium point, production is equal to grazing. The same property had also been evidenced by Blanchard *et al.* (2001) on another mudflat.

The implication of such a property is that there must be a loss of biomass – due to grazing and resuspension – between 2 successive diurnal emersions to pull the biomass away from the equilibrium value, so that production is allowed and can be realized during the next diurnal emersion if the weather conditions are appropriate. This brings the biomass back to its equilibrium point. Therefore, it turns out that the primary productivity of the system is highly dependent on the loss processes, and is fundamentally determined by a close coupling between the physical and biological processes.

This is illustrated by the fact that the gains or losses of biomass on the mudflat are significantly higher during Spring tides than during Neap tides (Figure 4 and 5). Schematically (Figure 7B), the coupling between high primary production and high resuspension rate (as during Spring tides) promotes a productive system since the

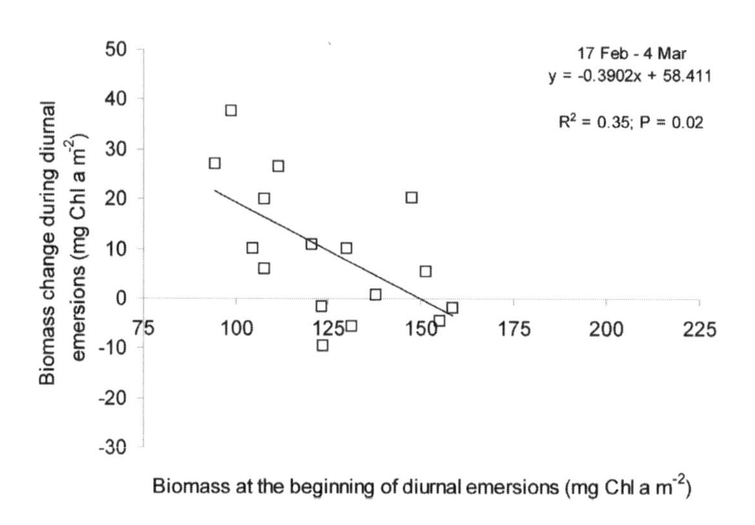

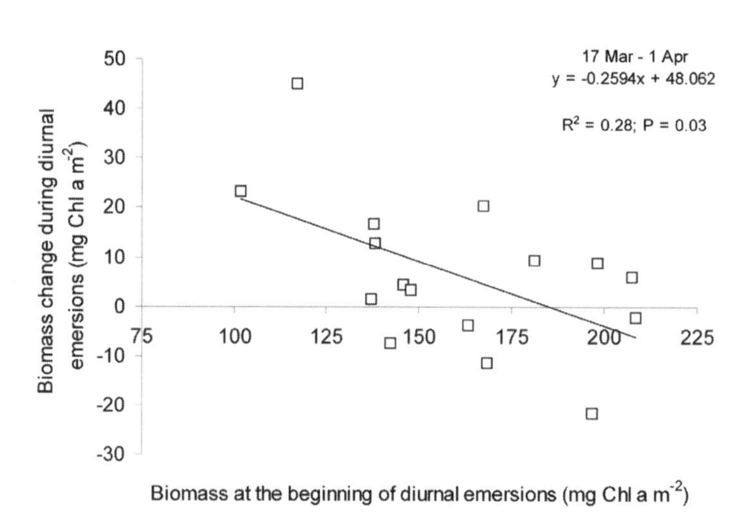

Figure 6. Biomass-dependent process. Relationship between the biomass change during diurnal emersions (mg Chl *a* m^{-2}) and the biomass level at the beginning of diurnal emersions (mg Chl *a* m^{-2}), for the February and March 16-day periods. Both inverse linear relations are statistically significant ($P<0.05$), and the coefficient of determination is indicated in each case.

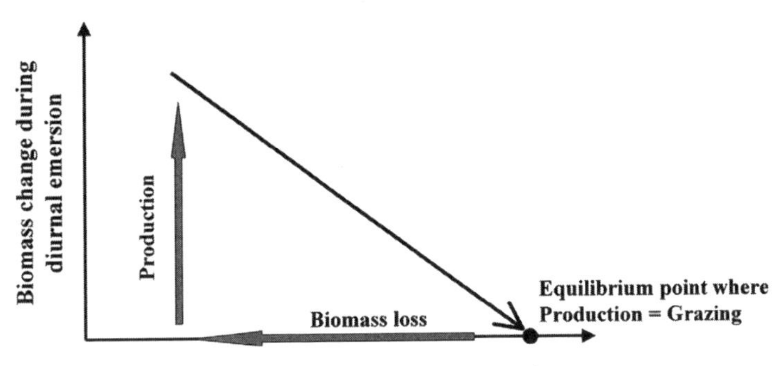

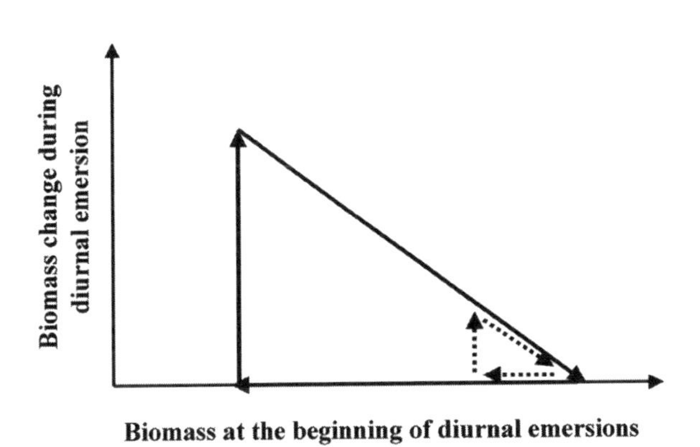

Figure 7. Schematic and conceptual representation of the dynamics of intertidal microphytobenthos primary production, through the coupling of physical and biological processes. A. Net biomass accumulation decreases as the quantity of biomass increases, thus indicating a convergence towards an equilibrium point where production is equal to grazing. There must be a loss of biomass (through grazing and resuspension) to pull it away from its equilibrium point and to allow it to grow again (through primary production). B. Two exemples to illustrate the dynamics of the primary production system (see text for explanations): (i) when there are good conditions for primary production and low resuspension (dashed arrows), the overall productivity is low because biomass remains in the vincinity of its equilibrium point; (ii) when there are good conditions for primary production and high resuspension (solid arrows), the overall productivity is higher because biomass is pulled away from its equilibrium point (loss processes enhance production).

biomass is pulled far away from its equilibrium point and is then brought back quickly (solid arrows). On the opposite, in situations with lower loss rates (even with good irradiance conditions), for instance during Neap tides, the overall productivity of the system is strongly reduced because biomass remains in the vicinity of its equilibrium point (dashed arrows). Intermediate situations can also be considered, for instance when poor conditions for primary production are associated with high loss rates: there would be a rapid and important decrease of the biomass in a few days. The biomass can come back to its initial level also within a few days when meteorological conditions improve.

Conclusion

The main outcome of the present study is that it confirms the existence of an equilibrium point towards which microphytobenthic biomass tends to converge. This is a key point in the understanding of the dynamics of the biomass since it brings a new insight in the interpretation of the short-term variability and proves to be of great ecological value. Indeed, taking into account this equilibrium point together with the coupling between physical and biological processes (succession of production and resuspension phases through the tidal cycle) leads to the elaboration of a conceptual scheme which strongly suggests the importance of loss processes on the system overall productivity. However, in the present state of our knowledge, we cannot separate grazing from resuspension, and we need to know how much of the production goes to grazing and how much to resuspension; the fate of the high microphytobenthic production is therefore a very important line of research for which we still have too little information. Finally, and more generally, the analysis of the biomass dynamics emphasizes the difficulty of upscaling isolated single measurements, because of the strong variability of the biomass at hourly and daily scales.

References

BLANCHARD, G.F., J.-M. GUARINI, C. BACHER and V. HUET. 1998. Contrôle de la dynamique à court terme du microphytobenthos intertidal par le cycle exondation-submersion. C. R. Acad. Sc. Paris **321**: 501-508.

BLANCHARD, G.F., J.-M. GUARINI, F. ORVAIN and P.-G. SAURIAU. 2001. Dynamic behaviour of benthic microalgal biomass in intertidal mudflats. J. Exp. Mar. Biol. Ecol. **264**: 85-100.

BLANCHARD, G.F., B. SIMON-BOUHET and J.-M. GUARINI. 2002. Properties of the dynamics of intertidal microphytobenthic biomass. J. Mar. Biol. Ass. UK **82**: 1027-1028.

COLIJN F. and V.N. DE JONGE. 1984. Primary production of microphytobenthos in the Ems-Dollard Estuary. Mar. Ecol. Prog. Ser. **14**: 185-196.

DE JONGE V.N. and J.E.E. VAN BEUSEKOM. 1992. Contribution of resuspended microphytobenthos to total phytoplankton in the Ems Estuary and its possible role for grazers. Neth. J. Sea Res. **30**: 91-105.

DE JONGE V.N. and J.E.E. VAN BEUSEKOM. 1995. Wind- and tide-induced resuspension of sediment and microphytobenthos from tidal flats in the Ems Estuary. Limnol. Oceanogr. **40**: 766-778.

GUARINI J.-M., G.F. BLANCHARD, PH. GROS, D. GOULEAU and C. BACHER. 2000a. Dynamic model of the short-term variability of microphytobenthic biomass on temperate intertidal mudflats. Mar. Ecol. Prog. Ser. **195**: 291-303.

GUARINI J.-M., G.F. BLANCHARD and PH. GROS. 2000b. Quantification of the microphytobenthic primary production in European intertidal mudflats – a modelling approach. Cont. Shelf Res. **20**: 1771-1788.

GUARINI J.-M., G.F. BLANCHARD and P. RICHARD. 2004. Modelling the dynamics of the microphytobenthic biomass and primary production in European intertidal mudflats. This book.

MACINTYRE H.L., R.J. GEIDER and D.C. MILLER. 1996. Microphytobenthos : the ecological role of the « secret garden » of unvegetated, shallow-water marine habitats. I. Distribution, abundance and primary production. Estuaries **19**: 186-201.

MILLER D.C., R.J. GEIDER and H.L. MACINTYRE. 1996. Microphytobenthos : the ecological role of the « secret garden » of unvegetated, shallow-water marine habitats. II. Role in sediment stability and shallow-water food webs. Estuaries **19**: 202-212.

PATERSON, D.M. 1989. Short-term changes in the erodibility of intertidal cohesive sediments related to the migratory behavior of epipelic diatoms. Limnol. Oceanogr. 34: 223-234.

PINCKNEY, J. and R.G. ZINGMARK. 1991. Effects of tidal stage and sun angles on intertidal benthic microalgal productivity. Mar. Ecol. Prog. Ser. **76**: 81-89.

SERÔDIO, J., J.M. DA SILVA and F. CATARINO. 1997. Non destructive tracing of migratory rhythms of intertdal benthic microalgae using in vivo chlorophyll a fluorescence. J. Phycol. **33**: 542-553.

UNDERWOOD, G.J.C. and J. KROMKAMP. 1998. Primary production by phytoplankton and microphytobenthos in estuaries. In Estuaries. Advances in Ecological Research. D.B. Nedwell and D.G. Raffaelli. Eds. Academic Press, 93-153.

Lawrence B. Cahoon

Upscaling primary production estimates: Regional and global scale estimates of microphytobenthos production

Abstract

Large-scale assessment of microphytobenthic production is a desirable goal for ecosystem modeling and prediction efforts. Several problems confound our ability to extrapolate from published measures to regional and global scale estimates, however. There are no adopted standards for measurement of microphytobenthic biomass and production. Most measurements have been obtained in relatively shallow habitats in temperate habitats, with few measurements from deeper, polar or tropical locations, and large regions of the world lack any measurements. Variability in published estimates of biomass and production, as well as estimates of photo-physiological parameters, is uncomfortably large and limits utility of mean values of any of these. Newer techniques for large scale estimation of biomass and production hold promise, but require ground-truthing with robust in situ measurements. Ultimately, the diverse nature of microphytobenthic assemblages themselves may account for much of the observed variability, and require a more synthetic understanding of the ecology of these organisms before we can generate large-scale estimates of their biomass and production with confidence.

The importance of microphytobenthic production in shallow marine ecosystems, including estuaries and coastal waters, has been well established by many studies throughout the world (see reviews by MacIntyre *et al.* 1996; Miler *et al.* 1996; Cahoon 1999; Underwood and Kromkamp 1999). Microphytobenthic production can equal or exceed phytoplankton production, supports significant secondary production, alters the properties of shallow sediments, and plays a key role in nutrient cycling in these ecosystems. Extrapolation of microphytobenthic production estimates from point or local scales to regional or even global scales of measurement has obvious implications for ecosystem modeling and management. However, efforts to do so lag behind similar efforts to extrapolate phytoplankton production on such large scales, e.g., Longhurst *et al.* (1995).

The problem of estimating microphytobenthos production on regional and global scales encompasses several challenging issues. Two papers have generated global estimates of microphytobenthic production, one of approximately 0.34 Gt (= 10^9 metric tons) C yr^{-1}, based on an average productivity estimate of 50 g C m^{-2} yr^{-1} for the depth interval 0-50 m (Charpy-Robaud and Sournia 1990), the other of approximately 0.5 Gt C yr^{-1}, based on regionally- and depth-weighted production estimates from approximately 108 studies (Cahoon 1999). Large-scale regional estimates, i.e., on the scale of the Baltic Sea, have not been attempted. There have been efforts to estimate microphytobenthic production in more restricted waters, such as individual estuaries (Thornton *et al.* 2002) or fjords (Glud *et al.* 2002), but extension of results from these studies to larger scales has not been attempted with much confidence. Virtually all individual studies of microphytobenthic biomass and production have been limited in their scope of time, space, or methodology, making extension of results to larger scales of time and space problematic.

There are some interesting parallels with the problems of estimating phytoplankton biomass and production at large scales, including the need for development of standard or at least comparable measurement techniques, generation of large sets of reliable and representative measurements, and adoption of methods for extracting predictive physiological information from remotely sensible properties. However, synoptic estimation of the biomass and productivity of the microphytobenthos at large spatial scales of space or time presents significant additional complications inherent in the organisms, their habitats, measurement techniques, and the logistic constraints these necessarily impose.

The ecological characteristics of and differences between estuarine phytoplankton and microphytobenthos have been reviewed thoroughly elsewhere (MacIntyre *et al.* 1996; Miller *et al.* 1996; Underwood and Kromkamp 1999) and do not require extensive elaboration here. A brief summary of the relevant considerations must include the effects of sediment substrates on microphytobenthos biomass and taxonomic composition (Cahoon 1999; Cahoon *et al.* 1999), attenuation of light by both the overlying water column and the sediment itself, considering also the changes in light fields during the emersion/immersion cycle in the intertidal zone (Serôdio and Catarino 1999), patchiness of microphytobenthos at various spatial scales and in response to different factors (Azovsky *et al.* 2000), and variability in physiological rates driven by temperature, light, environmental stress, and behavioral responses (Barranguet *et al.* 1998; Underwood *et al.* 1999; Serôdio and Catarino 2000). These factors complicate efforts to model microphytobenthic primary productivity, although limited attempts have been made with some success (Guarini *et al.* 2000*a*, *b*; Thornton *et al.* 2002).

Lack of standard methodology complicates comparisons among results from different studies of microphytobenthic biomass and production. About 50% of the reported production estimates from estuarine and coastal habitats world-wide have been derived from oxygen exchange measurements. Radiotracer techniques (14-C uptake) have been used in approximately 40% of similar studies. The remainder employed various other methods, including estimates of primary production from relationships between light intensity and chlorophyll biomass (Fuji *et al.* 1991; MacIntyre and Cullen 1996), CO_2 exchange measured by infrared gas analysis (Wilkinson 1981; Schories and Mehlig 2000), and oxygen profiles measured by microprobes (Glud *et al.* 2002). More recently fluorescence-based measures, such as PAM fluorometry (Kromkamp *et al.* 1998; Barranguet and Kromkamp 2000; Serôdio *et al.* 2001; Glud *et al.* 2002; Serôdio 2003) have been employed, although considerable methodological problems remain to be resolved (Perkins *et al.* 2002). Some cross-comparisons of methods have found reasonable agreement (Revsbech *et al.* 1981; Glud *et al.* 2002), but others have shown significant differences between simultaneous measurements using different methods (Perkins *et al.* 2002).

Large-scale geographical coverage of microphytobenthic production measurements is very patchy, with many potentially important areas of the world under-sampled or not sampled at all (Cahoon, 1999; Table 1). Furthermore, the total number of studies worldwide is strikingly small when one considers the numbers of measurements of phytoplankton biomass and production in contrast, e.g., Longhurst *et al.* 1995; Agawin *et al.* (2000). Most studies have been conducted in Europe and North America, for obvious reasons of funding, logistics, and local relevance. In contrast, estuarine and coastal habitats in many portions of the world, particularly polar and tropical regions, have been poorly studied. Large areas with few, if any, studies of microphytobenthic production include the Russian and Canadian Arctic, the Indian Ocean basin, most of the South and Central American coasts, and most of south and east Asia. Consequently, although estimates of microphytobenthos biomass and production in temperate zones may be relatively representative, our knowledge of these parameters in polar and tropical zones is based on an uncomfortably small number of studies in surprisingly few locations (Cahoon 1999 and more recent studies cited here), making accurate global estimates difficult.

Microphytobenthos production and biomass estimates are most confidently established in estuarine and other shallow (<20 m depth) habitats in temperate (30°-60° latitude) waters. On the order of 66 published studies of microphytobenthos production and biomass in such habitats yield average values of approximately 50-100 g C m^{-2} yr^{-1} and 80-130 mg chl*a* m^{-2}, respectively, with lower values from a smaller number of measurements at greater depths. The problem of extrapolating even reasonably approximate estimates of microphytobenthic production in estuaries is illustrated by examining the data for European and North American studies. Using estimates from 26 published studies (cited in Cahoon 1999 or later references; Meyercordt *et al.* 1999; Herman *et al.* 2001) a mean value for annual estuarine production (as reported or as calculated in Cahoon 1999) integrated over all depths reported was 97 g C m^{-2} yr^{-1} with a standard deviation of 102 in European estuaries (including the Baltic and Mediterranean Seas). Similarly, results from 34 studies in U.S. and Canadian estuaries yielded a mean estimate of annual microphytobenthic production of 104 g C m^{-2} yr^{-1}, with a standard deviation of 93 (Cahoon 1999). The substantial variability in these estimates from the most well-studied regions of the world indicate important sources of variation. Fewer (~19) and more spatially clustered studies in tropical (0°-30° latitude) waters yield generally higher values, mean = 527 g C m^{-2} yr^{-1} with a standard deviation of 856, and biomass values of 90-350 mg chl*a* m^{-2}, with lower values at deeper depths. Even fewer studies (~6) have been conducted in polar (60°-90° latitude) waters, yielding production estimates lower than elsewhere, mean = 24 g C m^{-2} yr^{-1} with standard deviation of 20, but relatively high biomass estimates (320-450 mg chl*a* m^{-2}), if sparse and probably unrepresentative. The paucity of microphytobenthos production and biomass data from large regions of the world, including areas where these are likely to be highly significant, makes most regional and global estimates uncertain. Although additional studies similar to those that have been published might usefully increase the confidence of these estimates, improvement would be incremental and subject to the vagaries listed above.

The range of habitats in which microphytobenthic biomass and production have been studied is also very biased. Approximately half of all studies have been conducted in the intertidal zone (Table 1), all but a very few of these in relatively sheltered estuarine ecosystems. The substrates most frequently described in these studies were muddy, and characterized by epipelic forms, in contrast to the microphytobenthic assemblages found on sandier substrates, e.g., Guarini *et al.* (2000a, b). Very few studies have examined microphytobenthos biomass and production in truly exposed habitats (Steele and Baird 1968; Souza and David 1996). As noted above, the intertidal zone imposes significantly different physical constraints on microphytobenthos than subtidal habitats, including enhanced risk of displacement, stronger variation in exposure to light, temperature, and salinity extremes during immersion/emersion cycles, and enhanced exposure to ultraviolet radiation. Therefore it is important to consider the differences in the species composition and physiological responses that may characterize the intertidal and subtidal microphytobenthos in deriving broader scale estimates of biomass and production, e.g., Thornton *et al.* (2002).

Most studies of microphytophytobenthic production have spanned relatively small depth ranges. Only 18 studies have measured production at depth ranges exceeding 5 m, and of these only 9 have spanned a range >5 m including the intertidal zone (References in Cahoon 1999; Kühl *et al.* 2001; Glud *et al.* 2002). Some of these show interesting and potentially useful relationships between depth and production, e.g., Plante-Cuny (1978), Sundbäck (1986), Charpy-Robaud (1988), but they are scattered across global regions and habitat types. Failure to establish sufficiently robust depth relationships for microphytobenthic production limits the reliability of depth-integrated and, therefore, spatially integrated production estimates. However, depth may not be an appropriate proxy for light flux, complicating extrapolation of P-E relationships in shallow waters into deeper waters. Inherent variability in microphytobenthos and their controlling factors, including patchiness, light fields, substrates, community types, and physiological responses, further complicate extrapolation of production measurements and derived models.

Knowledge of the basic photosynthetic physiology of microphytobenthos can provide powerful tools for estimating production with knowledge of basic parameters, i.e., light intensity, biomass (chlorophyll a), and temperature, although microphytobenthos present a more complicated situation than phytoplankton. Some empirical relationships have been described in the literature, e.g., Santos *et al.* (1997), Thornton *et al.* (2002). However, review of published studies ranging from empirical field studies to modeling studies reveals large variations in estimates of basic photosynthetic parameters of microphytobenthos (Table 2). Mean values from individual published studies of the saturating light intensity, E_k, range over 3 orders of magnitude, with the lowest values derived from studies of microphytobenthos in polar regions (Palmisano *et al.* 1985; Rivkin and Putt 1987; Kühl *et al.* 2001). Estimates of the maximum, biomass-normalized photosynthetic rate, P^{β}_{max}, also span approximately

Table 1. Spatial distribution of intertidal and subtidal studies measuring microphytobenthic production by all methods as of 2003. Data are numbers of published studies, from Cahoon (1999) and more recent References cited here.

Basin Sub-basin	Subtidal	Intertidal
Atlantic Ocean	14	31
Gulf of Mexico/Caribbean	8	2
Mediterranean	4	–
Black Sea	–	–
Baltic Sea	3	–
Antarctic Ocean	2	–
Arctic Ocean	4	–
Russian Arctic	–	–
Canadian Arctic	–	–
Indian Ocean	2	1
Persian Gulf	–	–
Pacific Ocean	7	7
Bering Sea	–	–
East Asia	–	2
Gulf of California	3	–
SE Asia-Australia	1	1
New Zealand		1

three orders of magnitude, with some evidence of variation between radiotracer-derived estimates and those derived from dissolved oxygen exchange (DOE) methods (Wolfstein and Hartig 1998). Estimates of the slope of the production-irradiance relationship, α, vary by much less, just somewhat more than one order of magnitude. Finally, estimates of biomass-normalized production, P^{β}, essentially all field measurements, vary by perhaps as much as two orders of magnitude. Much of the variation among the values of these several parameters must inevitably result from the different methodologies and artifacts involved in the respective studies, including different techniques for measuring production, chlorophyll *a*, and light intensity. Some of the variation obviously arises from the differing times and locations at which studies were conducted; some of the ranges of values in Table 2 reflect these differences within individual studies. However, substantive differences among the basic photosynthetic physiological characteristics of microphytobenthos themselves cannot be ruled out. Thus, there is no firm basis for using arbitrarily chosen values of these physiological parameters to upscale estimates of microphytobenthic production from one habitat or season to larger scales of space and time. Such efforts must still rely on empirical measurements to offer much confidence.

Table 2. Values of microphytobenthic photosynthetic parameters reported in published literature: E_k (saturating light intensity, *u*mol photons m^{-2} s^{-1}), P^{β}_{max} (maximum, biomass-normalized photosynthetic rate, mg C mg $chla^{-1}$ h^{-1}), α (slope of P-E relationship, mg C mg $chla^{-1}$ h^{-1} (*u*mol photons m^{-2} $s^{-1})^{-1}$), P^{β} (biomass-normalized photosynthetic rate, mg C mg $chla^{-1}$ h^{-1}).

Reference	E_k	P^{β}_{max}	α	P
Steele & Baird 1968	–	–	–	0.14-1.78
Admiraal & Peletier 1980	–	0.1-13	–	–
Hargrave *et al.* 1983	–	0.1-7	–	–
Rasmussen *et al.* 1983	160-360	–	0.013-0.028	0.16-0.57
Colijn and de Jonge 1984	–	0.43-0.49	–	–
Palmisano *et al.*1985	11	0.21	0.022	–
Mills & Wilkinson 1986	300->750	3.74	–	–
Rivkin & Putt 1987	6	–	0.08-0.10	0.53-0.60
Blanchard & Montagna 1992	108-215	2.98-20.0	0.01-0.16	–
Brotas & Catarino 1995	–	–	0.0021	–
MacIntyre & Cullen 1995	–	–	0.035-0.08	~1-12
Barranguet *et al.* 1998	150-450	3-12.7	0.008-0.042	–
Kromkamp *et al.* 1998	450-1200	–	–	–
Wolfstein & Hartig 1998	56-297	0.66-1.59^{14C} 1.70-4.10^{DOE}	0.011-0.057	–
Uthicke & Klumpp 1998	482-975	–	–	–
Meyercordt & Meyer-Reil 1999	27-367	4.04-54.2	0.017-0.339	–
Meyercordt *et al.* 1999	26-240	4.8-17.0	0.026-0.188	0.5-8.80
Barranguet & Kromkamp 2000	200-400	2-18	0.015-0.035	–
Goto *et al.* 2000	110	0.95	0.008540.4-0.8	
Guarini *et al.* 2000*b*	460	11.2	0.024	–
Miles & Sundbäck 2000	–	–	–	0.26-0.52
Kühl *et al.* 2001	4.6-6.9	–	–	–
Perkins *et al.* 2001	750	0.20	–	–
Glud *et al.* 2002	33	–	–	–
Guarini *et al.* 2002	16-30	2.3-4.7	0.033-0.165	–
Perkins *et al.* 2002	160-420	–	0.021-0.028	–

A significant and growing number of recent studies have examined new approaches to estimation of microphytobenthos distribution, biomass, and production. Optical sensor technologies offer potentially powerful methods to make rapid or even synoptic, large-scale observations of microphytobenthos through reflectance and fluorescence methods. Paterson *et al.* (1998) reported on the use of newer airborne and satellite mounted sensors to quantify microphytobenthic parameters in intertidal habitats. Roelfsema *et al.* (2002) have described use of Landsat imagery to map benthic microalgae in the shallow subtidal zone of a tropical reef ecosystem. Radiation sensors also offer the ability to determine depth, incident PAR, and temperature, other key parameters likely to control production. New approaches, such as use of second-derivative analysis of hyperspectral reflectance data (Stephens *et al.* 2003) and new algorithms for estimating bottom depth from reflectance data (Stumpf *et al.* 2003) may help resolve several obvious limitations of remote sensing methods. However, several significant sources of variation and resolution problems remain to be resolved, including effects of the microtopography of the bottom (Carder *et al.* 2003), confounding of reflectance signals by fluorescence (Mazel and Fuchs 2003) and extracellular polymeric secretions (Decho *et al.* 2003), and the inherent optical properties of the overlying water column (Boss and Zaneveld 2003). Finally, the inherent variability of microphytobenthos must be constrained and, perhaps most important, appropriate ground-truthing techniques must be developed and applied before useful and confident estimates of regional and global scale microphytobenthos production can be derived.

One final perspective must be offered about the philosophy of our approach to the study of microphytobenthos in estuaries and other habitats. Although we tacitly recognize the ecological, physiological, and taxonomic diversity implied in our use of the concept of microphytobenthos as communities of organisms, our methods frequently do not scale appropriately. Those studies that have focused on organism-scale factors and processes have often revealed patterns of variability that confound attempts to generalize to larger scales without incurring broad confidence limits. It is useful to consider the analogy of attempting to quantify the basic properties of the great diversity of terrestrial plant communities. Given the likelihood that microphytobenthos communities are similarly diverse at regional and global scales, the uncertainties in characterizing their biomass and productivity based on limited numbers of measurements in a relatively small number of arguably unrepresentative locations using different techniques that generate highly variable estimates of basic physiological parameters in response to inherently variable environmental factors are not surprising at all.

References

Admiraal, W., Peletier, H. 1980. Influence of seasonal variations of temperature and light on the growth rate of cultures and natural populations of intertidal diatoms. Mar. Ecol. Prog. Ser. **2**: 35-43.

Agawin, N.S.R., C.M. Duarte, and S. Agusti. 2000. Nutrient and temperature control of the contribution of picoplankton to phytoplankton biomass and production. Limnol. Oceanogr. **45**: 591-600.

Azovsky, A.I., Chertoprood, M.V., Kucheruk, N.V., Rybnikov, P.V., and Sapozhnikov, F.V. 2000. Fractal properties of spatial distribution of intertidal benthic communities. Mar. Biol. **136**: 581-590.

Barranguet, C., and Kromkamp, J. 2000. Estimating primary production rates from photosynthetic electron transport in estuarine microphytobenthos. Mar. Ecol. Prog. Ser. **204**: 39-52.

Barranguet, C., Kromkamp, J., and Peene, J. 1998. Factors controlling primary production and photosynthetic characteristics of intertidal microphytobenthos. Mar. Ecol. Prog. Ser. **173**: 117-126.

Blanchard, G.F., and Montagna, P.A. 1992. Photosynthetic response of natural assemblages of marine benthic microalgae to short- and long-term variations of incident irradiance in Baffin Bay, Texas. J. Phycol. **28**: 7-14.

Boss, E., and Zaneveld, J.R.V. 2003. The effect of bottom substrate on inherent optical properties: Evidence of biogeochemical processes. Limnol. Oceanogr. **48**: 346-354.

BROTAS, V., and CATARINO, F. 1995. Microphytobenthos primary production of the Tagus estuary (Portugal). Neth J. Aquat. Res. **29**: 333-339.
CAHOON, L.B. 1999. The role of benthic microalgae in neritic ecosystems. Oceanogr. Mar. Biol. Ann. Rev. **37**: 47-86.
CAHOON, L.B., NEARHOOF, J.E., and TILTON, C.L. 1999. Sediment grain size effect on benthic microalgal biomass in shallow aquatic ecosystems. Estuaries **22**: 735-741.
CARDER, K.L., LIU, C.-C., LEE, Z., ENGLISH, D.C., PATTEN, J., CHEN, F.R., IVEY, J.E., and DAVIS, C.O. 2003. Illumination and turbidity effects on observing faceted bottom elements with uniform Lambertian albedos. Limnol. Oceanogr. **48**: 355-363.
CHARPY-ROBAUD, C.J. 1988. Production primaire des fonds meubles du lagon de Tikahau (atoll des Tuamtotu, Polynésie française). Oceanol. Acta **11**: 241-248.
CHARPY-ROBAUD, C.J., and SOURNIA, A. 1990. The comparative estimation of phytoplanktonic, microphytobenthic, and macrophytobenthic primary production in the oceans. Mar. Microb. Food Webs **4**: 31-57.
COLIJN, F., and DE JONGE, V.N. 1984. Primary production of microphytobenthos in the Ems-Dollard Estuary. Mar. Ecol. Prog. Ser. **14**: 185-196.
DECHO, A.W., KAWAGUCHI, T., ALLISON, M.A., LOUCHARD, E.M., REID, R.P., STEPHENS, F.C., VOSS, K.J., WHEATCROFT, R.A., and TAYLOR, B.B. 2003. Sediment properties influencing upwelling spectral reflectance signatures: The biofilm gel effect. Limnol. Oceanogr. **48**: 431-443.
FUJI, A., WATANABE, H., OGURA, K., NODA, T., and GOSHIMA, S. 1991. Abundance and productivity of microphytobenthos on a rocky shore in southern Hokkaido. Bull. Fac. Fish. Hokkaido Univ. **42**: 136-146.
GLUD, R.N., KÜHL, M., WENZHÖFER, F., and RYSGAARD, S. 2002. Benthic diatoms of a high Arctic fjord (Young Sound, NE Greenland): importance for ecosystem primary production. Mar. Ecol. Prog. Ser. **238**: 15-29.
GOTO, N., MITAMURA, O., and TERAI, H. 2000. Seasonal variation in primary production of microphytobenthos at the Isshiki intertidal flat in Mikawa Bay. Limnol. **1**: 133-138.
GUARINI, J.-M., BLANCHARD, G.F., and GROS, PH. 2000a. Quantification of the microphytobenthic primary production in European intertidal mudflats – a modeling approach. Cont. Shel Res. **20**: 1771-1788.
GUARINI, J.-M., BLANCHARD, G.F., GROS, PH., GOULEAU, D., and BACHER, C. 2000*b*. Dynamic model of the short-term variability of microphytobenthic biomass on temperate intertidal mudflats. Mar. Ecol. Prog. Ser. **195**: 291-303.
GUARINI, J.-M., CLOERN, J.E., EDMUND, J., and GROS, P. 2002. Microphytobenthic potential productivity estimated in three tidal embayments of the San Francisco Bay: A comparative study. Estuaries **25**: 409-417.
HARGRAVE, B.T., PROUSE, N.J., PHILLIPS, G.A., and NEAME, P.A. 1983. Primary production and respiration in pelagic and benthic communities at two intertidal sites in the upper Bay of Fundy. Can J. Fish. Aquat. Sci. **40 (Suppl. 1)**: 229-243.
HERMAN, P.M.J., MIDDELBURG, J.J., and HEIP, C.H.R. 2001. Benthic community structure and sediment processes on an intertidal flat: results from the ECOFLAT project. Cont. Shelf Res. **21**: 2055-2071.
KROMKAMP, J., BARRANGUET, C., and PEENE, J. 1998. Determination of microphytobenthos PSII quantum efficiency and photosynthetic activity by means of variable chlorophyll fluorescence. Mar. Ecol. Prog. Ser. **162**: 45-55.
KÜHL, M., GLUD, R.N., BORUM, J., ROBERTS, R., and RYSGAARD, S. 2001. Photosynthetic performance of surface-associated algae below sea ice as measured with a pulse-amplitude-modulated (PAM) fluorometer and O_2 microsensors. Mar. Ecol. Prog. Ser. **223**: 1-14.
LONGHURST, A., SATHYENDRENATH, S., PLATT, T., and CAVERHILL, C. 1995. An estimate of global production in the ocean from satellite radiometer data. J. Plank.. Res. **17**: 1245-1271.
MACINTYRE, H.L., and CULLEN, J.J. 1995. Fine-scale vertical resolution of chlorophyll and photosynthetic parameters in shallow-water benthos. Mar. Ecol. Prog. Ser. **122**: 227-237.
MACINTYRE, H.L., and CULLEN, J.J. 1996. Primary production by suspended and benthic microalgae in a turbid estuary: time-scales of variability in San Antonio Bay, Texas. Mar. Ecol. Prog. Ser. **145**: 245-268.
MACINTYRE, H.L., GEIDER, R.J., and MILLER, D.C. 1996. Microphytobenthos: The ecological role of the 'secret garden' of unvegetated, shallow water marine habitats. I. Distribution, abundance, and primary production. Estuaries **19**: 186-201.
MAZEL, C.H., and FUCHS, E. 2003. Contribution of fluorescence to the spectral signature and perceived color of corals. Limnol. Oceanogr. **48**: 401.

MEYERCORDT, J., and MEYER-REIL, L.-A. 1999. Primary production of benthic microalgae in 2 shallow coastal lagoons of different trophic status in the southern Baltic Sea. Mar. Ecol. Prog. Ser. **178**: 179-191.

MEYERCORDT, J., GERBERSDORF, S., and MEYER-REIL, L.-A. 1999. Significance of pelagic and benthic primary production in two shallow coastal lagoons of different degrees of eutrophication in the southern Baltic Sea. Aq. Microb. Ecol. **20**: 273-284.

MILES, A., and SUNDBÄCK, K. 2000. Diel variation in microphytobenthic productivity in areas of different tidal amplitude. Mar. Ecol. Prog. Ser. **205**: 11-22.

MILLER, D.C., GEIDER, R.J., and MACINTYRE, H.L. 1996. Microphytobenthos: The ecological role of the 'secret garden' of unvegetated, shallow-water marine habitats. II. Role in sediment stability and shallow-water food webs. Estuaries **19**: 202-212.

MILLS, D.K., and WILKINSON, M. 1986. Photosynthesis and light in estuarine benthic microalgae. Bot. Mar. **29**: 125-129.

PALMISANO, A.C., SOOHOO, J.B., WHITE, D.C., SMITH, G.A., STANTON, G.R., and BURCKLE, L.H. 1985. Shade adapted diatoms beneath Antarctic sea ice. J. Phycol. **21**: 664-667.

PATERSON, D.M., DOERFFER, R., KROMKAMP, J., MORGAN, G., and GIESKE, W. 1998. Assessing the biological and physical dynamics of intertidal sediment ecosystems: A remote sensing approach. Pp. 377-390 in Barthel, *et al.* (eds.) Third European marine science and technology conference (MAST conference), Lisbon, 23-27 May 1998: Project synopses Vol. I.

PERKINS, R.G., OXBOROUGH, K., HANLON, A.R.M., UNDERWOOD, G.J.C., and BAKER, N.R. 2002. Can chlorophyll fluorescence be used to estimate the rate of photosynthetic electron transport within microphytobenthic biofilms? Mar. Ecol. Prog. Ser. **228**: 47-56.

PERKINS, R.G., UNDERWOOD, G.J.C., BROTAS, V., SNOW, G.C., JESUS, B., and RIBEIRO, L. 2001. Responses of microphytobenthos to light: primary production and carbohydrate allocation over an emersion period. Mar. Ecol. Prog. Ser. **223**: 101-112.

PLANTE-CUNY, M.R. 1978. Pigments photosynthétiques et production primaire des fonds meubles néritiques d'une region tropicale (Nosy-Bé, Madagascar). Trav. Doc. L'ORSTOM **96**, Paris.

RASMUSSEN, M.B., HENRIKSEN, K., and JENSEN, A. 1983. Possible causes of temporal fluctuations in primary production of the Danish Wadden Sea. Mar. Biol. **73**: 109-114.

REVSBECH, N.P., JØRGENSEN, B.B., and BRIX, O. 1981. Primary production of microalgae in sediments measured by oxygen microprofile, $H^{14}CO_3$ fixation, and oxygen exchange methods. Limnol. Oceanogr. **26**: 717-730.

RIVKIN, R.B., and PUTT, M. 1987. Photosynthesis and cell division by Antarctic microalgae: Comparison of benthic, planktonic, and ice algae. J. Phycol. **23**: 223-229.

ROELFSEMA, C.M., PHINN, S.R., and DENNISON, W.C. 2002. Spatial distribution of benthic microalgae on coral reefs determined by remote sensing. Coral Reefs **21**: 264-274.

SANTOS, P.J.P., CASTEL, J., and SOUZA-SANTOS, L.P. 1997. spatial distribution and dynamics of microphytobenthos biomass in the Gironde estuary (France). Oceanol. Acta **20**: 549-556.

SCHORIES, D., and U. MEHLIG. 2000. CO_2 gas exchange of benthic microalgae during exposure to air: a technique for rapid assessment of primary production. Wetlands Ecol. Manage. **8**: 273-280.

SERÔDIO, J. 2003. A chlorophyll fluorescence index to estimate short-term rates of photosynthesis by intertidal microphytobenthos. J. Phycol. **39**: 33-46.

SERÔDIO, J., and CATARINO, F. 1999. Fortnightly light and temperature variability in estuarine intertidal sediments and implications for microphytobenthos primary productivity. Aquat. Ecol. **33**: 235-241.

SERÔDIO, J., and CATARINO, F. 2000. Modelling the primary productivity of intertidal microphytobenthos: time scales of variability and effects of migratory rhythms. Mar. Ecol. Prog. Ser. **192**: 13-30.

SERÔDIO, J., MARQUES DA SILVA, J., and CATARINO, F. 2001. Use of *in vivo* chlorophyll *a* fluorescence to quantify short-term variations in the productive biomass of intertidal microphytobenthos. Mar. Ecol. Prog. Ser. **218**: 45-61.

SOUSA, E.C.P.M., and DAVID, C.J. 1996. Daily variation of microphytobenthos photosynthetic pigments in Aparecida Beach – Santos (23°58'48' S, 46°19'00' W), Sao Paulo, Brazil. Rev. Brazil. Biol. **56**: 147-154.

STEELE, J.H., and I.E. BAIRD. I.E. 1968. Production ecology of a sandy beach. Limnol. Oceanogr. **13**: 14-25.

STEPHENS, F.C., LOUCHARD, E.M., REID, R.P., and MAFFIONE, R.A. 2003. Effects of microalgal communities on reflectance spectra of carbonate sediments in subtidal optically shallow marine environments. Limnol. Oceanogr. **48**: 535-546.

STUMPF, R.P., HOLDERIED, K., SINCLAIR, M. 2003. Determination of water depth with high-resolution satellite imagery over variable bottom types. Limnol. Oceanogr. **48**: 547-556.

SUNDBÄCK, K. 1986. What are the benthic microalgae doing on the bottom of Laholm Bay? Ophel. Suppl. **4**: 273-286.

THORNTON, D.C.O., DONG, L.F., UNDERWOOD, G.J.C., and NEDWELL, D.B. 2002. Factors affecting microphytobenthic biomass, species composition and production in the Colne Estuary (UK). Aq. Microb. Ecol. **27**: 285-300.

UNDERWOOD, G.J.C., and J.C. KROMKAMP. 1999. Primary production by phytoplankton and microphytobenthos in estuaries. Adv. Ecol. Res. **29**: 93-153.

UNDERWOOD, G.J.C., NILSSON, C., SUNDBÄCK, C., WULFF, A. 1999. Short-term effects of UVB radiation on chlorophyll fluorescence, biomass, pigments, and carbohydrate fractions in a benthic diatom mat. J. Phycol. **35**: 656-666.

UTHICKE, S., and D.W. KLUMPP. 1998. Microphytobenthos community production at a near-shore coral reef: seasonal variation and response to ammonium recycled by holothurians. Mar. Ecol. Prog. Ser. **169**: 1-11.

WILKINSON, V. 1981. Production ecology of microphytobenthos populations in Manukau Harbour. M.S. thesis, Univ. Auckland, New Zealand.

WOLFSTEIN, K., and HARTIG, P. 1998. The Photosynthetic Light Dispensation System: application to microphytobenthic primary production measurements. Mar. Ecol. Prog. Ser. **166**: 63-71.

Rodney M. Forster and Jacco C. Kromkamp

Estimating benthic primary production: scaling up from point measurements to the whole estuary

Abstract

Good predictions of benthic primary production are needed in order to guide management plans for the estuarine environment, and to understand more about estuarine ecology. This article discusses the advantages and disadvantages of three different methods for estimating net and gross rates of microphytobenthic primary production. Calculations using different modeled profiles of microalgal chlorophyll in the sediment were applied to data from a mid-intertidal site in the Westerschelde estuary. Gross primary production was closely correlated to the daily photon dose received during emersion periods, but there were considerable differences between the three modeling approaches. A model in which chlorophyll was evenly distributed throughout the upper 2 mm of the sediment surface resulted in lower gross production than models in which vertical migration of cells to the sediment surface was simulated. Net production was shown to be highly dependent on the selected value for dark respiratory losses. Respiration was important because much of the benthic chlorophyll is not situated in the photic zone, and because periods of darkness at night and during tidal immersion were longer than the periods of light. Biomass-normalised gross primary production was estimated for a range of different high and mid-shore sites throughout the Westerschelde estuary, and was linearly related to the daily photon dose. Following this, a method is proposed for the mapping of primary production at the whole estuary scale.

Introduction

Estuaries and wetlands are judged to be valuable to mankind because of the high rates of useful biogeochemical activity that they support (Turner *et al.* 2000; Woodward and Wui 2001). However, in many developed countries the total area of estuarine habitat, including intertidal sediments and salt marshes are diminishing, due to land reclamation and port development. For example, in the British Isles, a national loss of mudflats of 25% has been calculated, with some estuaries showing 80% to 100% losses in area (UK Biodiversity Action Plan 2002). Consequently, there may have been major detrimental changes of ecosystem function in the form of lower capacity for nutrient cycling and food production. There is a clear need from a management point of view to be able to better quantify the biological processes occurring in estuaries, in order to guide development plans and if necessary to design replacement habitats of equal value. From a scientific standpoint, better characterization of the key ecosystem biochemical pathways – of which net primary production and denitrification could be considered the most important – is a step towards understanding the ecology and evolutionary adaptations of marine intertidal organisms. The aim of this article is to consider some existing methods for estimating primary production of intertidal sediments, and in particular to identify areas where information is lacking.

Microphytobenthic photosynthesis takes place in a thin layer called the photic zone at the sediment surface. Rates of photosynthesis are affected by many factors, varying from the position of the cells with respect to the gradient of light, the sediment surface temperature, and the availability of sufficient nutrients to build enzymes. Researchers cannot measure all of these factors accurately, and the financial limitations of many monitoring programs may mean that often none can be routinely measured!

Photosynthetic rate measurements usually involve the incubation or measurement of a small amount of sample material, selected from a low number of sites, at a limited number of times in the year. The distribution of MPB can be variable across space and time, and obtaining a representative number of samples can be difficult. Interpolation and kriging techniques are necessary to scale up from point samples to larger scales (Guarini *et al.* 1998). In this respect, estimation of MPB biomass from remote sensing can be very useful.

So what are the options for routine estimation of primary production of mudflats? The tendency in the scientific literature has been for increasing layers of complexity to be added to models, which requires more and more parameters to be measured. Coastal zone managers have taken an alternative approach, for example by using annual averages of MPB biomass at a site as an indicator of the potential level of primary production (de Jong and de Jonge 1995). A similar approach, in which the photosynthetic biomass, the daily dose of light, and the optical depth are combined, seems to work well for phytoplankton in estuaries (Cole and Cloern 1987). Here, we test the results of different models using a detailed series of measurements in the Westerschelde estuary.

Methods and modelling

Experimental dataset

Environmental conditions, microphytobenthic biomass and photosynthetic activity were measured throughout the Westerschelde estuary during 2002 and 2003. For this work, a subset of data was selected for the period May 2003, from an intertidal site, Appelzak, located close to the Belgian Netherlands border. The site was located at the mid-tide level, with mud and fine sands dominating the sediment. Emersion times were calculated for each day by using measured water heights within the estuary and were checked on two occasions by direct observation of the moment when the site was inundated by the flood and exposed again by the ebb. The measured emersion times were found to agree with the predicted times. Due to the strong attenuation of irradiance by the water column of this part of the estuary, photosynthesis was considered to be zero during tidal immersion.

Daily values for air temperature, wind, rainfall and sunshine hours were collated from records at a local meteorological station. Maximum air temperature was used as a proxy for sediment surface temperature in the primary production model (for a more advanced method see Guarini *et al.* 1997). The mean irradiance (PAR; µmol photons m^{-2} s^{-1}) was recorded at hourly intervals by a Licor Li-192 sensor located at the Netherlands Institute of Ecology in Yerseke. Hourly PAR values at the sediment surface were calculated using look-up tables for all periods of emersion during the study month.

Algal biomass (chlorophyll a) was measured in the top 2 mm of the sediment by using the contact coring technique (Ford and Honeywill 2002). All pigment determinations were made using 90% acetone extraction and HPLC separation. Photosynthesis-irradiance curves were measured on suspensions of microphytobenthos in filtered estuary water using radiocarbon uptake. Carbon fixation rates were normalized to the algal biomass of the suspension, and simultaneous least-squares fitting was used to fit the cardinal parameters of photosynthesis-irradiance curves (P_{max} and α, Figure 1). The effect of temperature on microphytobenthic photosynthesis was not

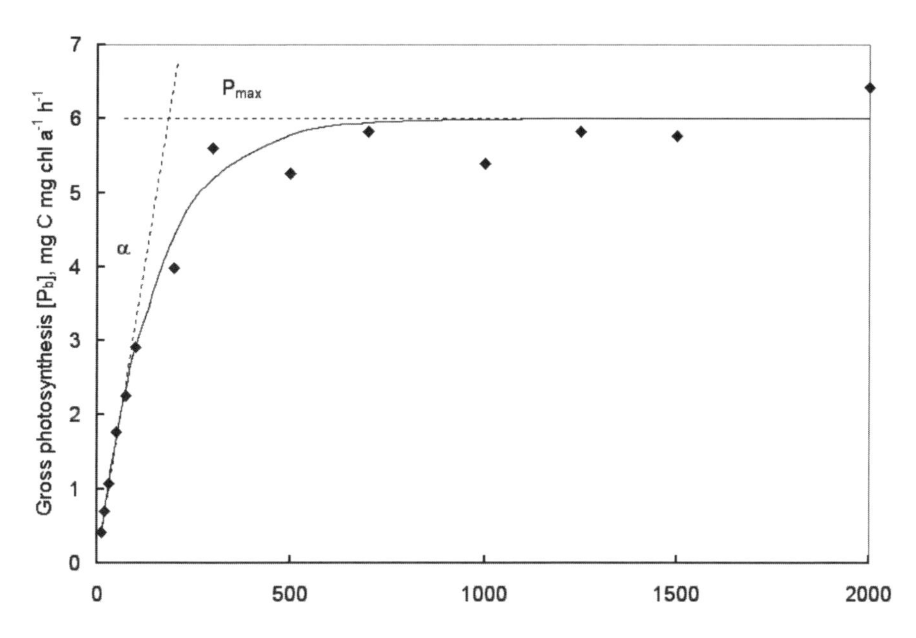

Figure 1. The relationship between gross photosynthetic carbon fixation and irradiance for microphytobenthic cells suspended in seawater. P_{max} is the maximum rate of photosynthesis at light saturation, and α is the efficiency of light use at subsaturating irradiance.

measured in this study, but has been examined in detail by Blanchard *et al.* (1996, 1997) and by Morris and Kromkamp (2003). These results showed clearly that α is independent of temperature, and that P_{MAX} has a unimodal relationship with temperature. The temperature response equations and coefficients of Blanchard *et al.* (1997) were used to convert P_{max} from the measurement temperature (20 °C) to values that would occur at the SST of the site.

As no published data is available on the respiratory loss rates of MPB, we assumed that this would vary in proportion to metabolic activity. Accordingly, R was set to a fixed percentage of the temperature-adjusted P_{max}, which should be related to the growth rate and activity of the MPB cells (Geider 1992). Excretion of 'EPS' and other loss processes were beyond the scope of this modelling exercise.

Approaches to modeling primary production

A number of different production models have been described in the more recent literature. All of these models feature a functional separation between the total MPB biomass, and a subfraction of the biomass that is exposed to light at the sediment surface. This fraction is usually termed the photosynthetically-active biomass, or PAB (e.g. Serôdio *et al.* 1997). There is widespread agreement that the ratio between total biomass and PAB is determined by the migratory behaviour of diatom cells, and that this ratio is to some extent predictable (Pinkney and Zingmark 1991). Probably the most thoroughly tested of the current generation of benthic production models is that of Blanchard and colleagues. Their model has been tested against real observations of MPB populations in mesocosm experiments and field data, and its development can be seen in these papers: Blanchard *et al.* (2001), Guarini *et al.* (2000a,b), and elsewhere in this volume. The basis of this model is shown in Figure 2 – when the conditions are right (e.g. during daylight, during low tide) there is a movement of MPB to the surface. The surface of the sediment is seen as a box, or pool, which can be filled up with cells. Once full, there is no room for any further increases in PAB. Only the cells in this box can photosynthesise and add to that day's primary production. Photosynthetic rates in the PAB box are calculated from photosynthetic rate parameters measured on suspensions of MPB (Figure 1). This model corresponds very well to the latest electron microscope images of sediments surfaces (Herlory *et al.* 2004), in which a layer of cells is clearly seen lying on top of the sediment. The amount of chlorophyll in the box is fixed at 25 mg m^{-2}, and any chlorophyll in excess of this amount is deemed non-photosynthetically active. However, the total chlorophyll content in the upper layers of the sediment will often greatly exceed this value, for example we have recorded values in Westerschelde sediments of up to 400 mg chl a m^{-2} in the top 2 mm of the sediment.

Macintyre *et al.* (1995) and Barranguet *et al.* (1998) used a different approach to model daily primary production at their intertidal sites. These authors used essentially the same techniques for modelling sediments as others have used for measuring primary production in a mixed water column. In these sediment optical models, an attenuation coefficient determines the penetration of irradiance into the sediment. Algal cells are uniformly distributed with depth, with rates of photosynthesis per unit

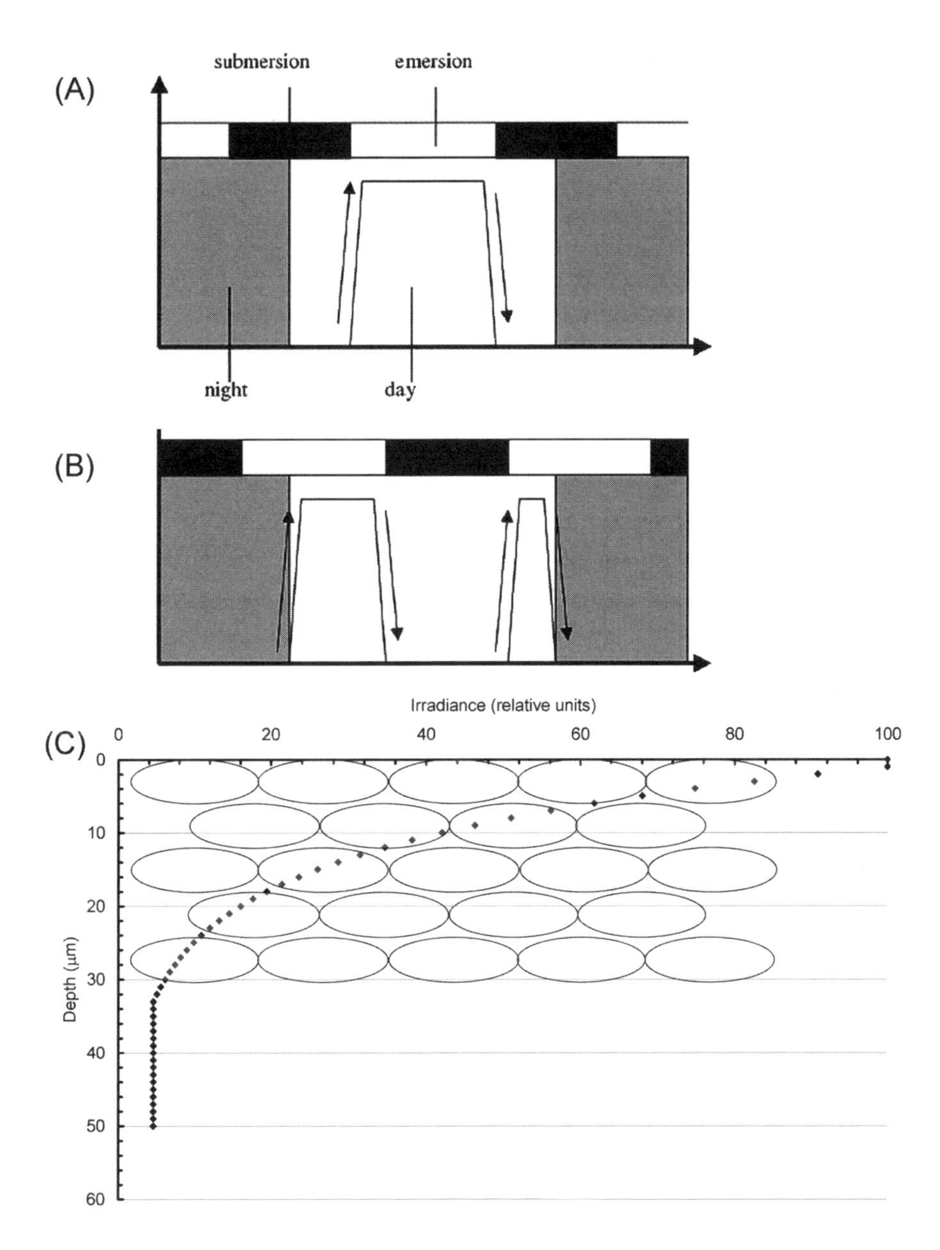

Figure 2. Conceptual model of the changes in photosynthetically-active biomass (PAB) in microphytobenthos. (A) Cells migrate to the photic zone at the surface of the sediment during an emersion period centred around noon (B) Upward and downward migration of cells during two short daylight emersion periods in early morning and late afternoon (modified after Guarini *et al.* 2000a) (C) Close packing of diatom cells at the sediment surface and the associated decrease in irradiance within the sediment.

chlorophyll again defined by laboratory measurements on dilute suspensions. The interaction of irradiance and photosynthetic parameters, together with the amount of chlorophyll allows the rate of photosynthesis to be calculated at each depth. Integration over depth and time gives the daily gross primary carbon fixation.

However, there are other data that argue against the use of the 'surface pool' or 'uniform profile' models. If the surface of the sediment is frozen, then sectioned with a very fine technique, called cryo-slicing, which gives layers of sediment only 0.1 mm thick, then a quantitative distribution of chlorophyll with depth can be made. Details of this method can be found in papers by de Brouwer and Stal (2001), Kelly *et al.* (2001) and Honeywill *et al.* (2002). The results showed that not all of the chlorophyll was at the surface, nor was there a uniform distribution with depth (except at night, after downward migration). In fact, the most common result of cryo-slicing was an exponentially-decreasing gradient of chlorophyll with depth.

Gross and net primary production calculations

Which of these descriptions of the microphytobenthic environment is correct, and what are the implications for modelling primary production? Forster and Kromkamp (2004) gave a description of irradiance conditions within cohesive, muddy sediments in relation to different profiles of chlorophyll. Primary production values for three of the possible chlorophyll profiles given in that paper are calculated here – these are the UNI profile (corresponding to a uniform distribution of chlorophyll with depth as in Macintyre *et al.* 1995, Barranguet *et al.* 1998), the SML profile (a surface monolayer of cells, Guarini *et al.* 2000a), and the KY profile (an exponentially decreasing gradient of chlorophyll with depth, Kelly *et al.* 2001).

Daily gross primary production was calculated for the Appelzak field site during May 2003 for each of the profiles, using a depth interval of 10 μm and a time step of 10 min. Additional calculations were made for the integral net primary production by subtraction of respiratory losses, and the effect of using different values for respiration was tested.

Results

The daily photon dose at the sediment surface during emersion periods varied from 6 to 32 mol photon m^{-2}, driven by meteorological and tidal conditions. Changes in irradiance appeared to be the main cause of variability in daily gross primary production (P_g) (Figure 3). Day-to-day variability in P_g was high, for example see the large differences between the 5th and 6th of May 2003. Production rates were highest at the beginning and end of the month, corresponding to periods when low water was centred on the local solar noon. Mean P_g at the mid-intertidal site varied from 270 mg C m^{-2} d^{-1} to 490 mg C m^{-2} d^{-1} according to the type of model used (Table 1). The UNI and SML models give similar, lower values of P_g, whereas KY gave consistently higher values.

The net daily primary production (P_n) models were run with a dark respiration rate that was equivalent to 5% of the maximum photosynthetic rate. These runs also

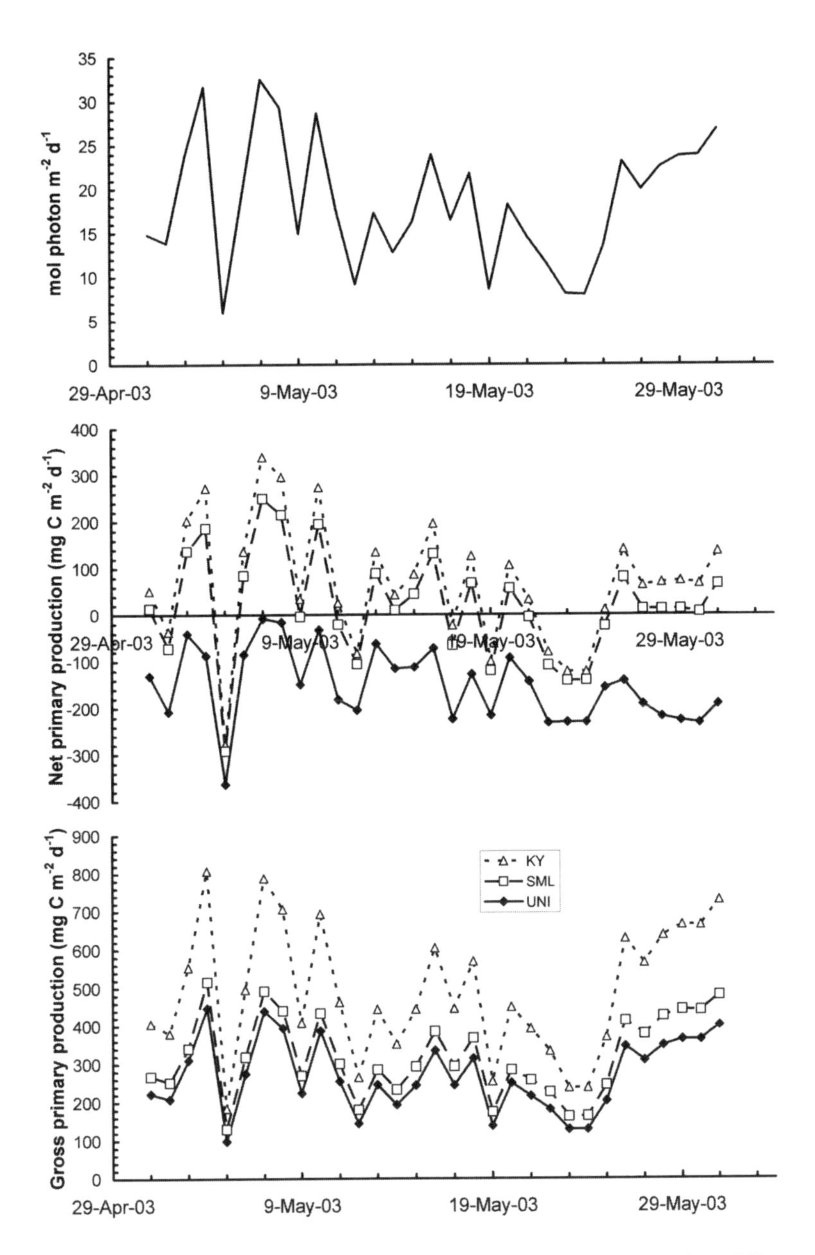

Figure 3. Calculations of primary production at Appelzek in May 2003 using different modeling approaches. (A) Daily photon dose received at the intertidal site (B) Net primary production using the three different chlorophyll profiles (C) Gross primary production.

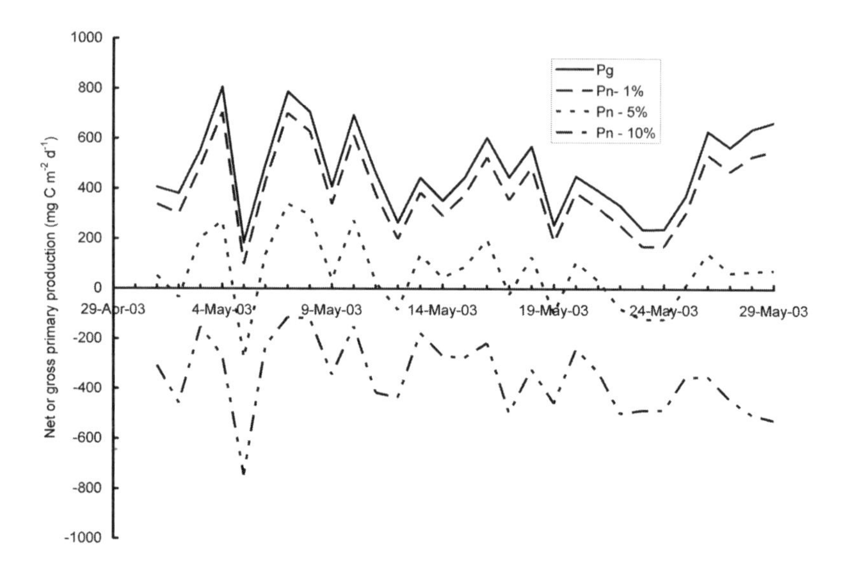

Figure 4. The effect of dark respiration rate on calculated daily net primary production at Appelzak during May 2003. Net primary production at respiration rates of 1%, 5% and 10% of the maximum photosynthetic capacity are shown in comparison to the gross primary production.

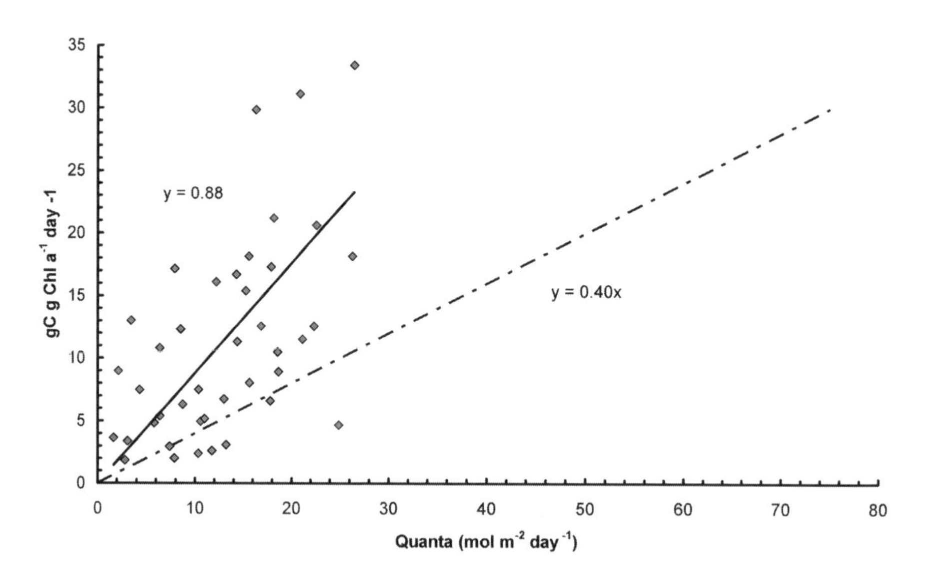

Figure 5. The relationship between biomass-normalised gross daily production and irradiance for microphytobenthic samples from the Westerschelde estuary collected during 2002 and 2003. The linear regression for MPB (y=0.88x) is compared with a similar estimate of light use efficiency for phytoplankton (y=0.40x) made by Falkowski *et al* (1994).

showed the same ordering of values as for P_g, with UNI<SML<KY, but now SML and KY showed the best agreement. The UNI model predicted negative rates of P_n for each day of May 2003, whereas negative production occurred only on days with low incident irradiance in the other two models. The biweekly periodicity caused by the spring-neap tidal cycle was less obvious in the graph of P_n. Net primary production was only weakly correlated with photon dose, whereas gross production was more closely related to the variability in incident irradiance.

The calculation of P_n was highly sensitive to the respiration rate, as shown in Figure 4 where the KY model calculations are shown for three different respiratory loss rates. Increasing the respiration rate successively from 1% to 5% then 10% greatly decreased the net production, and dampened the sensitivity of production to irradiance.

The previous examples were calculated for a site with a constant algal biomass, and constant photosynthetic parameters during the month of observations. Calculations of P_g were made for samples from other areas in the Westerschelde estuary, where a range of chlorophyll concentrations and photosynthetic parameters were encountered. The KY chlorophyll profile was used to model gross production at each site. After normalisation to the amount of biomass, there was a linear dependence of P_g on the daily photon dose (Figure 5). The slope of the relationship, which describes the light use efficiency of an intact microphytobenthic assemblage, was higher than the efficiency previously calculated for phytoplankton in various water columns (Falkowski *et al.* 1994).

Discussion

A model can only be as accurate as its underlying data and algorithms, and it is important to remain aware of the potential for multiplication of errors as the complexity of the model increases. In the case of microphytobenthic production estimates, whilst there are some factors that can be accurately quantified (e.g. incident irradiance), there are others for which little information is available (e.g. respiration rates). The input parameters that can be most accurately measured or predicted are the irradiance and temperature at the sediment surface during emersion, the total microalgal biomass, and the photosynthetic activity of a sample in suspension. The latter can be estimated using a variety of techniques such as uptake of radiocarbon label, oxygen evolution, or PAM fluorescence (see chapters by Glud, and Kromkamp and Forster elsewhere in this volume).

The more difficult to measure parameters are those that define the actual exposure of benthic microalgal pigments to light. A quantitative description of the chlorophyll-depth profile in the sediment can be used if available, together with a description of the attenuation of irradiance in the sediment. Alternatively, the detection of chlorophyll fluorescence from cells at the sediment surface can be used as a proxy for the level of PAB; this method has been pioneered and validated in the work of Serôdio and colleagues (Serôdio *et al.* 1997, 2001; Serôdio 2003). All three model scenarios tested here had equal amounts of microalgal chlorophyll in the upper 2 mm of sediment – the only difference was in the shape of the chlorophyll-depth profiles. The exponentially-decreasing chlorophyll profile (KY) gave the highest gross and net production. In this profile, more of the algal chlorophyll is exposed to the light than in the uniform profile (UNI), where the deeper strata of chlorophyll receive no light. The SML profile also had chlorophyll placed in the optimum position to capture photons, but the pool size of PAB was in this case smaller than for the KY profile, hence the lower production rates per unit area of sediment.

A further uncertainty in the calculation of primary production occurs when models such as the ones used here are unconstrained by the supply of dissolved inorganic carbon, the substrate for photosynthesis. Situations can arise when the DIC demand from a photosynthesizing biofilm will greatly exceed the pool of DIC in the porewater and overlying surface water. This is likely to occur when dense surface biofilms are exposed to prolonged periods of high light. The relatively weak isotopic fraction observed between ^{13}-C and ^{12}-C which has been observed in virtually all studies of MPB to date suggests that DIC limitation may be widespread. Inhibition of photosynthesis by high porewater concentrations of oxygen may also occur. In both case, the observed photosynthetic rate would be lower than the maximum rate determined on a dilute suspension. Preventing this overestimation of primary production will require the additional parameterization of concentration gradients and diffusion coefficients in the sediment photic zone (shown previously by Ludden *et al.* 1985). It is likely that the differences between the different models will be reduced if DIC limitation is taken into account, as this will affect the KY and SML models more than the UNI model.

Net primary production is perhaps the most relevant variable to consider when working on the ecology of tidal flats. P_n can be closely related to the growth rate of MPB if the conversion factor from units of carbon to chlorophyll is known. However, the calculation of P_n is highly dependent upon the dark respiration value used. Respiration rates are difficult to measure, and are known to be both irradiance- and temperature-dependent. Therefore, quantification of respiration rates as well as other loss or organic carbon such as exudation, should be a priority area for research. From the modeling of P_n made here, for a mid-tidal site with equal emersion and immersion periods, and also from studies in which we have modeled the growth of biofilms in mesosocms experiments, we suggest that respiration (and other losses) are not likely to be more than 5% of the maximum photosynthetic rate. A respiration rate as low as 1.5% of P_{max} was required in order to balance a growth model of phytoplankton in the Westerschelde estuary (Kromkamp and Peene 1995). It is feasible that respiration rates similar to, or lower than this could occur in MPB due to oxygen limitation in the deeper sediment layers during immersion.

The good correlations observed between biomass-normalised gross production and daily photon dose across a range of intertidal sites show that general rules can be applied to benthic production models. The higher light use efficiency observed for MPB than for phytoplankton is a direct result of diatom vertical migration, which maximizes photon capture by placing cells at the surface of the sediment. Phytoplankton in a mixed water column must compete with water, cDOM and other attenuating compounds, thus water columns have a lower photosynthetic efficiency compared to

Table 1. A summary of the modelling results for net and gross primary production at Appelzak during May 2003. For each unit of production, the daily mean and monthly sum are shown. The three chlorophyll profiles, KY, SML and UNI are described in the text.

		P_n Sum	Mean	P_g Sum	Mean
		mg C m^{-2}	mg C m^{-2} d^{-1}	mg C m^{-2}	mg C m^{-2} d^{-1}
Model	KY	2076	67	15182	490
	SML	570	18	9906	320
	UNI	-4723	-152	8370	270

cohesive sediments. Given data on the photon dose and biomass for a given site, it should be possible to predict P_g with reasonable accuracy. The photon dose can be calculated for any site if a digital elevation model of the estuary together with surface irradiance measurements and tidal curves are available. Synoptic maps of microphyto-benthic biomass can be obtained from remote sensors such as Landsat, so it should be possible in the near future to routinely map primary production at the estuary scale.

Acknowledgements

This work was funded by the European Union 5th Framework project 'EVK3-CT-2001-00052 A system of hierarchical monitoring methods for assessing changes in the biological and physical state of inter-tidal areas (HIMOM)'. This is NIOO publication number 3544.

References

Barranguet, C., J.C. Kromkamp, and J. Peene 1998. Factors controlling primary production and photosynthetic characteristics of intertidal microphytobenthos. Mar. Ecol. Prog. Ser **173**: 117-126.

Blanchard, G.F., J.M. Guarini, P. Gros, and P. Richard 1997. Seasonal effect on the relationship between the photosynthetic capacity of intertidal microphytobenthos and temperature. J. Phycol. **33**: 723-728.

Blanchard, G.F., J.M. Guarini, F. Orvain, and P.G. Sauriau 2001. Dynamic behaviour of benthic microalgal biomass in intertidal mudflats. Journal Of Experimental Marine Biology And Ecology **264**: 85-100.

Blanchard, G.F., J.M. Guarini, P. Richard, P. Gros, and F. Mornet 1996. Quantifying the short-term temperature effect on light-saturated photosynthesis of intertidal microphytobenthos. Mar. Ecol. Prog. Ser **134**: 309-313.

Cole, B.E., and J.E. Cloern 1987. An empirical model for estimating phytoplankton productivity in estuaries. Mar. Ecol. Prog. Ser **36**: 299-305.

de Brouwer, J.F. C., and L.J. Stal 2001. Short-term dynamics in microphytobenthos distribution and associated extracellular carbohydrates in surface sediments of an intertidal mudflat. Mar. Ecol. Prog. Ser **218**: 33-44.

de Jong, D.J., and V.N. de Jonge 1995. Dynamics and distribution of microphytobenthic chlorophyll-a in the western Scheldt estuary (SW Netherlands). Hydrobiologia **311**: 21-30.

FALKOWSKI, P.G., R. GREEN, and Z. KOLBER. 1994. Light utilization and photoinhibition of photosynthesis in marine phytoplankton, p. 409–434. *In* [eds.] N.R. Baker and J.R. Bowyer, Photoinhibition of photosynthesis from molecular mechanisms to the field. BIOS.

FORD, R.B., and C. HONEYWILL 2002. Grazing on intertidal microphytobenthos by macrofauna: is pheophorbide a a useful marker? Mar. Ecol. Prog. Ser **229**: 33-42.

FORSTER, R.M., and J.C. KROMKAMP 2004. Modelling the effects of chlorophyll fluorescence from subsurface layers on photosynthetic efficiency measurements in microphytobenthic algae. Mar. Ecol. Prog. Ser **284**: 9-22.

GEIDER, R.J. 1992. Respiration: taxation without representation, p. 333-360. *In* [eds.], P.G. Falkowski and A.D. Woodhead Primary Productivity and Biogeochemical Cycles in the Sea. Plenum.

GUARINI, J.M., G.F. BLANCHARD, C. BACHER, P. GROS, P. RIERA, P. RICHARD, D. GOULEAU, R. GALOIS, J. PROU, and P.-G. SAURIAU 1998. Dynamics of spatial patterns of microphytobenthic biomass: inferences from a geostatistical analysis of two comprehensive surveys in Marennes-Oleron bay (France). Mar. Ecol. Prog. Ser **166**: 131-141.

GUARINI, J.M., G.F. BLANCHARD, and P. GROS 2000a. Quantification of the microphytobenthic primary production in European intertidal mudflats – a modelling approach. Cont. Shelf Res. **20**: 1771-1788.

GUARINI, J.M., G.F. BLANCHARD, P. GROS, D. GOULEAU, and C. BACHER 2000b. Dynamic model of the short-term variability of microphytobenthic biomass on temperate intertidal mudflats. Mar. Ecol. Prog. Ser **195**: 291-303.

GUARINI, J.M., G.F. BLANCHARD, P. GROS, and S. J. HARRISON 1997. Modelling the mud surface temperature on intertidal flats to investigate the spatio-temporal dynamics of the benthic microalgal photosynthetic capacity. Mar. Ecol. Prog. Ser **153**: 25-36.

HERLORY, O., J.M. GUARINI, P. RICHARD, and G.F. BLANCHARD 2004. Microstructure of microphytobenthic biofilm and its spatio-temporal dynamics in an intertidal mudflat (Aiguillon Bay, France). Mar. Ecol. Prog. Ser **282**: 33-44.

HONEYWILL, C., D. M. PATERSON, AND S. E. HAGERTHEY 2002. Determination of microphytobenthic biomass using pulse- amplitude modulated minimum fluorescence. Eur. J. Phycol. **37**: 485-492.

KELLY, J.A., C. HONEYWILL, and D.M. PATERSON 2001. Microscale analysis of chlorophyll-a in cohesive, intertidal sediments: the implications of microphytobenthos distribution. J. Mar. Biol. Ass. U K **81**: 151-162.

KROMKAMP, J., and J. PEENE 1995. Possibility of net phytoplankton primary production in the turbid Schelde estuary (SW Netherlands). Mar. Ecol. Prog. Ser **121**: 249-259.

LUDDEN, E., W. ADMIRAAL, and F. COLIJN 1985. Cycling of carbon and oxygen in layers of marine microphytes; a simulation model and its eco-physiological implications. Oecologia **66**: 50-59.

MACINTYRE, H.L., and J.J. CULLEN 1995. Fine-scale vertical resolution of chlorophyll and photosynthetic parameters in shallow-water benthos. Mar. Ecol. Prog. Ser **122**: 227-237.

MORRIS, E.P., and J.C. KROMKAMP 2003. Influence of temperature on the relationship between oxygen and fluorescence-based estimates of photosynthetic parameters in a marine benthic diatom (*Cylindrotheca closterium*). Eur. J. Phycol. **38**: 133-142.

PINCKNEY, J., and R. ZINGMARK 1991. Effects of tidal stage and sun angles on intertidal benthic microalgal productivity. Mar. Ecol. Prog. Ser **76**: 81-89.

SERÔDIO, J. 2003. A chlorophyll fluorescence index to estimate short-term rates of photosynthesis by intertidal microphytobenthos. J. Phycol. **39**: 33-46.

SERÔDIO, J., J.M. DA SILVA, and F. CATARINO 1997. Nondestructive tracing of migratory rhythms of intertidal benthic microalgae using in vivo chlorophyll a fluorescence. J. Phycol. **33**: 542-553.

SERÔDIO, J., J.M. DA SILVA, and F. CATARINO 2001. Use of in vivo chlorophyll a fluorescence to quantify short- term variations in the productive biomass of intertidal microphytobenthos. Mar. Ecol. Prog. Ser **218**: 45-61.

TURNER, R.K., J.C.J.M. VAN DEN BERGH, T. SODERQVIST, A. BARENDREGT, J. VAN DER STRAATEN, E. MALTBY, and E.C. VAN IERLAND 2000. Ecological-economic analysis of wetlands: scientific integration for management and policy. Ecological Economics **35**: 7-23.

WOODWARD, R.T., and Y.S. WUI 2001. The economic value of wetland services: a meta-analysis. Ecological Economics **37**: 257-270.

UK BIODIVERSITY ACTION PLAN (2002) online at www.ukbap.org.uk/reporting.

VI. Microphytobenthos and benthic-pelagic exchange

Graham J. C. Underwood and Mark Barnett

What determines species composition in microphytobenthic biofilms?

Abstract

Despite significant advances over the past few years in our understanding of biofilm physiology and functioning, we still have a very patchy and unclear picture of what influences the species composition of biofilms. Much of our understanding of biofilm functioning comes from studies that have not treated species composition as a significant factor. Conversely, many studies concerned with species distribution have not encompassed our understanding of the dynamic nature of biofilms. Nutrients are a strong contender for an environmental variable that influences species composition. Yet, much of the available data comes from mechanistic floristic surveys that do not really inform as to the driving mechanisms determining species composition. So does the local nutrient environment really select for particular taxa? Can we separate out the effects of nutrients, sediment particle size, species physiology and other, perhaps, unmeasured, covariables that help to define the synecological niche? In this paper we give a brief review of some of the main areas that influence species composition, and look at approaches where these different issues have been addressed.

Introduction

Microphytobenthos play an important role in the ecology of many coastal habitats. Abundant biofilms are present on intertidal sand and mudflats (Underwood and Kromkamp 1999), within saltmarsh creeks and amongst saltmarsh plants (Sullivan 1999), and, where water clarity is sufficient to permit light penetration, on subtidal sediments (Sundbäck and Jönsonn 1988; Underwood 2002). The primary production of microphytobenthos ranges between 50-875 g C m^{-2} a^{-1} and the activity of the films on the sediment surface influences nutrient transformations within the sediments (nitrification, denitrification) and nutrient exchange across the sediment-water interface (Underwood and Kromkamp 1999; Thornton *et al.* 1999; Dong *et al.* 2000; Rysgaard-Petersen 2003).

Microphytobenthos consist of a number of phylogenetic groups, with diatoms, cyanobacteria and euglenophytes being the most common. Diatoms are ubiquitous and are usually dominant in terms of biomass, but there are situations when other photosynthetic microorganisms are more abundant (Cahoon 1999). This paper will generally discuss patterns in diatom abundance, where there is a reasonable literature. Our knowledge of the dynamics of the other algal groups in microphytobenthos is patchy in comparison, and more research is needed on the role of other taxonomic groups (Sullivan and Currin 2000). The diatom composition of microphytobenthic mats changes both seasonally and spatially within estuaries. Spatial changes are usually related to the salinity (and associated nutrient) gradients within estuaries. From such studies some general patterns concerning species distribution are revealed (Sullivan 1999; Underwood and Kromkamp 1999). Yet, intercomparison between studies is often difficult due to differences in the major taxa found in different estuaries, or differences in nomenclature and identification by different authors. The high level of synonymy in benthic diatom taxonomy and the problems of standardisation of taxonomy between different sets of literature also confound these problems.

There is a requirement to understand the niches of estuarine diatoms. It has been known since the 1970's that the species within a biofilm do not function identically. Round (1979) found differences in the pattern of migration of different species within biofilms in Mass. Harbour. Different taxa of microphytobenthos change their position within the biofilm in response to a number of variables, of which probably the best studied is irradiance. *Euglena* and diatoms rapidly change position at the surface of sediments as light intensities increase and decrease (Paterson *et al.* 1998; Perkins *et al.* 2002; Kingston 1999). Single cell PAM fluorescence imaging has shown that *Euglena* has significantly higher operating efficiencies than diatoms within the same biofilm at higher irradiances (Oxborough *et al.* 2000), and such a strategy for *Euglena* fits with its observed pattern of occurrence, being commoner during the summer months (Underwood 1994). Different taxa migrate to the surface of a biofilm at different times throughout a diel tidal exposure and different species exhibit different photosynthetic efficiencies at the same irradiance (Perkins *et al.* 2002; Underwood *et al.* 2005). This indicates that species may function quite differently. By accepting the premise that the species composition influences biofilm function (Underwood *et al.* 2005), it becomes important to be able to predict what species we might find in a particular environment. Can this be done with any certainty?

Accurate description of the niche of organisms has been a problem in biology for a long period. Charles Darwin, in his published *Journal of the Voyages of the Beagle* (first pub 1839), wrote,

> 'we do not steadily bear in mind, how profoundly ignorant we are of the conditions of existence of every animal, nor do we always remember, that some check is constantly preventing the too rapid increase of every organism being left in a state of nature. ...we feel so little surprise at one, of two species closely allied in habits, being rare and the other abundant in the same district, or again, that one *(species)* should be abundant in one district and another, filling the same place in the economy of nature, should be abundant in another district, differing very little in its conditions. If asked how this is, one immediately replies that it is determined by some slight difference in climate, food or the number of enemies: yet how rarely, if ever, we can point to the precise cause and manner of action of the check. We are, therefore, driven to the conclusion, that causes generally quite unappreciable by us, determine whether a given species shall be abundant or scanty in numbers.' *(Darwin 1860).*

Despite the major advances in our understanding of biofilm functioning in the last 20 years (since Admiraal, 1984), these cautionary words of Darwin are highly valid when applied to our current understanding of the distribution and abundance of microphytobenthic taxa.

Large scale differences in MPB species composition between estuaries

At what spatial scale do differences in biofilm species composition in estuaries occur? Diatom assemblages from the Severn, Colne and Blackwater estuaries, U.K. were compared. These studies took place over a number of years; Severn estuary 1990-1991, Blackwater 1992-1995, Colne estuary 1996-1998, and further details are given in Underwood (1994; 1997; 2000) and Thornton *et al.* (2002). The studies all had a similarity of approach and taxonomic standardisation permitting intercomparison (Table 1). Multivariate analysis was applied to relative abundance (RA) data and also to biomass-weighted RA data (RA of each taxon in a sample multiplied by the sediment Chl *a* concentration for that sample.

PCA of relative abundance data

Principal components 1-4 derived from relative abundance data for the diatom assemblages for the three estuaries explained 54.9% of the variation in the data set (Table 2). The data for each of the estuaries sampled generally clustered together, with the Colne and Blackwater (Northey Island) estuaries separated from the Severn estuary by PC1, with PC2 separating Colne from Blackwater samples (Figure 1). The major gradient (PC1) was due to high relative abundances of *Navicula pargemina*, *N. flanatica*, *Nitzschia epithemioides*, *C. signata* and *Rhaphoneis* found in the Severn estuary samples, compared to high RA's of *N. phyllepta*, *N. salinarum*, *Plagiotropis vitrea* and other species found in the Colne-Blackwater complex (Figure 1). Lower shore samples from the Blackwater estuary showed a greater overlap with Colne estuary mudflat samples than those from the upper saltmarsh. Colne mudflat samples tended to have higher RA of *Plagiotropis, N. gregaria*, *Staurophora amphioxys* and *N. rostellata*, while the saltmarsh sites in the Blackwater estuary were characterised by higher RAs of *Nitzschia frustulum, N. sigma, C. closterium* and *Amphora marina* (Figure 1).

Principal component 1, derived from species abundance data, was significantly negatively correlated (Table 2) with sediment ash free dry weight, chl *a*, and total and colloidal carbohydrate concentrations (all environmental variables that represent a gradient of increasing organic content, nutrient availability and microalgal biomass in the sediments), and negatively correlated with salinity. The same environmental variables (with the addition of temperature) were also negatively correlated with PC2. Thus *N. phyllepta*, *N. frustulum*, *Cylindrotheca closterium* and *N. salinarum* show greater relative abundances in more eutrophic sites and at higher temperatures, while *C. signata*, *N. epithemiodes* and *N. pargemina* also favour warmer conditions, but have higher relative abundances in less eutropic conditions. The following diatom taxa, *Raphoneis minutissima*, *N. flanatica*, *Coscinodiscus* may thus be representative of lower temperatures, higher salinity and lower microphytobenthic biomass and nutrient conditions.

Table 1. Median (minimum-maximum) relative abundance (RA) values of benthic diatom species (taxa occurring at least once at over 1% RA) and median (min-max) values for environmental variables for 3 different U.K. estuaries.

Taxon	**code in Figs1-4**	**Severn**	**Blackwater**	**Colne**
Navicula phyllepta Kütz.	NPHY	2 (0-64)	35 (5-75)	17 (0-97)
N. pargemina Underwood & Yallop	NPAR	23 (0-97)	5 (5-75)	ND
N. flanatica Grun.	NFLA	0 (0-58)	5 (5-35)	5 (0-64)
N. salinarum Grun.	NSAL	0 (0-5)	5 (5-75)	0 (0-51)
N. cincta Ehrenb.	NCIN	0 (0-2)	15 (5-75)	0 (0-12)
N. rostellata Kütz	NROST	0.5 (0-20)	5 (5-75)	2 (0-48)
N. digitoradiata (Greg.) A. Schmidt	NDIGIT	0 (0-1)	5 (5-35)	0 (0-13)
N. gregaria Donkin	NGREG	2 (0-59)	5 (5-75)	3 (0-57)
N. diserta Hustedt	NDIS	ND	5 (3-35)	0 (0-10)
Fallacia pygmaea (Kütz.) Stickle & D.G. Mann	FPYG	0 (0-1)	ND	0, (0-10)
Amphora salina W. Smith	ASAL	0 (0-2)	5 (5-35)	0 (0-4)
A. marina W. Smith	AMAR	ND	5 (5-35)	ND
Nitzschia sigma (Kütz) W. Smith	NSIG	0 (0-3)	5 (5-75)	ND
Nitzschia frustulum (Kütz) Grun.	NFRU	1 (0-65)	15 (5-75)	1, (0-9)
Nitzschia epithemiodes Grun.	NEPI	4 (0-82)	ND	ND
Amphiprora paludosa W. Smith	AMP	0 (0-23)	ND	ND
Cylindrotheca closterium (Ehrenb.) Reimann & Lewin	CCLOS	0 (0-76)	15 (5-75)	ND
Cylindrotheca signata Reimann & Lewin	CSIG	1 (0-94)	ND	ND
Staurophora amphioxys (Greg.) D.G. Mann	STAURO	0 (0-50)	ND	0 (0-31)
Pleuosigma angulatum (Quekett) W. Smith.	PANG	0 (0-21)	5 (5-75)	0 (0-7)
Gyrosigma limosum Sterrenburg et Underwood	GLIM	0 (0-19)	ND	0 (0-82)
Plagiotropis vitreae (W.Smith) Cleve.	PLAG	ND	ND	5, (0-100)
Raphoneis minutissima Hustedt	RAPH	4 (0-37)	ND	0, (0-35)
Cymatosira belgica (Grun.) Van Heurck.	CBELG	ND	ND	0, (0-25)
Coscinodiscus sp.	COSC	2 (0-26)	ND	ND
water content %		60 (31-79)	68, (34-84)	ND
ADFW %		10 (3-20)	15 (4-43)	ND
salinity ‰		29 (15-39)	33 (16-210)	33 (8-34)
temp °C		16 (1-28)	19 (4-30)	13 (4-24)
Chl *a* (μg g^{-1})		11 (1-206)	80 (9-295)	40 (3-219)
total carbohydrate (mg g^{-1})		19.6 (2-46)	30.1 (45-131)	2.7 (0.5-14)
colloidal carbohydrate (μg g^{-1})		250 (0-5334)	3055 (584-6345)	35 (4-2568)

Table 2. The percent variation explained by principal components (PC) 1-4 of the relative abundance (RA) and biomass-weighted relative abundance (RA^B) of microphytobenthic diatom assemblages from the Severn, Colne and Blackwater estuaries and the correlation between each PC and environmental variables. Values in **bold** are significant at $p < 0.05$ or less.

	% var.	AFDW	salinity	Temp	Chl *a*	Total carbo.	colloidal carbo.
RA							
PC1	27.6	**-0.46**	**-0.32**	0.07	**-0.57**	**-0.34**	**-0.70**
PC2	10.8	**-0.36**	**-0.32**	**-0.34**	**-0.42**	**-0.30**	**-0.59**
PC3	9.0	0.01	-0.08	**0.40**	**0.27**	**0.12**	**0.14**
PC4	7.5	0.10	0.10	**0.12**	0.06	0.01	0.05
$\mathbf{RA^B}$							
PC1	22.8	**0.603**	**0.318**	0.085	**0.731**	**0.538**	**0.797**
PC2	10.8	-0.043	**-0.113**	**0.263**	**0.333**	**0.158**	0.092
PC3	9.8	**0.357**	**0.344**	**0.118**	**0.175**	**0.198**	**0.367**
PC4	5.8	**0.133**	**0.180**	0.024	**0.211**	**0.126**	0.107

The clustering of samples by estuary evident in PC1 and PC2 was not present with principal components 3 and 4. PC3 and 4 explained an additional 16.5% of the variation in the data set, and samples from the 3 estuaries showed substantial overlap. Temperature, sediment Chl *a* content and total and colloidal carbohydrate content all correlated positively with PC3, and temperature with PC4. Thus the main gradient across Figure 2 (from lower left to upper right) is a seasonal gradient of increasing temperature and biomass. The species vectors can be placed in 3 groups, low biomass-temperature winter taxa (*Raphoneis* , *N. flanatica*, *Coscinodiscus*, *N. frustulum*, a mid temperature-biomass range of taxa (*N. cincta*, *N. sigma*, *N. pargemina*, *P. angulatum*, *Stauroneis* and *Plagiotropis*) and summer taxa (*N. phyllepta*, *N. gregaria*, *C. signata* and *Nz. epithemioides*). This major gradient present in species composition is seen across all the three estuaries investigated, as can be seen from the scatter of samples from each estuary across the graph.

PCA of biomass-modified RA data

Modification of the relative abundance data by a biomass factor resulted in a less pronounced separation of samples with PC1 and PC2 (Figure 3) into clusters for each estuary (as was seen in Fig 1), and an increase in scatter for all estuarine data. PC1 and 2 explained 33.6% of the data set, with the Blackwater and Colne samples intermingled, and some overlap between the Severn, Colne and Blackwater data. The incorporation of biomass into the analysis factor removed the low biomass taxa (*Raphoneis*, *Coscinodiscus*, *Cymatosira* etc.) from the list of significant species relating to PC 1 and 2, with increasing significance of a number of commonly occurring

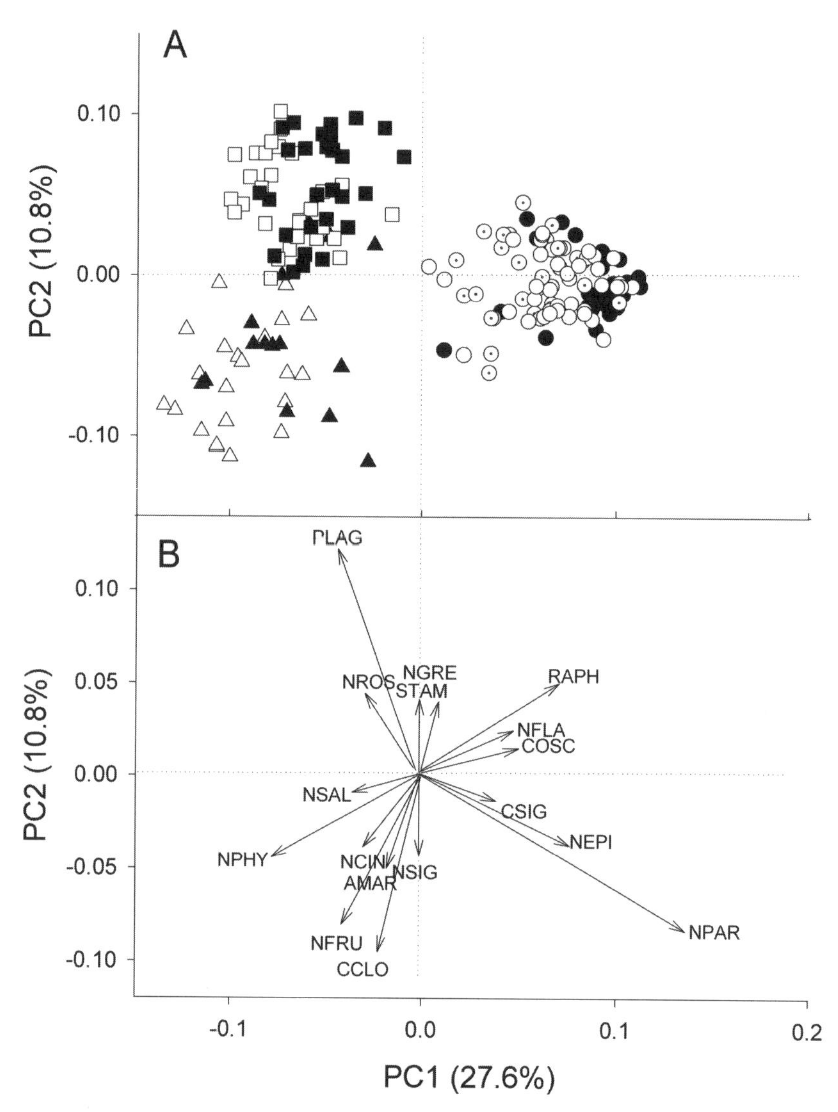

Figure 1. (A). Principal components analysis scatter plot of components 1 (PC1) and 2 (PC2) derived from relative abundance values of microphytobenthic diatom species composition in samples from sites in 3 U.K. estuaries. (■) Colne estuary sites 1 and 2, (□) Colne estuary sites 3 and 4 (Thornton *et al.* 2002), (▲) Blackwater Estuary upper marsh sites (△) Blackwater estuary lower marsh and mudflat sites (Underwood 1997), (○) Severn estuary Aust, (⊙) Severn estuary, Portishead, (●) Severn estuary, Sand Bay (Underwood 1994). (B). Vectors showing the direction of increasing relative abundance of particular diatom species. Only species with vectors significantly correlated ($p < 0.05$) with either PC1 or PC2 are shown. For species codes see Table 1.

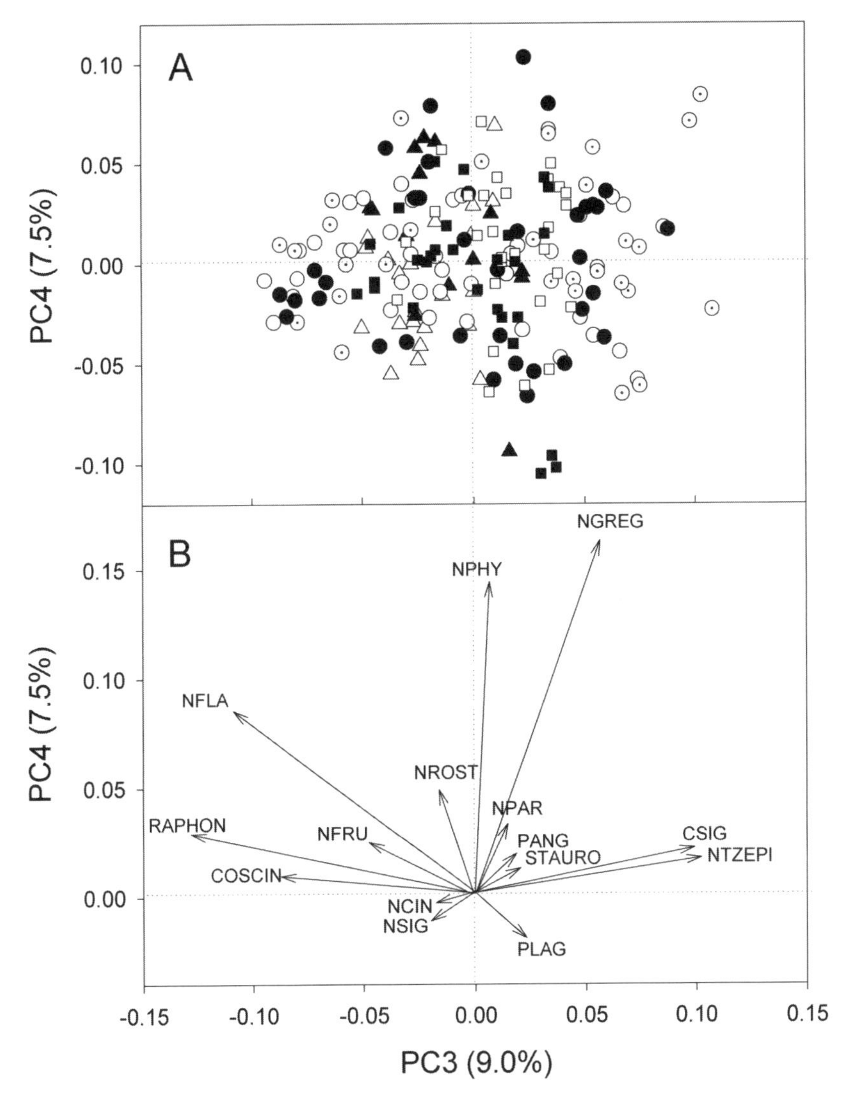

Figure 2. (A). Principal components analysis scatter plot of components 3 (PC3) and 4 (PC4) derived from relative abundance values of microphytobenthic diatom species composition in samples from sites in 3 U.K. estuaries; (■, □) Colne, (▲, △) Blackwater, and (○, ⊙, ●) Severn estuaries. For detailed key see Figure 1. (B). Vectors showing the direction of increasing relative abundance of particular diatom species. Only species with vectors significantly correlated ($p < 0.05$) with either PC3 or PC4 are shown. For species codes see Table 1.

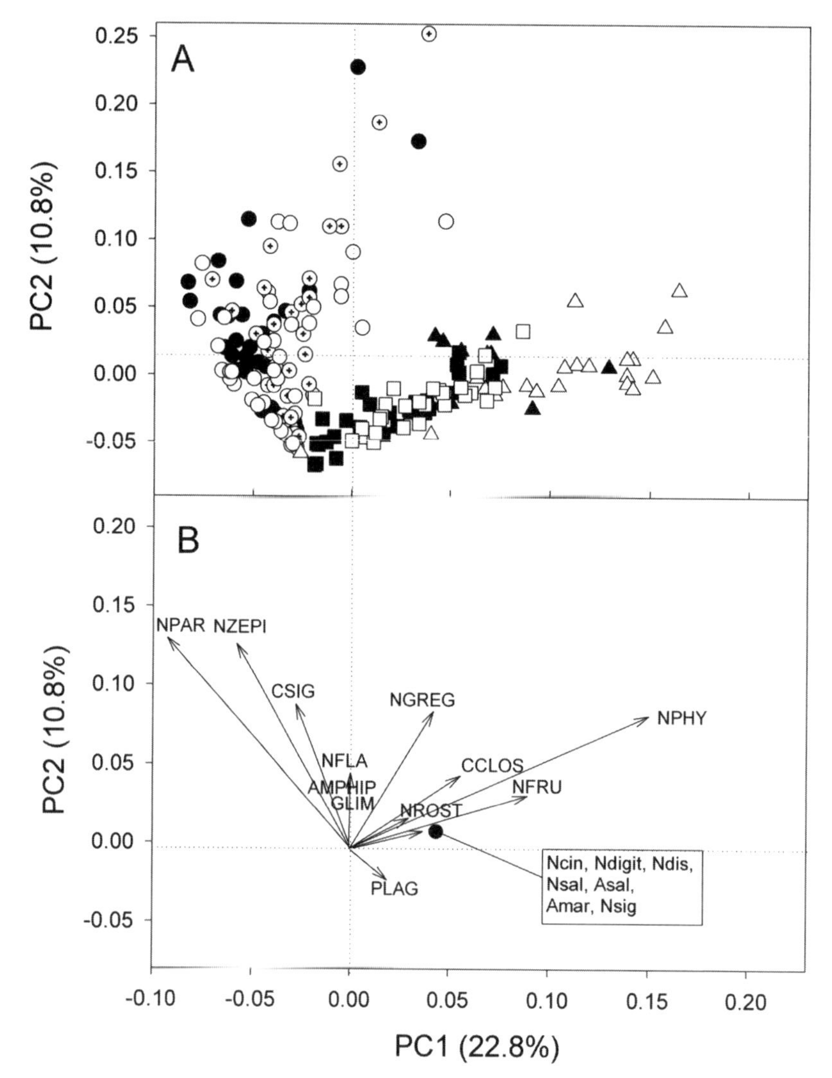

Figure 3. (A). Principal components analysis scatter plot of components 1 (PC1) and 2 (PC2) derived from biomass-weighted relative abundance values of microphytobenthic diatom species composition in samples from sites in 3 U.K. estuaries; (■, □) Colne, (▲, △) Blackwater, and (○, ⊙, ●) Severn estuaries. For detailed key see Figure 1. (B). Vectors showing the direction of increasing relative abundance of particular diatom species. Only species with vectors significantly correlated ($p < 0.05$) with either PC1 or PC2 are shown. For species codes see Table 1.

(though often in low relative abundance) taxa that were present when the microphytobenthic biomass was higher (e.g. *Navicula cincta*, *N.rostellata*, *N. digitoradiata*, *N. diserta*, *Amphora salina*, *A. marina*, *Nitzschia sigma*). These taxa are common members of the assemblages, but usually do not appear within the set of important taxa due to low relative abundance. PC1 was significantly correlated with the same set of biomass/organic richness as in Figure 1 (Table 2), while PC2 was negatively correlated with temperature and positive with Chl *a* and total carbohydrate concentrations. Thus the effect of biomass-normalisation on PC1 and PC2 was to reduce the separation into estuary-specific clusters, widen the scatter of the data set, remove taxa associated with very low biomass and increase the significance of commonly found taxa yet that usually have a fairly low relative abundance.

PC3 and 4 explained a further 15.6% of the biomass-adjusted data, and separated out the upper saltmarsh samples from the other mudflat samples, and separated the Colne from the Blackwater samples (Figure 4). PC3 was positively correlated with biomass and organic content variables, as well as salinity. Taxa responding positively to this gradient (*Amphora marina*, *Cylindrotheca closterium*, *Nitzschia sigma* and *N. frustulum*) were common in the saltmarsh assemblages sampled at Northey Island. Separation of samples along PC4 was also correlated with a gradient of increasing salinity and organic content, with lower estuary samples tending to have a higher PC4 score. These two gradients of salinity place *N. phyllepta* and *N. gregaria* as taxa abundant in lower salinity samples, and *Pleurosigma angulatum* and *N. rostellata* and the cluster of 'saltmarsh' taxa more characterstic of higher salinity samples (Figure 4).

By incorporating a biomass-factor into these analyses, some of the taxa that were acting to separate out estuaries become less important (for example, winter taxa and tychoplankonic diatoms in the Severn estuary), and the UK East coast estuarine sites show more overlap. But there are still large differences in the flora of these sites, with some species characteristic of high biomass conditions in one estuary not common in another. For example, in the Severn, *N. pargemina, Nz. epithemioides, C. signata* while on the east coast, *N. phyllepta, N. gregaria, C. closterium* and *Nz. frustulum* are dominant taxa. This is despite the ranges of environmental data measured for these three estuaries showing significant overlap (Table 1) so it is not immediately obvious what factor may be causing differences in species composition. Monthly nutrient data for the Severn and Blackwater sites were not measured during those studies, but area-normalised nutrient loads for all 3 estuaries have been calculated (Table 3; Nedwell *et al.* 2002). These data would suggest greater similarities between the Severn and Blackwater estuaries, not a pattern observed in the floristic data. This would suggest that nutrient loads don't appear a good predictor at this very broad geographical scale.

This example illustrates the care that must be taken in drawing conclusions about nutrient status and estuarine diatom flora. The Severn estuary epipelic diatom flora (see also Oppenheim 1991) does appear to differ from that found in UK southern North Sea estuaries. The flora of the latter estuaries is quite similar to the flora of Dutch and Flemish estuaries (Colijn and Dijkema 1981; Peletier 1996; Sabbe 1993). Whether this is evidence for regional floristic differences is a hypothesis that needs

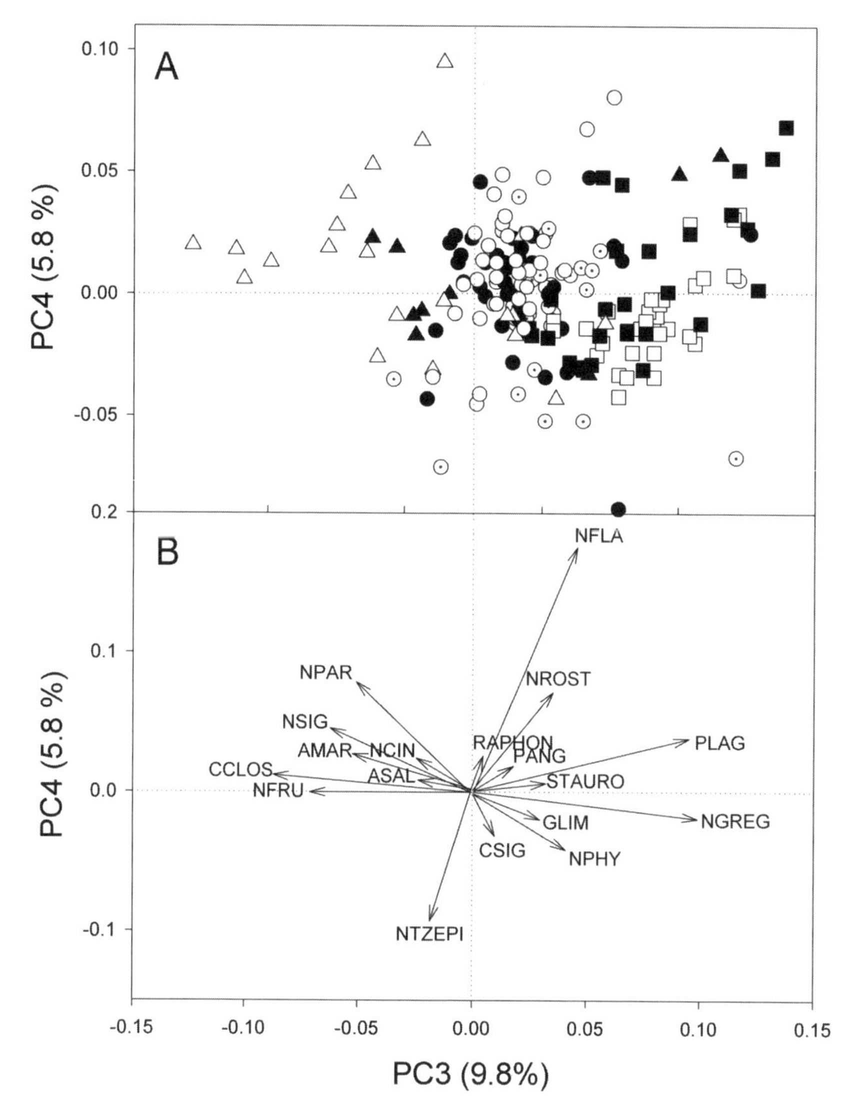

Figure 4. (A). Principal components analysis scatter plot of components 3 (PC3) and 4 (PC4) derived from biomass-weighted relative abundance values of microphytobenthic diatom species composition in samples from sites in 3 U.K. estuaries; (■, □) Colne, (▲, △) Blackwater, and (○, ⊙, ●) Severn estuaries. For detailed key see Figure 1. (B). Vectors showing the direction of increasing relative abundance of particular diatom species. Only species with vectors significantly correlated ($p < 0.05$) with either PC3 or PC4 are shown. For species codes see Table 1.

Table 3. Estuary area-normalised loads of nitrate and ammonium (Mmol N km^{-2} y^{-1}) for the Blackwater "Colne and Severn" Estuaries in 1995-1996 (from Nedwell *et al.* 2002).

Load (Mmol N km^{-2} y^{-1})	N-NO_3^-	N-NH_4^+
Blackwater	4	0.06
Colne	2	1
Severn	5	0.6

investigating. Interestingly, the Severn estuary also has a benthic fauna different from other UK estuaries, attributed to the physical dynamics of the estuary (Warwick *et al.* 1991). Is disturbance a selective force for the dominant diatom taxa in the Severn estuary?

Many benthic marine diatom species appear to have a cosmopolitan distribution (Witkowski *et al.* 2000), though some genera do show higher diversity in some biogeographic areas (e.g. *Mastogloia* in the tropics). However, great care needs to be taken in drawing conclusions concerning global or even regional distribution patterns. Many regions are undersampled, and taxonomic difficulties are present in poorly described floras. For example, *Navicula flanatica* is a common European mudflat taxon that tends to be more dominant in assemblages during the cooler months of the year (Colijn and Dijkema 1981; Admiraal *et al.* 1984; Underwood 1994; Thornton *et al.* 2002). Yet, apparently the same species (identical in the morphological features used to identify diatoms) is common within mangrove and muddy intertidal sediments in Fiji, where temperature do not drop below 25 °C, and often exceeds 40°C (Underwood, unpublished observations). Combined taxonomic, physiological and genetic studies are needed to clarify our definitions of species and to determine intraspecific variability in niche widths.

Within estuary patterns of species distribution

Salinity and nutrients

As well as broad, regional patterns of species difference, changes in the relative numbers of different taxa are very evident along much shorter gradients (Figure 5, Table 4). Within estuaries, the estuarine gradient is, characterised usually by a decrease in nutrient concentrations and increasing salinity (Underwood and Kromkamp 1999; Thornton *et al.* 2002). Salinity has been implicated as a major driver of species distribution (Van Dam *et al.* 1994; Watt 1998), and there do appear to be distinct brackish water and marine species. In the stable salinity gradient of the Baltic Sea, species occur in narrow salinity bands (Snoeijs 1993; Snoeijs and Vilbaste 1994; Snoeijs and Potapova 1995; Snoeijs and Kasperovičiene 1996; Snoeijs and Balashova 1998; Stachura-Suchoples 2001), but other estuarine taxa have a wide tolerance to salinity (Admiraal 1984; Cox 1995). In meso- and macrotidal estuaries, there is wide fluctuation in the

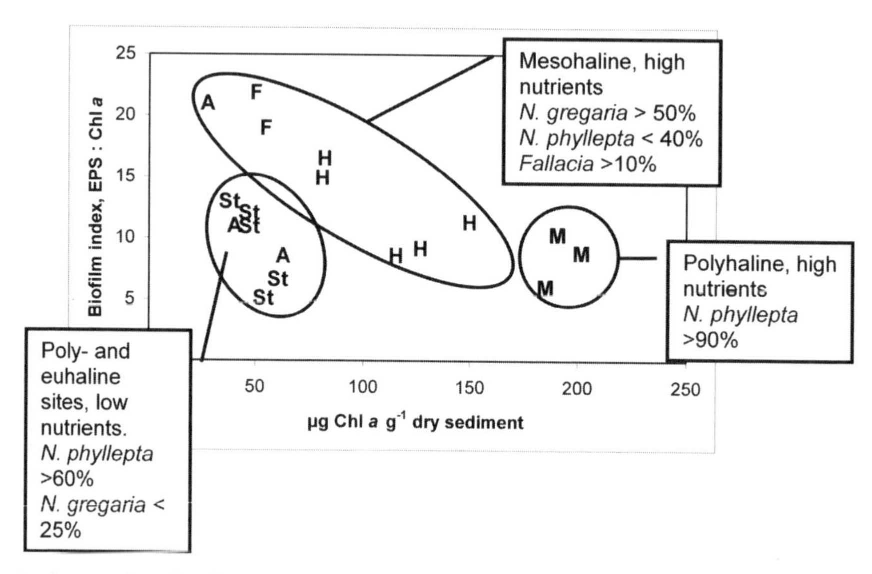

Figure 5. Scatter plot of sediment Chl *a* content and biofilm index (ratio of EPS: Chl *a* content) from 5 sites (F, St, A, H, M) in the Colne-Blackwater estuary complex, indicating major species composition of the different biofilms. For site details see Table 4.

Table 4. Chemical conditions at 5 sites in the Colne (F, S, A, H) and Blackwater (M) estuaries, Essex.

(µM)	Fingringhoe (F)	St Osyth (St)	Alresford (A)	Hythe (H)	Maldon (M)
Overlying water					
Salinity (‰)	12	32	28	7	22
T.Ox.N.	274 ± 45.1	11.0 ± 1.18	26.0 ± 2.12	649 ± 48.3	297 ± 46.5
Phosphate (µM)	17.7 ± 2.54	0.17 ± 0.07	1.37 ± 0.16	34.0 ± 2.12	31.9 ± 6.72
Ammonia (µM)	14.2 ± 1.60	22.7 ± 3.61	9.76 ± 1.20	79.6 ± 16.3	52.8 ± 3.77
Silicate (µM)	125 ± 22.4	15.3 ± 0.88	30.5 ± 3.20	247 ± 8.21	131 ± 37.2
Porewater					
Salinity (‰)	7.9 ± 0.33	35.5 ± 0.16	33.5 ± 0.81	4.0 ± 0.0	15.8 ± 0.20
T.Ox.N.	7.31 ± 1.95	1.01 ± 0.90	6.26 ± 0.77	13.8 ± 8.59	6.24 ± 6.24
Phosphate (µM)	65.6 ± 4.36	12.0 ± 0.86	30.5 ± 5.36	112 ± 14.6	68.0 ± 22.0
Ammonia (µM)	29.3 ± 6.29	15.1 ± 2.40	152 ± 32.8	630 ± 57.5	232 ± 36.7
Silicate (µM)	338 ± 15.7	18.0 ± 3.39	56.1 ± 13.5	643 ± 105	418 ± 138

salinity regime over the spring-neap tide cycle and also annually. Such variation is likely to select for species tolerant to changing salinities. Patterns in salinity also co-vary with nutrient concentrations and loads (Underwood *et al.* 1998; Ribeiro *et al.* 2003), so it is difficult to separate out the effect of each variable, though multivariate analyses can be used (Oppenheim 1991; Thornton *et al.* 2002).

Experimental approaches can be used to manipulate environmental variables and determine the response of different populations of species. Underwood *et al.* (1998) manipulated ammonium concentrations in tidal mesocosms and found that certain taxa showed significant treatment-related changes in population density. Ammonium concentrations may play an important role in determining species composition due to the toxic effects of ammonia (Admiraal 1977) under conditions of high pH. Similarly sulphide probably has a selective effect on assemblage composition (Sullivan 1999). Concentrations of these ions are high in organically-enriched sediments, and species such as *Fallacia (Navicula) pygmeae* and *Navicula salinarum* appear particularly tolerant of sulphide and ammonia (Peletier 1996; Sullivan 1999; Table 4; Figure 5). A series of nutrient enrichment experiments carried out on saltmarshes on the eastern seaboard of the USA also found shifts in species composition due to nitrogen enrichment (Sullivan 1999), while a major reduction in nutrient loads in the Ems-Dollard estuary over a 10-year period have resulted in significant changes in the diatom flora (Peletier 1996).

Culturing experiments have also shown significant differences in growth rates of estuarine species across a range of nitrate, ammonium and salinity conditions with evidence of separation of the autecological niches of *Navicula phyllepta*, *N. salinarum*, *N. perminuta* and *Cylindrotheca closterium* across this matrix of conditions (Underwood and Provot 2000). These experimental data agree with known distributions of some taxa. For example, experimental data suggests that *N. phyllepta* is a mesotrophic species, favouring nitrate and ammonium concentrations in the range of 100-400 μM, which matches the occurrence of assemblages dominated by this taxon in the Colne estuary (Thornton *et al.* 2002). The relative abundance of *Navicula phyllepta* also shows an inverse relationship with that of *N. gregaria* across a wide nutrient range (Figure 6). Both taxa are common in a range of estuaries and are commonly found together.

Grazing

A very under-researched area is the impact that grazing might have on assemblage species composition. Grazing has been shown to have very significant impacts of the species composition of epiphytic and epilithic assemblages, both in freshwater habitats and on rocky shores, with smaller, more adpressed taxa having an advantage under grazed conditions. However, though there are numberous studies that have demonstrated that grazing can significantly reduce biomass of estuarine microphytobenthos (e.g. Sullivan and Currin 2000), detailed grazing studies concentrating of species composition are generally lacking. Grazing experiments with *Nereis diversicolor* and *Corophium volutator* have shown a shift in species composition, compared to ungrazed controls (Smith *et al.* 1996), and recently it has been shown that certain

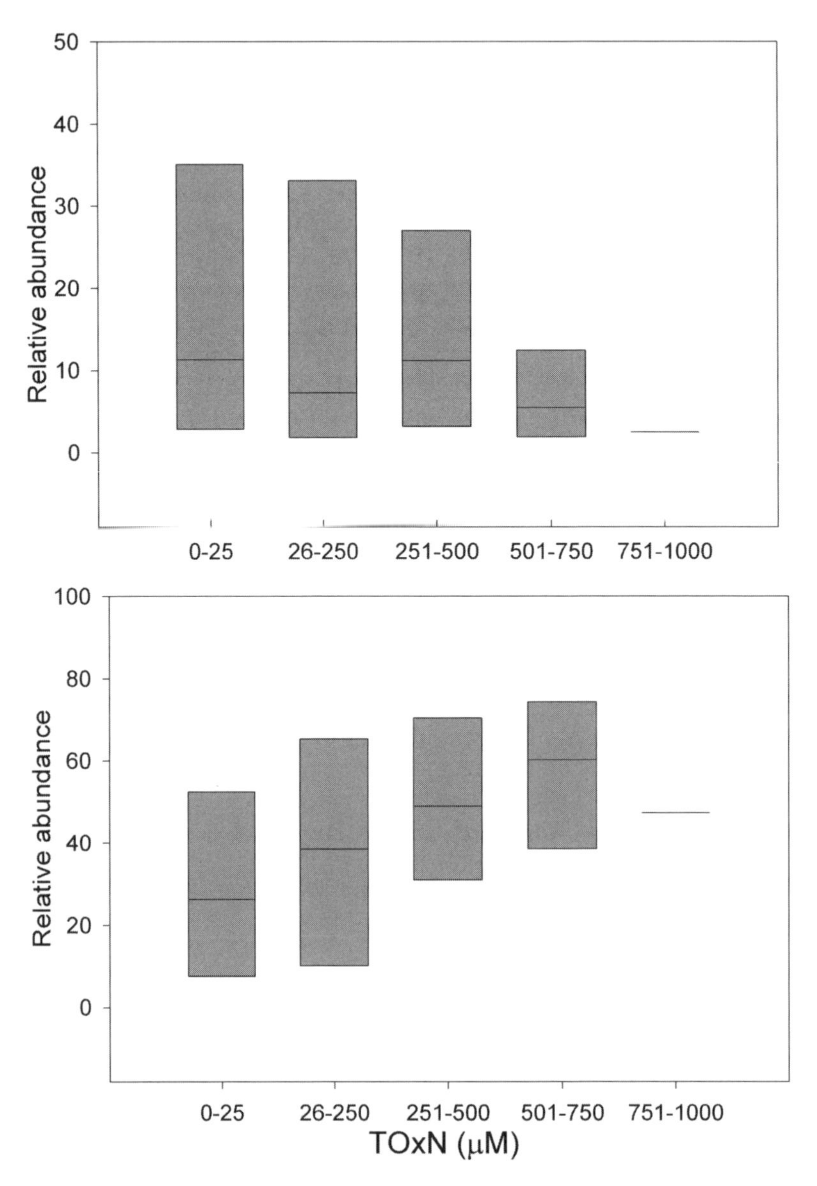

Figure 6. Box plots (median value, 25 and 75% quartiles) of the relative abundance of (A) *Navicula gregaria* and (B) *Navicula phyllepta* across a range of total oxidised nitrogen (T.Ox.N) concentrations in the Colne, Great Ouse, Conwy and Orwell estuaries, U.K.

ciliate species select actively different taxa of epipelic diatoms (Hamels *et al.* 2004). Whether these grazing interactions are sufficiently intense to result in substantial species shifts *in situ* remains to be determined (i.e., the relative importance of top-down versus bottom-up effects in determining microphytobenthic species composition).

Sediment type

Benthic diatoms are often separated into epipsammic and epipelic taxa, depending on their mode of attachment and motility in sediments. Epipsammic species are usually very small, virtually non-motile, and live closely attached to depressions and crevices in sand grains by pads or short stalks. Epipelic diatoms are non-attached and are highly motile (Round *et al.* 1990). In the natural environment, the distinction between these two growth forms is blurred, as sands and muds are found together and an assemblage can consist of both epipelic and epipsammic taxa (Hamels *et al.* 1998). Some large, motile naviculoids, e.g. *Navicula rostellata*, *N. peregrina* and *N. arenaria*, are found in sandy muds, and sand can support dense populations of motile *Hantzschia* species (Underwood 2002; Kingston 1999) as well as the smaller, more typical epipsammic taxa. Where sands are scoured and moved by currents, the density of epipelon is reduced (Sabbe 1993; Hamels *et al.* 1988; Vilbaste *et al.* 2000,) and epipsammic species dominate the assemblage. In estuaries, the location of sandflats and mudflats will depend partly on hydrodynamics (Dyer 1998; Flemming 2000) and in partially-mixed estuaries (the common NW European system), sandier sediments will be more abundant towards the mouth of the estuary (Viles and Spencer 1995). Thus spatial changes in diatom flora along estuarine gradients may also reflect increasing sediment grain size, as well as increasing salinity and a dilution of riverine nutrient inputs.

A cautionary example

The temptation to assume that variables measured in a study are the causative factors in determining environmental patterns is very strong, as Darwin (1860) (*op.cit.*) pointed out. Yet, it is not always clear what the real causative factors are. As an example, data on microphytobenthic species composition, nutrient conditions, photophysiological responses of microphytobenthos and behavioural characteristics of different biofilms from various shallow water habitats in Fiji are summarised in Figure 7. Signficant differences in species community composition were found between different intertidal and subtidal sites, with sites of similar physical characteristics clustering (by diatom species) together, even though geographically separate. Microphytobenthos showed a significant gradient of light acclimation and potential maximum rates of photosynthesis across these sites, which also correlated with a significant gradient in water column nutrient concentrations (see Underwood 2002 for details). Clear patterns of tidally-linked endogenous rhythms of vertical migration were present in intertidal biofilms, but not recorded in subtidal biofilms, and there were also significant differences in carbohydrate and relative EPS content across these gradients. It is not possible to determine which, if any, of these variables is

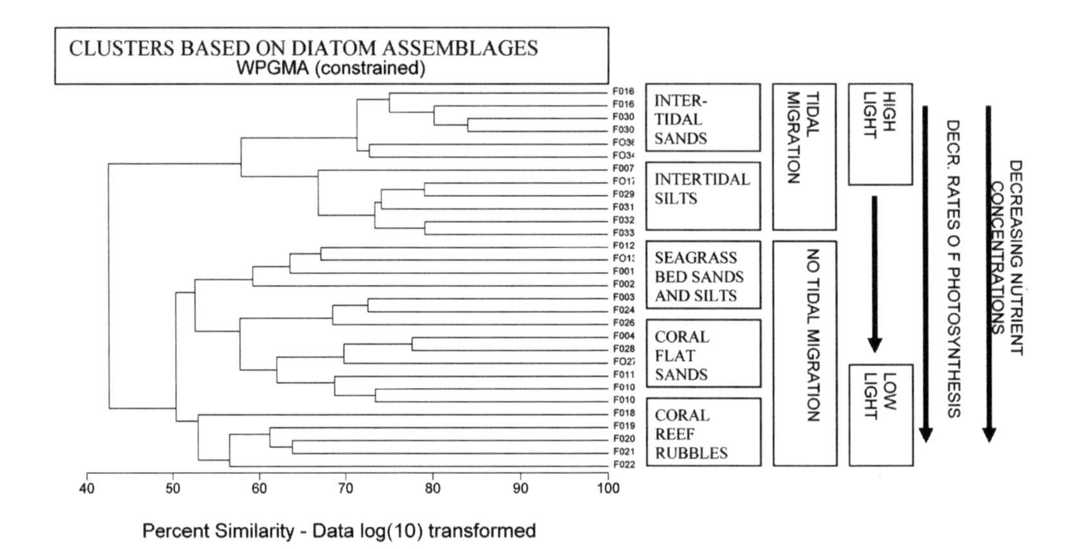

Figure 7. Dendrogram of sample clusters based on percentage similarity in diatom species composition for 30 benthic assemblages from intertidal and shallow subtidal sites in Fiji. The different site and biofilm characteristics (vertical migration, photophysiology etc.) for these clusters are indicated (Underwood 2002).

causative of the patterns in species distribution observed. What is interesting is that differences in ecologically-important properties of biofilms, for example potential primary production and EPS concentrations) are also related to differences in species composition (Cahoon 1999). It might be that the species profile changes biofilm properties? This suggests that greater attention needs to be paid to the species composition of biofilms. While more descriptive studies on spatial and seasonal patterns in species composition will be useful, manipulative studies are also needed to determine species-specific responses to environmental variables. Combination of these data will allow more accurate understanding of what determines species composition in estuarine microalgal biofilms.

References

ADMIRAAL, W. 1977. Salinity tolerance of benthic estuarine diatoms as tested with a rapid polarographic measurment of photosynthesis. Mar. Biol. **39**: 11-18.

CAHOON, L.B. 1999. The role of benthic microalgae in neritic ecosystems. Oceanogr. Mar. Biol. Ann. Rev. **37**: 47-86.

COLIJN, F., and DIJKEMA, K.S. 1981. Species composition of benthic diatoms and the distribution of Chlorophyll a on an intertidal flat in the Dutch Wadden sea. Mar. Ecol. Prog. Ser. **4**: 9-21.

COX, E.J. 1995. Morphological variation in widely distributed diatom taxa: taxonomic and ecological implications. 13th International Diatom Symposium 1994. pp 335 – 345, Biopress Ltd. Bristol.

DARWIN, C. 1860. *Naturalist's voyage. Journal of researches into the geology and natural history of the various countries visited during the voyage of H.M.S. Beagle round the world.* J.M.Dent and Co., London.

DONG, L.F., THORNTON, D.C.O., NEDWELL, D.B. and UNDERWOOD, G.J.C. 2000. Denitrification in sediments of the River Colne estuary, England. Mar. Ecol. Prog. Ser. **203**: 109-122.

DYER. K.R. 1998. The typology of intertidal mudflats. *In* Black, K.S., Paterson, D.M. and Cramp, A. (Eds.) *Sedimentary processes in the intertidal zone.* Geological Society, London. Special Publications **139**: 11-24.

HAMELS, I., SABBE, K., MUYLAERT, K., BARRANGUET, C., LUCAS, C., HERMAN, P. and VYVERMAN, W. 1998. Organisation of microbenthic communities in intertidal estuarine flats, a case study from the Molenplaat (Westerschelde Estuary, The Netherlands). Europ. J. Protistol. **34**: 308-320.

HAMELS, I., MUSSCHE, H., SABBE, K., MUYLAERT, K. and VYVERMAN, W. 2004. Evidence for constant and highly specific active food selection by benthic ciliates in mixed diatom assemblages. Limnol. Oceanogr. **49**: 58-68.

FLEMMING B.W. 2000. A revised textural classification of gravel-free muddy sediments on the basis of ternary diagrams. Cont. Shelf. Res. **20**: 1125-1137.

KINGSTON, M.B. 1999. Effect of light on vertical migration and photosynthesis of *Euglena proxima* (Euglenophyta). J. Phycol. **35**: 245-253.

NEDWELL, D.B., DONG, L.F., SAGE, A.S. and UNDERWOOD, G.J.C. 2002. Variations of the nutrient loads to the mainland U.K. estuaries: correlation with catchment areas, urbanisation and coastal eutrophication. Est. Coastal Shelf Sci. **54**: 951-970.

OPPENHEIM, D.R. 1991. Seasonal changes in epipelic diatoms along an intertidal shore, Berrow flats, Somerset. J. Mar. Biol. Assoc. U. K., **71**: 579-596.

OXBOROUGH, K., HANLON, A.R.M., UNDERWOOD, G.J.C., & BAKER N.R. 2000. *In vivo* estimation of the photosystem II photochemical efficiency of individual microphytobenthic cells using high-resolution imaging of chlorophyll *a* fluorescence. Limnol. Oceanogr. **45**: 1420-1425.

PATERSON, D.M., WILTSHIRE, K.H., MILES, A., BLACKBURN, J., DAVIDSON, I., YATES, M.G., MCGROTY, S. and EASTWOOD, J.A. 1998. Microbiological mediation of spectral reflectance from intertidal cohesive sediments. Limnol. Oceanogr. **43**: 1027-1221.

PELETIER, H. 1996. Long-term changes in intertidal estuarine diatom assemblages related to reduced input of organic waste. Mar. Ecol. Prog. Ser. **137**: 265-271

PERKINS, R.G., OXBOROUGH, K., HANLON, A.R.M., UNDERWOOD, G.J.C. and BAKER, N.R. 2002. Can chlorophyll fluorescence be used to estimate the rate of photosynthetic electron transport within microphytobenthic biofilms? Mar. Ecol. Prog. Ser., **228**v 47-56.

RIBEIRO, L., BROTAS, V., MASCARELL, G. and COUTÉ, A. 2003. Taxonomic survey of the microphytobenthic communities of two Tagus estuary mudflats. Acta Oecologica **24**: S117-S123.

ROUND, F.E. 1979. Occurrence and rhythmic behaviour of *Tropidoneis lepidoptera* in the epipelon of Barnstable Harbor, Massachusetts, USA. Mar. Biol. **54**: 215-217.

RYSGAARD-PETERSEN, N. 2003. Coupled nitrification – denitrification in autotrophic and heterotrophic estuarine sediments: on the influence of benthic microalgae. Limnol. Oceanogr. **48**: 93-105.

SABBE, K. 1993. Short-term fluctuations in benthic diatom numbers on an intertidal sandflat in the Westerschelde Estuary (Zeeland, The Netherlands). Hydrobiol. **269/270**: 275-284.

SMITH, D., HUGHES, R.G, and COX, E.J. 1996. Predation of epipelic diatoms by the amphipod *Corophium volutator* and the polychaete *Nereis diversicolor*. Mar. Ecol. Prog. Ser. **145**: 53-61.

SNOEIJS, P. 1993. Intercalibration and distribution of diatom species in the Baltic Sea. Vol. 1, The Baltic Marine Biologists Publication No. 16b, Opulus Press, Uppsala.

SNOEIJS, P. and BALASHOVA, N. 1998. Intercalibration and distribution of diatom species in the Baltic Sea. Vol. 5, The Baltic Marine Biologists Publication No. 16e, Opulus Press, Uppsala.

SNOEIJS, P. and KASPEROVIČIENE, J. 1996. Intercalibration and distribution of diatom species in the Baltic Sea. Vol. 4, The Baltic Marine Biologists Publication No. 16d, Opulus Press, Uppsala.

SNOEIJS, P. and POTAPOVA, M. 1995. Intercalibration and distribution of diatom species in the Baltic Sea. Vol. 3, The Baltic Marine Biologists Publication No. 16c, Opulus Press, Uppsala.

SNOEIJS, P. and VILBASTE, S. 1994. Intercalibration and distribution of diatom species in the Baltic Sea. Vol. 2, The Baltic Marine Biologists Publication No. 16a, Opulus Press, Uppsala.

STACHURA-SUCHOPLES, K. 2001. Bioindicative values of dominant diatom species from the Gulf of Gdańsk, Southern Baltic Sea, Poland. In. Lange-Bertalot-Festschrift (Jahn, R., Kociolek, J.P., Witkowski, A. and Compère, Eds.) 477-490. Ganter, Ruggell. Koeltz Scientific Books. Königstein, Germany.

SULLIVAN, M.J. 1999. Applied diatom studies in estuarine and shallow coastal environments. In *The Diatoms: Applications for the Environmental and Earth Sciences* (Stoermer, E.F. and Smol, J.P., editors). 334-351. Cambridge University Press, Cambridge.

SULLIVAN, M.J. and CURRIN, C.A. 2000. Community structure and functional dynamics of benthic microalgae in salt marshes. In, *Concepts and Controversies in Tidal Marsh Ecology* (Weinstein, M.P. and Kreeger, D.A., Eds.). pp. 81-106. Kluwer Academic Publishers, Dordrecht.

SUNDBÄCK, K. and JÖNSONN, B. 1988. Microphytobenthic productivity and biomass in sublittoral sediments of a stratified bay, southeastern Kattegat. J. exp. Mar. Biol. Ecol. **122**: 63-81.
THORNTON, D.C.O., DONG, L.F., UNDERWOOD, G.J.C., and NEDWELL, D.B. 2002. Factors affecting microphytobenthic biomass, species composition and production in the Colne estuary (UK). Aquat. Microb. Ecol., **27**: 285-300.
THORNTON, D.C.O., UNDERWOOD, G.J.C., and NEDWELL, D.B. 1999. Effect of illumination and emersion period on the exchange of ammonium across the estuarine sediment-water interface. Mar. Ecol. Prog. Ser. **184**: 11-20.
UNDERWOOD, G.J.C. 1994. Seasonal and spatial variation in epipelic diatom assemblages in the Severn estuary. Diatom Res. **9**: 451-472.
UNDERWOOD, G.J.C. 1997. Microalgal colonisation in a saltmarsh restoration scheme. Est. Coastal Shelf Sci. **44**: 471-481.
UNDERWOOD, G.J.C. 2000. Changes in microalgal species composition, biostabilisation potential and succession during saltmarsh restoration. In *British Saltmarshes Eds*. Sherwood B.R., Gardiner B.G., and Harris T. publ. Linnean Society of London / Forrest Text. Cardigan. pp 418. Chapter 9, p 143-154.
UNDERWOOD, G.J.C. 2002. Adaptations of tropical marine microphytobenthic assemblages along a gradient of light and nutrient availability in Suva Lagoon, Fiji. Eur. J. Phycol. **37**: 449-462.
UNDERWOOD, G.J.C. and KROMKAMP, J. 1999. Primary production by phytoplankton and microphytobenthos in estuaries. Adv. Ecol. Res. **29**: 93-153.
UNDERWOOD, G.J.C, PERKINS, R.G., CONSALVEY, M., HANLON, A.R.M., OXBOROUGH, BAKER, N.R. and PATERSON, D.M. 2005. Patterns in microphytobenthic primary productivity: species-specific variation in migratory rhythms and photosynthesis in mixed species biofilms. Limnol. and Oceanogr. **50**: 755-767.
UNDERWOOD, G.J.C, PHILLIPS, J. and SAUNDERS, K. 1998. Distribution of estuarine benthic diatom species along salinity and nutrient gradients. Eur. J. Phycol. **33**: 173-183.
UNDERWOOD, G.J.C. and PROVOT, L. 2000. Determining the environmental p References of four estuarine epipelic diatom taxa – growth across a range of salinity, nitrate and ammonium conditions. Eur. J. Phycol. **35**, 173–182.
VAN DAM, H., MERTENS, A., and SINKELDAM J. 1994. A coded checklist and ecological indicator valies or freshwater diatoms from the Netherlands. Neth. J. Aquat. Ecol. **28,** 117-133.
VILBASTE S., SUNDBÄCK, K., NILSSON. C. and TRUU, J. 2000. Distribution of benthic diatoms in the littoral zone of the Gulf of Riga, the Baltic Sea. Eur. J. Phycol. **35**: 373-385.
VILES, H., and SPENCER, T. 1995. *Coastal Problems, geomorphology, ecology and society at the coast.* Edward Arnold, London. pp. 350.
WARWICK, R.M., GOSS-CUSTARD, J.D., KIRBY, R., GEORGE, C.L., POPE, N.D., and ROWDEN, A.A., 1991. Static and dynamic environmental factors determining the community structure of estuarine macrobenthos in SW Britain: Why is the Severn estuary different? J. Appl. Ecol. **28**: 329-345.
WATT, D.A. 1998. Estuaries of contrasting trophic status in KwaZulu-Natal, South Africa. Est. Coastal Shelf Sci. **47**: 209-216.
WITKOWSKI, A, LANGE-BERTALOT, H. and METZELTIN D. 2000. *Diatom flora of marine coasts*. Iconographia diatomologica v. 7. Ganter, Ruggell. Koeltz Scientific Books. Königstein, Germany.

James L. Pinckney

System-scale nutrient fluctuations in Galveston Bay, Texas (USA)

Abstract

The purpose of this review is to identify and characterize the major estuarine nutrients, illustrate the annual to biweekly fluctuations of water column nutrients, and provide insights into the major factors that regulate the spatio-temporal distributions of nutrients in a typical subtropical estuary in the Gulf of Mexico, Galveston Bay, Texas. Decadal and annual estimates of total N loading highlight the importance of interannual variation and long-term trends, which are useful for developing nutrient input management strategies and assessing the effectiveness of mitigation programs. However, long-term averaging of large spatial areas dampens the frequency and magnitude of seasonal events and system perturbations. Shorter timescale measurements were undertaken to determine event-based nutrient fluctuations in Galveston Bay. Water samples were collected and hydrographic profiles were recorded at 9 stations at biweekly intervals from 1999 to 2002 along a fixed transect. Sample analyses included nutrient concentrations, phytoplankton biomass, and phytoplankton community composition. Data were analyzed using spatio-temporal contour plots. The results showed that nutrient concentrations in this estuary reflected a balance between river discharge, benthic regeneration, and water temperature (season). Major perturbations, such as the passage of tropical storm Allison, upset this balance and profoundly changed the biogeochemistry of the estuary. After this 'flushing' event, the system seemed to slowly return to a 'normal' pattern of nutrient distributions. Documenting nutrient fluctuations in estuaries is relatively straightforward provided the funding, resources, and time are available. Understanding the relative importance of the different processes driving these fluctuations on a system scale is a much more formidable task. Mechanistic studies of benthic and planktonic biogeochemical processes, coupled in space and time, are a necessary prerequisite for developing reliable models to simulate, and possibly predict, estuarine nutrient fluctuations.

Introduction

Estuaries are semi-enclosed coastal bodies of water which have a free connection with the sea and within which seawater is measurably diluted by freshwater. The freshwater sources for most estuaries are streams, rivers, and possibly groundwater.

In the continental US, estuaries comprise more than 80% of the coastline along the Atlantic Ocean and Gulf of Mexico and more than 10% of the Pacific coast (Bricker *et al.* 1999). Most estuaries are very efficient at retaining dissolved and particulate matter (Hobbie 2000). In this respect, estuaries are often thought of as filters or traps that are located between the land and the sea. Because these systems are so efficient at retaining dissolved chemicals and particulates, they are very susceptible to nutrients and pollutants that are washed into the estuary (*NRC* 2000). These chemicals typically have long residence times in estuaries and tend to accumulate over time. High rates of nutrient inputs from land runoff stimulate primary production and thereby increase the rate of organic matter loading in the estuary. Excessive nutrient loading can lead to primary production that exceeds the assimilative capacity of the system and consequently leads to eutrophication (Nixon 1995). The purpose of this review is to identify and characterize the major estuarine nutrients, illustrate the annual to biweekly fluctuations of water column nutrients, and provide insights into the major factors that regulate the spatio-temporal distributions of nutrients in a typical subtropical estuary in the Gulf of Mexico, Galveston Bay, Texas

Overview of estuarine nutrients

As the conduit between the land and sea, estuaries are exposed to a variety of dissolved chemical species. Some of these compounds are used by resident biota and promote growth or provide energy for physiological processes such as photosynthesis and are considered nutrients. In estuaries, the most important nutrients for primary production are inorganic and organic nitrogen compounds, phosphate, silicate, and dissolved organic matter.

Nitrogen (N) is a key constituent for life on Earth and this element occurs in many different chemical pools and states in the biosphere. All living organisms require N for the synthesis of amino acids and proteins for growth. Most of the N in marine waters is in the form of N_2 (dissolved dinitrogen gas) at a concentration of 1 mM (Pilson 1998). Concentrations of N_2 are relatively uniform in seawater but vary as function of the temperature-salinity dependent solubility of the gas (Pilson 1998). Nitrous oxide (N_2O) and nitric oxide (NO) are trace constituents in seawater, mostly due to microbial activity (Capone 2000). Nitrate (NO_3^-), nitrite (NO_2^-), and ammonium (NH_4^+), collectively termed dissolved inorganic nitrogen (DIN), are the most abundant non-gaseous N species in estuarine waters (Sharp 1991; Alongi 1998). Nitrate is highly mobile and is usually the dominant form of N in runoff, riverine input, groundwater discharge, and atmospheric deposition. Nitrite is a minor component of the total DIN pool, but concentrations may be high at redox interfaces. Ammonium concentrations can vary widely in estuaries depending on season, location, and depth in the estuary. Furthermore, NH_4^+ concentrations are usually high in hypoxic/anoxic environments, near sewage/wastewater outfalls, in agricultural runoff, and in areas of high benthic biomass (e.g., oyster reefs, clam beds, etc.) (Dame *et al.* 1989). Ammonium is usually the dominant form of DIN in the porewater of estuarine sediments. In estuaries, water column concentrations of DIN usually range from 0 to 100 μM (Hobbie 2000).

Dissolved organic nitrogen (DON) is a mixture of organic molecules, such as amino acids, nucleic acids, simple proteins, and urea, that contain N (Dortch 1990). However, the chemical identities of many of the compounds found in DON remain uncharacterized but most of the DON seems to be composed of refractory biopolymers (Ogawa *et al.* 2001). DON is a major N pool in estuarine systems but the biological utilization of DON as a source of N depends greatly on the amount of labile N in the total DON pool (Ward and Bronk 2001). DON sources in estuaries are the same as DIN sources but the ratios of DIN to DON vary depending on the source.

Like N, phosphorus (P) is an important chemical for life and is a major element in organic matter. In aerobic environments, phosphorus occurs almost exclusively as orthophosphate, which is any salt of H_3PO_4 (phosphoric acid). The dissociation products are $H_2PO_4^-$, HPO_4^{2-}, and PO_4^{3-} and the major ion in seawater is HPO_4^{2-} (Pilson 1998). Polyphosphates are formed when H_2O is removed by cellular processes, but these compounds are not commonly found in seawater except near sewage outfalls. Organophosphates are phosphate esters derived from living cells. Phosphate (as orthophosphate) is the most abundant form of P in estuarine and coastal waters (Alongi 1998; Pilson 1998). Phosphates readily adsorb onto particulates; under aerobic conditions, phosphate adsorbs onto oxyhydrides, calcium carbonates, and clay mineral particles, where phosphate is substituted for silicate in the lattice structure of clays (Pilson 1998). Phosphates also tend to form insoluble compounds with certain metals and readily precipitate with cations such as Ca^{2+}, Al^{3+}, and Fe^{3+} (francolite, hydroxyapatite) (Sharp 1991). In anaerobic environments, the combination of bacteria and H_2S results in the reduction of ferric iron (Fe^{3+}) to ferrous iron (Fe^{2+}) and the release of dissolved phosphate. Ferric oxyhydroxides dissolve and solubilize phosphate. Thus interstitial porewaters usually have high phosphate concentrations except in carbonate sands (Milliman 1974; Alongi 1998). Some of the dissolved phosphate may reprecipitate at the oxic/anoxic interface and sediments may be a source of phosphate if the overlying waters are anoxic.

Silicon (Si), an important element for cell wall formation in diatoms, silicoflagellates, radiolarians, and some sponges, is a common element in clay particles and is usually present in high concentrations in estuaries. In marine waters, the common particulate form is silica (SiO_2) and the dissolved form is silicate [$Si(OH)_4$] (Pilson 1998). Biogenic silica, produced by organisms, exists as opaline, a non-crystalline form of hydrated silica. Under conditions of high N-loading or during large diatom blooms, silicate may become a limiting nutrient for diatoms (Fisher *et al.* 1992; Glibert *et al.* 1998).

Dissolved organic matter (DOM) is one of the largest organic carbon pools on earth and plays a central role in the biogeochemistry of a variety of elements in estuarine ecosystems (Farrington 1992). DOM is a mixture of a range of sizes of organic molecules which are usually characterized in terms of molecular weight fractions. In estuarine habitats, organic matter occurs in both particulate (POM; plant debris, detritus, phytoplankton) and dissolved forms (DOM; humics, mucopolysaccharides, peptides, lipids). The distinction between POM and DOM is arbitrary and depends on the methods used to separate these two fractions (e.g., filtration, ultrafiltration, centrifugation, dialysis). In addition, the rates of supply of different forms of DOM are highly variable and are predicated by the land use characteristics of the watershed (rural, agricultural, urban, etc.), hydrology, and climatology (Guo *et al.* 1999). Sources of DOM can come from two major pathways. Allochthonous DOM originates outside the estuary and is transported into the estuary from watershed runoff, stream/wetland inflow, or anthropogenic point sources. Autochthonous DOM is generated within the system, mostly through photosynthesis by primary producers or by benthic regeneration of OM. In estuarine habitats, the dominant primary producers of autochthonous DOM are phytoplankton, benthic microalgae, epiphytes/periphyton, and submerged aquatic vegetation.

Riverine DOM is a mixture of inputs from watershed runoff, wastewater treatment facilities, and aquatic primary production (Burton and Liss 1976). Fluxes and concentrations of DOM (measured as DOC, dissolved organic carbon) into coastal waters can be substantial over annual time scales. For example, the average DOC concentration in large river systems in the Gulf of Mexico region ranges from 270 to 833 μM with loading rates of 0.1 to 209 x 10^{10} g C yr^{-1} (Guo *et al.* 1999). The pool of organic substances is dynamic. Some high molecular weight DOM can be rapidly recycled by bacteria (Amon and Benner 1994) while other compounds undergo photochemical degradation into more labile low molecular weight DOM (Kieber *et al.* 1989). In addition, the bioavailability of selected molecules in water appears to be altered from that observed in pure solution (Keil and Kirchman 1994), mainly because of the presence of DOM. Phytoplankton may secrete exopolymers to create conditions more favorable for growth, to regulate nutrient and trace metal availability, and for protection from toxic compounds (Moffett and Brand 1996; Ahner *et al.* 1997; Croot *et al.* 2000; Tang *et al.* 2001; Hung *et al.* 2001).

Different sources of DOM may have varying effects on phytoplankton communities. Many facultative autotrophs (mixotrophs), are able to supplement their cellular organic C requirements using DOM obtained from outside the cell (Bennett and Hobbie 1972; Droop 1974; Neilson and Lewin 1974; Hellebust and Lewin 1977; Gaines and Elbrächter 1987). DOM is also an important source of nutrients (e.g., glucose, urea, amino acids) for the growth and metabolism of heterotrophic bacteria in estuaries (Azam *et al.* 1983, 1993; Kirchman *et al.* 1991). The bacterial community may compete with phytoplankton for available inorganic nutrients, and, in the case of the presence of mixotrophic algae and/or heterotrophic flagellates, for the available DOM (Carlsson *et al.* 1998). These competitive interactions may be an important regulator of DOM transformations as well as a determinant of phytoplankton community structure and function (Sherr 1988).

Nutrient sources and sinks

New nutrients are nutrients that are supplied from outside a system while *regenerated* nutrients are derived from *in situ* chemical cycling within the estuary. The major sources of *new* nutrient inputs into estuaries are associated with freshwater flow into the estuary. Dissolved nutrients in river discharge constitute the primary nutrient source for many estuaries that receive significant freshwater input. Nutrient inputs from identifiable sources such as sewage outfalls, industrial process water, and other 'pipes' are point sources. Non-point sources are those which cannot be defined as discrete points such as agricultural runoff or groundwater. In some regions, atmospheric deposition, as wet deposition (rain) or dry deposition (dust), may supply a significant quantity of nutrients, especially N, to the surface of the estuary (Paerl 1995; Prospero *et al.* 1996). Another potential source of *new* N in estuaries may be N fixation by resident prokaryotes (Paerl 1990). Groundwater may be a significant nutrient source, but quantifying the input rates from this source is difficult in estuarine habitats and is virtually unknown for many estuaries (Tobias *et al.* 2001). In estuaries that are surrounded by watersheds low in nutrients, nutrients may be imported from coastal waters through tidal exchange and gravitational flow. The major sources of *regenerated* nutrients are the benthos and recycling in the water column. The relative importance of these two sources depends on the water volume and benthic surface area of the estuary. In shallow estuaries, benthic processes tend to dominate while water column processes are more important in deeper estuaries (Hobbie 2000).

Although estuaries function as nutrient 'traps', there are four major sinks that are responsible for nutrient removal. In most estuaries (except negative estuaries), the net flow of water is out of the estuary. Thus some quantities of nutrients are exported to coastal waters and lost from the system. Tidal exchange volume, flushing time, and residence time are determinants of the importance of this export mechanism. Inorganic nitrogenous compounds such as nitrate (NO_3^-), nitrite (NO_2^-), and ammonium (NH_4^+) are removed from the system by microbial processes. Nitrification and denitrification result in the oxidation of NH_4^+ to NO_2^- to NO_3^- and the reduction of NO_3^- to gaseous forms such as nitrous oxide (N_2O) or dinitrogen gas (N_2). These two process are coupled in marine sediments, with nitrifying bacteria supplying the NO_2^- and NO_3^- used by denitrifying bacteria. The efficient coupling of nitrification-denitrification results in the conversion of bioavailable N forms to gaseous N_2 and is a primary mechanism for N losses in coastal ecosystems. However, nitrification is an aerobic process while denitrication is an anaerobic process. Under anoxic conditions, which are commonly found below 1 cm in sediments, organic N is regenerated by microbial processes (ammonification) to produce NH_4^+. In the absence of oxygen, or when nitrification rates are less than NH_4^+ supply rates, the primary form on N released from sediments is NH_4^+. Thus the nitrogen species and fluxes from sediments are a function of the rate balances of diffusion and microbial processes. In estuaries with high sedimentation rates, organic matter and nutrients may be buried deep in sediments and effectively removed from the system. However, severe wind events (e.g., hurricanes) or dredging activities may uncover these deeper sediments, allowing the reintroduction of nutrients into the estuary. Finally, fisheries harvests, which result in the removal of biota from the estuary, are another important but rarely quantified nutrient export mechanism.

The spatio-temporal fluctuations of water column nutrients are essentially unique to each estuary. For this review, Galveston Bay, Texas, an estuary typical for the Gulf of Mexico, was selected as a case study to illustrate estuarine nutrient fluctuations.

Galveston Bay

Galveston Bay is one of 7 major bays along the Texas coast and is the second largest estuary in the Gulf of Mexico (Figure 1). The drainage basin includes the metropolitan areas of Dallas/Fort Worth and Houston, Texas. The human population surrounding the bay exceeds 5 million people. Galveston Bay has a high socioeconomic value and is responsible for approximately one-third of the state's fishing income. Commercially important species include shrimp, blue crabs, and oysters. For example, the average annual oyster harvest for the bay is ca. 1,800 metric tons with a value of 8 million US dollars (Robinson *et al.* 2000).

Physical description

Galveston Bay (29.5°N, 94.8°W), a coastal plain estuary, has a surface area of 1,554 km^2 and is surrounded by 526 km^2 of marshland. A deep (12-14 m), but narrow (120-160 m) ship channel that serves the Port of Houston bisects the bay. With the exception of oyster reefs and dredge spoil banks, the bathymetry is generally flat and regular with water depths ranging from 2 to 3 m. Compared with other estuaries in the US, Galveston Bay has a small water volume relative to the benthic surface area (Figure 2). The shallow water column and broad, flat bottom create a situation in which the role of the benthos as a modulator of water column processes is likely enhanced in comparison with other estuaries.

The major freshwater inputs into the estuary are from the Trinity (83%) and San Jacinto (8%) Rivers (NOAA 1989). The tidal range in the bay averages 40 cm, is primarily diurnal, and fosters the long hydraulic residence time of the estuary (40-88 days) (Santschi 1995). Most of the tidal exchange (ca. 80%) in Galveston Bay occurs through Bolivar Roads at the mouth of estuary while a small, manmade cut through Bolivar Peninsula (Rollover Pass) provides minor circulation in East Bay (NOAA 1989). Tidal exchange in West Bay occurs through San Luis Pass. The low tide water volume of Galveston Bay is 2.50 x 10^9 m^3 and the intertidal volume is 2.10 x 10^8 m^3 (Lowery 1998). Thus, only 7.7% of the total water volume of the bay is exchanged by tidal processes. The absence of a strong tidal influence reinforces the potential importance of benthic/pelagic exchange in governing nutrient concentrations in Galveston Bay. The long-term average salinity for Galveston Bay is 15.5 psu and can range from 13.7 to 18.0 during low and high riverine discharge conditions, respectively (NOAA 1989; Lowery 1998). Furthermore, winds are more important than tides for circulation in Galveston Bay and the predominant wind direction is southeasterly. Bay circulation is generally anticyclonic during periods of high freshwater inflow and cyclonic during low inflow (NOAA 1989).

Sediments in the bay are composed primarily of mud (silt and clay) and sandy mud (NOAA 1989). The upper bay is underlain by shallow piercement salt domes that contain oil and gas deposits and, consequently, there are a number of active oil wells and platforms in Trinity Bay. An extensive analysis of trace metal concentrations in the water column and sediments of Galveston Bay found metal levels similar to more pristine bays elsewhere (Morse *et al.* 1993). Trace metal concentrations in

Figure 1. Location map for Galveston Bay, Texas (29.5°N, 94.8°W). The bay transect is denoted by the solid line and station locations are marked with numbers (0.0 to 56.3) corresponding to the distance (km) upstream from the mouth of the estuary.

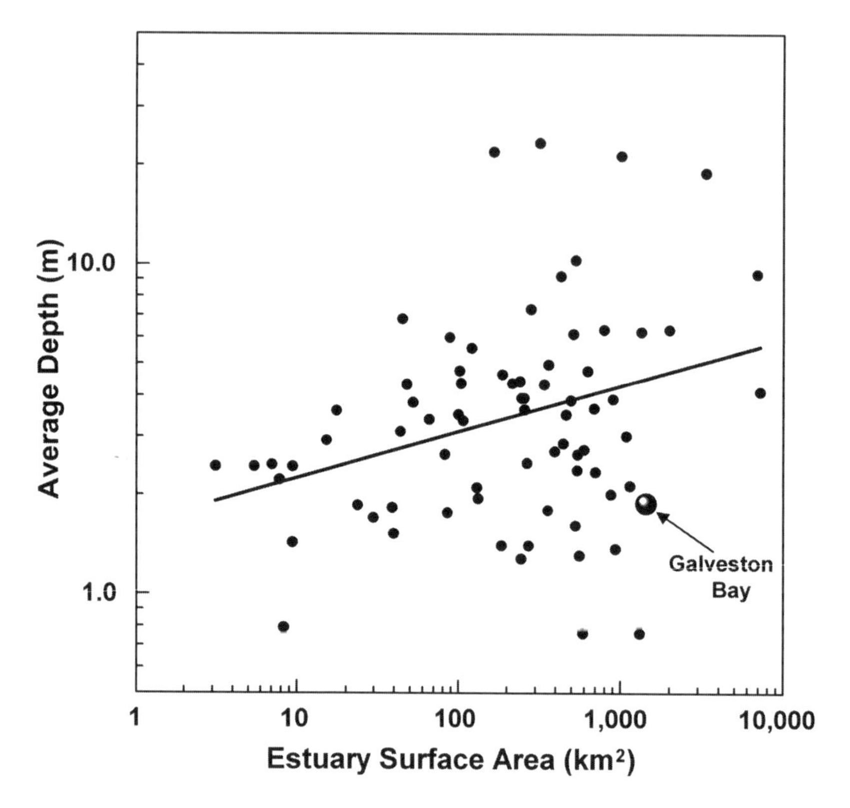

Figure 2. Plot of the average depth and surface area for 74 estuaries in the continental US. The solid line is a linear regression line for all points. Data from Bricker *et al.* 1999.

the water column were significantly correlated with concentrations of suspended particulate matter, suggesting that wind and tidal resuspension of sediments was the primary cause for elevated metal concentrations. In addition, many of the metals in the top 10 cm of sediment were coprecipitated with pyrite (Morse *et al.* 1993).

Long-term changes in nutrient concentrations

Improvements in municipal water treatment discharges into the San Jacinto River and the Houston Ship Channel have resulted in a shift from ammonium to nitrate as the primary source of N entering lower Galveston Bay (Crocker and Koska 1996). In the period from 1971 to 1991, there was an increasing trend in the loading of nitrate and nitrite from this source. However, overall water quality in the HSC improved markedly over the 20-year period (Crocker and Koska 1996). Criner and Johnican (2002), using data from the Texas Natural Resources and Conservation Commission (TNRCC), summarized the long term trends in water and sediment quality for Galveston Bay during the period from 1969 to 1999. Ammonium, nitrate-nitrite, and total phosphorus concentrations declined over the period. They also reported a concomitant decline in the monthly average chlorophyll *a* concentration (phytoplankton biomass). Santschi (1995) examined the Texas Water Commission database (1980-1989) to determine the seasonality of water column nutrient concentrations in Galveston Bay. Phosphate concentrations exhibited a recurring seasonal maximum in September which was attributed to benthic regeneration of phosphorus at the end of summer. Nitrate concentrations were inversely correlated with salinity and suggested that the Trinity River was the major source of nitrate for the estuary (Santschi 1995).

Annual N loading

Stanley (2001) summarized nitrogen loading rates for Galveston Bay from 1977 to 1990. Estimates of total nitrogen (NO_3^- + NO_2^- + TKN) loading were determined using data from USGS river flow gage stations (gaged flows), ungaged watershed flows, wastewater (return flows), and wet deposition directly to the bay surface (rain) (Figure 3). Groundwater inputs were assumed to be insignificant and not included in the calculations. Over the 13-year period, N loading attributed to wastewater and rain was relatively constant. In contrast, gaged and ungaged flows were the primary sources of N loading. The high interannual variability in loading rates was attributed to rainfall in the drainage basin (Brock 2001). Using these data, Brock (2001) estimated that the mean annual load of total nitrogen to Galveston Bay was 38,350 ± 11,488 sd x 10^3 kg N y^{-1} with a total N load per unit surface area

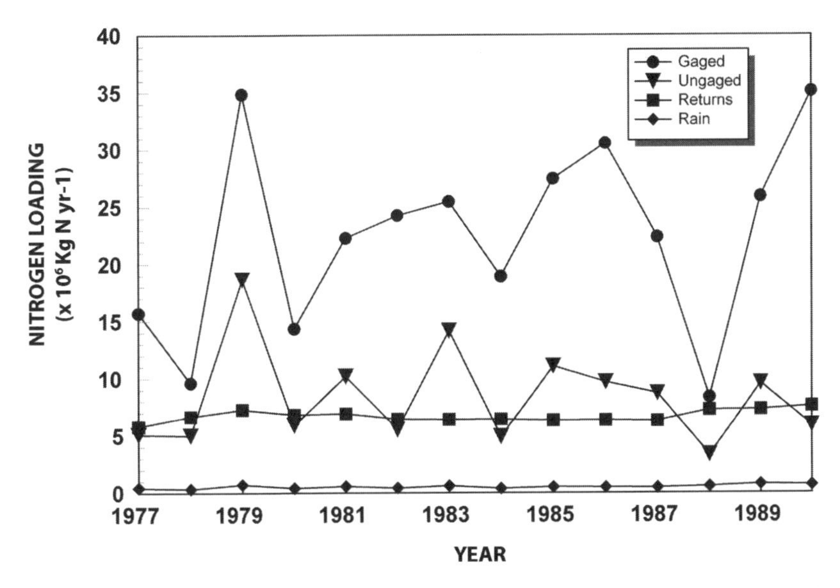

Figure 3. Annual estimates (1977-1990) of total nitrogen loading from 4 different sources for Galveston Bay. (Data from Stanley 2001).

of 0.633 ± 0.190 sd x 10^3 kg N km^{-2} y^{-1}. In comparison with four other large Texas estuaries, Galveston Bay received more than double the next highest area-specific N loading (Brock 2001). Atmospheric deposition of N directly to the estuary surface accounted for only 1.4% of the total N loading.

Using data from the National Atmospheric Deposition Program (NADP) for the period 1985 - 1996, Meyers *et al.* (2001) determined the total annual N deposition to the Galveston Bay watershed, which includes the open areas of the bay. The average annual loading rates (in kg N km^{-2} y^{-1}) were 0.0209 for wet deposition NH_4, 0.0188 for wet deposition NO_3, 0.0497 for wet deposition total N (NH_4 + NO_3 + DON), 0.0033 for dry deposition NH_4, 0.0328 for dry deposition NO_3, and 0.0394 for dry deposition total N (Meyers *et al.* 2001). Therefore the total estimated atmospheric nitrogen deposition (wet and dry) to Galveston Bay was 0.089 kg N km^{-2} y^{-1}.

Monthly N loading

D. Brock (Texas Water Development Board, Bays and Estuaries Program, Austin, Texas) graciously supplied monthly estimates of total N loading for Galveston Bay from 1977 to 1990 based upon modeling methods outlined in Brock *et al.* (1996). Monthly N loading events, like annual loading, was highly variable in both duration and magnitude (Figure 4). However, N loading was usually higher in the late winter-early spring months. The years 1989 and 1990 were 'wet' years with protracted river runoff and, consequently, high N loading rates. The higher resolution monthly rate

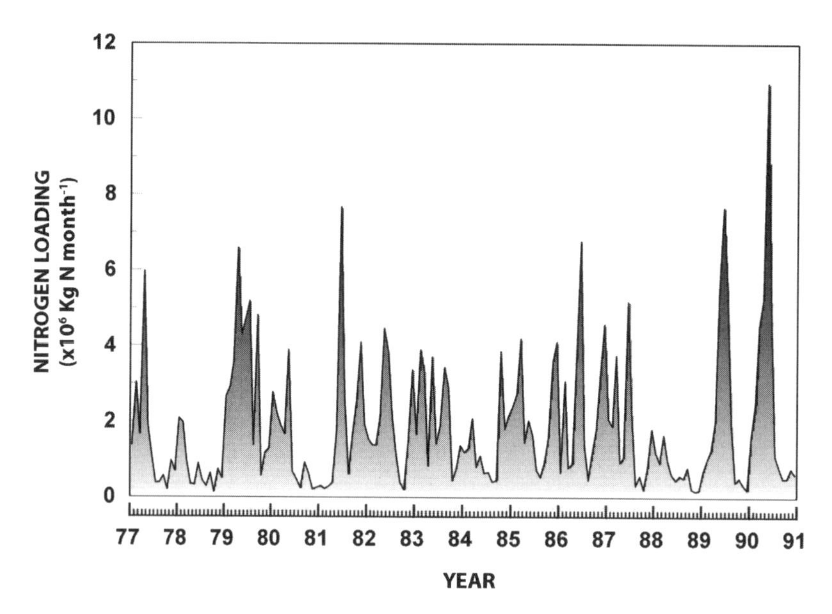

Figure 4. Monthly estimates (1977-1990) of total nitrogen loading for Galveston Bay. (Data courtesy of D. Brock, Texas Water Development Board).

estimates provide a mean annual loading estimate of 22,559 ± 8,368 sd x 10^3 kg N y^{-1}, which is much lower (59%) than the annual estimate given in Brock (2001). This discrepancy is attributed to the shorter-term averaging (monthly) which incorporates month to month variations in loading rates.

Benthic fluxes

Zimmerman and Benner (1994) undertook a study of denitrification, nutrient regeneration, and carbon mineralization in sediments collected from 5 stations in Galveston Bay. Denitrification rates, based on laboratory incubations of sediment cores, ranged from 0 to 47 μmol N_2 m^{-2} h^{-1} at temperatures ranging from 16.0 to 29.5°C. Summer (July) denitrification rates were the highest (30.2 μmol N_2 m^{-2} h^{-1}), followed by May (19.8) and March (5.4). Fluxes of nitrate + nitrite were from the sediment to the water (0 to 0.93 μmol N m^{-2} h^{-1}) at the lower estuary sites and from the water to the sediment (-0.21 to -1.21 μmol N m^{-2} h^{-1}) in the upper estuary. Ammonium fluxes (-0.29 to 3.21 μmol N m^{-2} h^{-1}) were highest in March and nearly always out of the sediments. Phosphate fluxes were small (<4 μmol P m^{-2} h^{-1}) and usually from the sediment to the overlying water. Overall, denitrification was responsible for ca. 75% of the benthic inorganic nitrogen flux in the spring and summer. In the winter, the majority (80%) of the nitrogen flux was in the form of ammonium. In the upper estuary, 24% of the N_2 production could be supported by the nitrate+nitrite influx into sediments and only 7% could be supported in the lower estuary. Thus the influx of water column nitrate and nitrite could not support the measured denitrification rates and illustrated a strong coupling of nitrification and denitrification in these sediments. Elemental analysis of pre- and post-incubation sediments showed preferential remineralization of nitrogen relative to carbon. Based on these measurements, Zimmerman and Benner (1994) estimated that denitrifiers were responsible for 37% of the total benthic carbon remineralization in the upper bay and 13% in the lower estuary and they attributed nearly one-third of the total sediment oxygen consumption to nitrification in this estuary. Overall, they estimated that denitrification removed only 14% of the total N loading into Galveston Bay.

Joye and An (1999) undertook a comprehensive 3-year study of denitrification at four locations in Galveston Bay. The average dentrification rate was 1.80 mmol N m^{-2} d^{-1} and ranged from 0.00 to 4.58, with lower rates in the winter and higher rates in the summer. The NO_3 flux into sediments supported, on average, 25% of the measured denitrification and suggested that the main source of NO_3 was from *in situ* nitrification. By extrapolating denitrification rates across the estuary, Joye and An (1999) estimated that the annual denitrification rate in Galveston Bay was 8.3 x 10^7 mol N $month^{-1}$ and N removal via denitrification averaged 52% of the total N-loading.

In a similar study in Trinity Bay, An and Joye (2001) used light and dark *in situ* benthic chambers to measure the diel variability in sediment oxygen demand and the flux of N_2. Winter N_2 fluxes (in mmol m^{-2} d^{-1}) were 1.00 (light) and 0.48 (dark) while summer rates were 2.2 (light) and 0.1 (dark). Ammonium fluxes (mmol N m^{-2} d^{-1}) were 0.00 (winter) and 0.82 (summer) while nitrate fluxes were -0.1 (winter) and 0.2 (summer). The daily rates for denitrification (mmol N m^{-2} d^{-1}) were 0.58 ± 0.04 sd for winter and 1.93 ± 0.1 sd for summer. These denitrification rates are four times larger than the rates reported by Zimmerman and Benner (1994). An and Joye (2001) attribute the disparity to the fact that their samples were incubated in both light and dark conditions while Zimmerman and Benner (1994) incubated their samples under dark conditions only. An and Joye (2001) found that denitrification rates were higher during the day and proposed that the higher rates were due to oxygen production by benthic microalgae, which enhanced the coupling between nitrification and denitrification. Using a simple model, An and Joye (2001) showed that when photosynthesis was inhibited, the rates of nitrification and denitrification decreased significantly. Based on the results of their study and modeling efforts, An and Joye (2001) estimated that more than 50% of the N load to the estuary is denitrified as compared to the estimate of 14% by Zimmerman and Benner (1994).

The findings of An and Joye (2001) highlight the importance of mixing events and turbulence on benthic/pelagic exchange of nutrients in Galveston Bay. During periods when sediments are resuspended, the overall effect is a reduction in the amount of light available for benthic microalgal photosynthesis. Thus denitrification rates would be slowed and the primary source of inorganic N would be NH_4 rather than NO_3. During calm weather periods, when turbidity is low, benthic microalgal photosynthesis may enhance the rates of nitrification-denitrification and reduce the input of dissolved inorganic nitrogen from sediments into overlying waters.

Warnken *et al.* (2000) undertook a study of benthic nutrient exchange in Trinity Bay from 1994 to 1996. Using both benthic chambers and porewater nutrient profiles, Warnken *et al.* (2000) calculated benthic and diffusive fluxes across the sediment water interface. Phosphate fluxes ranged from 0.08 to –0.53 mmol P m^{-2} d^{-1} while ammonium fluxes were always directed out of the sediments and ranged from –0.44 to –5.1 mmol N m^{-2} d^{-1}. Silicate fluxes were almost always out of the sediment and ranged from 1.7 to –1.2 mmol Si m^{-2} d^{-1}. A comparison of the rates and magnitudes of nutrient fluxes revealed that both direct measurements and diffusive flux methods agreed on the direction of fluxes, but differed in the magnitude. The diffusive flux method, based on porewater nutrient profiles, underestimated nutrient exchange rates relative to those determined by benthic chamber measurements. During the study period, turnover times for phosphate, silicate, and ammonium were 7-135 d, 4-56 d, and 0.3-10 d, respectively. These turnover times were significantly shorter than the average water residence time (1.5 yr.) for Trinity Bay during the study period. Benthic nutrient inputs of ammonium and phosphate were 1 to 2 orders of magnitude higher than inputs from the Trinity River and suggested that benthic regeneration was the primary source of these nutrients (NH_4 and PO_4) for Trinity Bay. Warnken *et al.* (2000) also found that benthic nutrient inputs were higher in the middle and outer regions of Trinity Bay. Elemental analysis of sediments showed lower C: N ratios for these areas and, therefore, suggested that the deposited autochthonous organic matter was of higher quality than other areas of the bay.

Short-term nutrient fluctuations in Galveston Bay

In May 1999, a biweekly sampling program was instituted to collect data on water quality, nutrient concentrations, and phytoplankton dynamics in Galveston and Trinity Bays. Samples for physical, chemical, and biological parameters were undertaken at biweekly to monthly intervals over a 33-month period (1999-2002) in Trinity Bay and Galveston Bay (Figure 1). The purpose of this program was to characterize short (monthly) to long-term (annual) changes in environmental variables, nutrients, and phytoplankton community structure and function.

Materials and methods

Eight to ten sampling stations were located along the long axis of Galveston Bay (Figure 1). Sampling at the stations consisted of vertical profiles of physical and chemical (O_2, pH, temperature, salinity, conductivity) parameters using a Hydrolab H20 water quality monitor. Irradiance (PAR) in the water column was determined using a LiCor 4π quantum sensor. The upper meter of the water column was sampled with a 1 m integrated water sampler (opaque PVC baler) and transported to a field laboratory for sample processing and preservation. Phytoplankton biomass, community composition and nutrient concentrations were determined for each sample. For nutrient analyses, water samples were filtered through combusted (350 °C, 2.5 h) Whatman GF/F (25 mm) filters, poured into acid-rinsed Nalgene bottles and stored at –20 °C until analyzed. Nitrate, nitrite, urea, phosphate and silicate were quantified according to Atlas *et al*. (1971) and ammonium was quantified according to Harwood and Kuhn (1970).

High performance liquid chromatography (HPLC) was used to determine chemosystematic photosynthetic pigments to estimate phytoplankton abundance. Aliquots (0.5 L) of seawater were filtered under a gentle vacuum (<50 kPa) onto 2.5 cm dia. glass fiber filters (Whatman GF/F), immediately frozen, and stored at –80 °C. For analyses, frozen filters were placed in 100 % acetone (1 ml), sonicated, and extracted at –20 °C for 18-20 h. Filtered extracts (375 µL) were injected into a Shimadzu HPLC equipped with a monomeric (Rainin Microsorb-MV, 0.46 x 10 cm, 3 µm) and a polymeric (Vydac 201TP, 0.46 x 25 cm, 5 µm) reverse-phase C_{18} column in series. A nonlinear binary gradient was used for pigment separations (Pinckney *et al*., 1996). Absorption spectra and chromatograms (440 nm) were acquired using a Shimadzu SPD-M10av photodiode array detector. Pigment peaks were identified by comparison of retention times and absorption spectra with pure standards, including chlorophylls *a*, *b*, β-carotene, fucoxanthin, lutein, canthaxanthin, echinenone, gyroxanthin, peridinin, alloxanthin, and zeaxanthin (DHI, Denmark). Other pigments were identified by comparison to extracts from phytoplankton cultures and quantified using the appropriate extinction coefficients (Jeffrey *et al*. 1997). Detailed protocols for the HPLC photopigment methods are posted at http: //www.biol.sc.edu/~jpinckney/lab_protocols.htm.

Results

The Trinity River is the primary source of freshwater and new nutrient inputs into Trinity Bay and Galveston Bay (Figure 5). From May 1999 to February 2002, river discharge (x 10^7 m^3 d^{-1}) ranged from 224.3 to 1.8 with a mean flow of 24.9 (Figure 5). Peak discharge occurred in June 2001 and was associated with tropical storm Allison, which dropped ca. 94 cm of rain on the watershed over a 5 day period (National Weather Service, unpubl. data). Dry years (1999 and 2000) resulted in low river discharge and 2001 was already a relatively wet year prior to the arrival of the tropical storm. The low river discharge in 1999 and 2000 resulted in high salinities in excess of 25 psu in upper Trinity Bay (Figure 5). Likewise, high discharge in 2001 displaced isohalines towards the mouth of Galveston Bay and the tropical storm lowered salinties to < 5 psu over much of the bay. Bay salinities closely tracked freshwater discharge from the Trinity River. The diffuse attenuation coefficient (k_d) ranged from <1 to >10 m^{-1} with a mean of 2.28 ± 1.68 sd (Figure 5). Water clarity was generally lower during periods of high river discharge. Resuspension of sediments during moderate wind events (5-10 m sec^{-1}) was a major contributor to turbidity and light attenuation in the water column. Using an average k_d value of 2.28 m^{-1}, the depth of the 1% light level (the theoretical limit for net photosynthesis) would be 2.02 m. Since the average depth of Galveston Bay is 2.4 m, it is likely that benthic microalgal photosynthesis was frequently light-limited, especially in Trinity Bay. Phytoplankton chlorophyll *a* (Chl *a*) averaged 10.4 ± 10.1 sd µg L^{-1} and ranged from 0.27 to 101.5 µg Chl *a* L^{-1} over the study period (Figure 5). Phytoplankton blooms were associated with periods of moderate to high river discharge and the largest bloom, composed primarily of the dinoflagellate *Prorocentrum minimum,* occurred in February 2001. The red tide dinoflagellate, *Karenia brevis*, was advected into Galveston Bay from coastal waters in September 2000 and a moderate bloom (28.3 µg Chl *a* L^{-1}) was detected in the lower bay.

High river discharge events, especially in 2001, resulted in high concentrations of nitrite (NO_2) and nitrate (NO_3) in Trinity and Galveston Bays (Figure 6). For a period of ca. 4 months, NO_3 levels were > 30 µmol N L^{-1} over much of the upper bay. The passage of tropical storm Allison (June 2001), and associated high runoff and flushing, reduced NO_3 concentrations to less than 5 µmol N L^{-1} within a few days. In general, NO_2 concentrations were higher in the lower bay while NO_3 was higher in the upper bay and was directly linked to river discharge, suggesting that the Trinity River was the primary source of NO_3. Occasionally, ammonium (NH_4) concentrations exceeded 14 µmol N L^{-1}, but these occurrences were limited to short durations at isolated stations in the bay and were not associated with hypoxic/anoxic events. In December 2001 and June 2002, NH_4 concentrations were generally high (> 2 µmol N L^{-1}) throughout the bay (Figure 6). Although some NH_4 inputs appear to be associated with the Trinity River (only during times of high discharge), there are other times when high NH_4 concentrations do not seem to be related to river flow (e.g., summer 2001). The high NH_4 concentrations in March-May 2000 occurred when concentrations of NO_3 were low and may be indicative of spring runoff from newly fertilized croplands in the watershed. The high NH_4 concentrations in the middle and

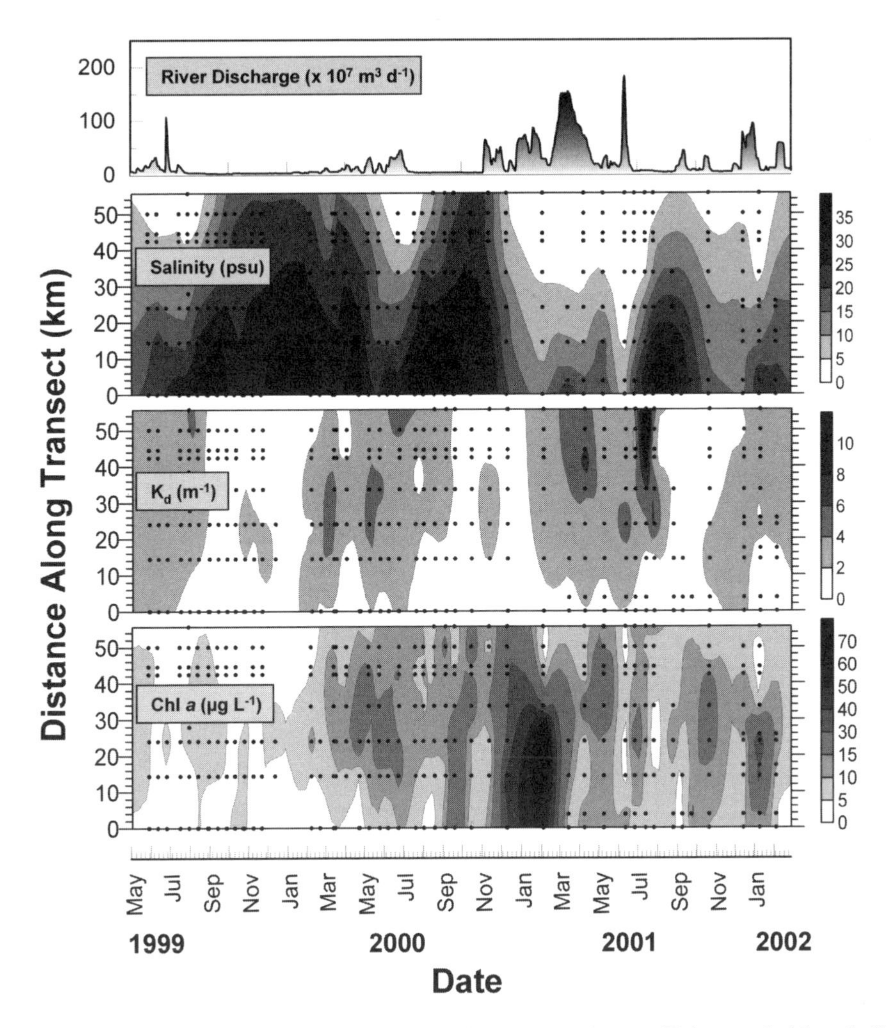

Figure 5. Spatio-temporal contour plots of salinity, diffuse attenuation coefficient, and chlorophyll *a* for Galveston Bay (1999-2002). The distance along transect refers to the distances given in Figure 1. The dots depict sampling dates and locations.

lower reaches of the bay may be attributable to benthic regeneration. The pattern for overall dissolved inorganic nitrogen (DIN = NO_2 + NO_3 + NH_4) was that concentrations were low from 1999 to mid-2000, then increased due to freshwater inflow from the Trinity River (Figure 6). Tropical storm Allison and associated heavy rainfall released a large amount of freshwater low in DIN (due to dilution) into Galveston Bay. By November 2001, the normal pattern of riverine DIN inputs resumed. The periods of high DIN concentrations were also correlated with phytoplankton biomass in the bay and indicates the tight coupling between DIN inputs and phytoplankton responses (Figs. 5, 6).

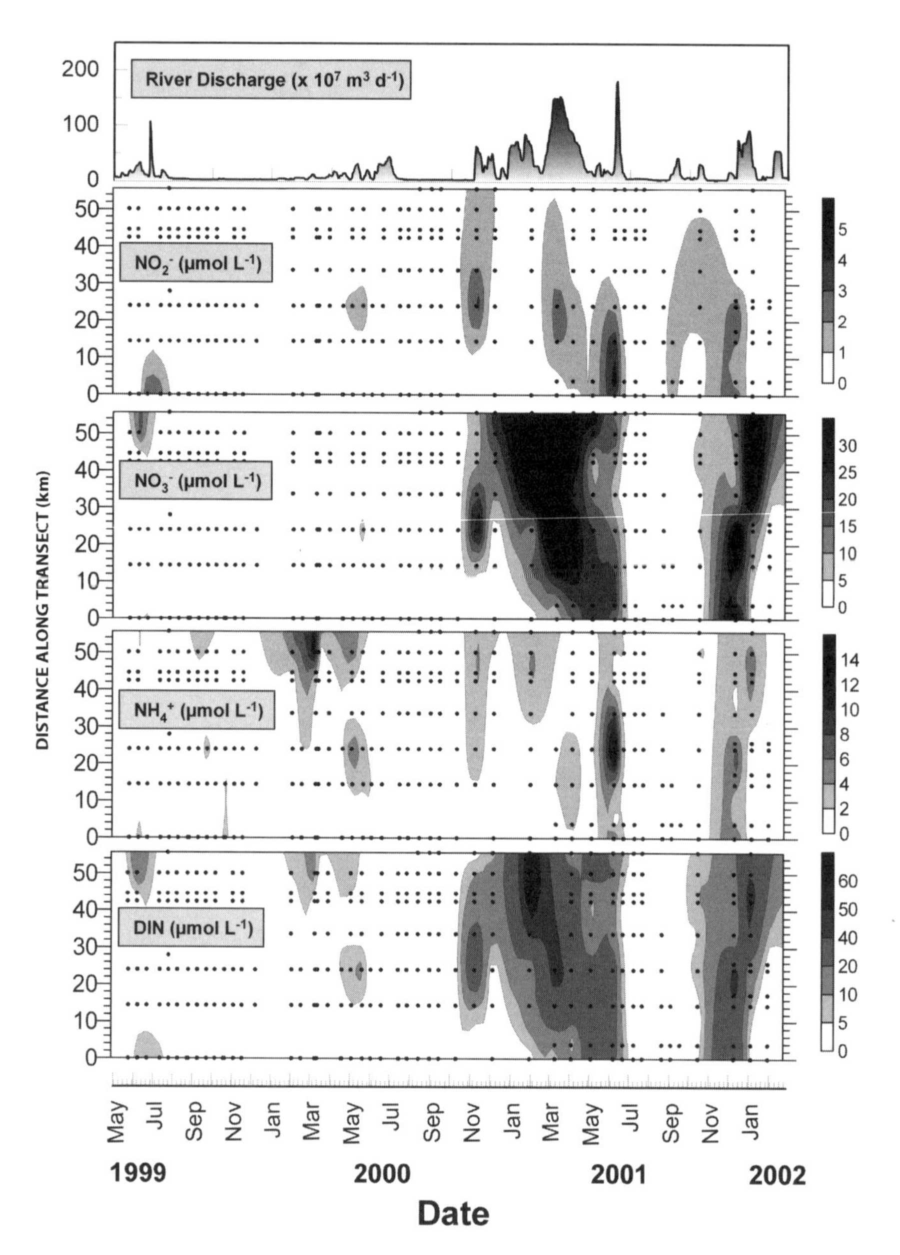

Figure 6. Spatio-temporal contour plots of nitrite, nitrate, ammonium, and total dissolved inorganic nitrogen for Galveston Bay (1999-2002). The distance along transect refers to the distances given in Figure 1. The dots depict sampling dates and locations.

Urea concentrations were generally low (< 2 μmol N L^{-1}) but reached a high of 30.3 μmol N L^{-1} in the lower bay in June 2001, following tropical storm Allison (Figure 7). Phosphate (PO_4) concentrations averaged 3.23 ± 2.38sd μmol P L^{-1} and ranged from 0 to 14.5 (Figure 7). Concentrations were generally higher in the summer months in the upper bay, with the exception of January 2002 when high PO_4 levels were detected in the lower bay. In general, PO_4 showed an inverse relationship with river discharge, suggesting that the major source of PO_4 may be benthic regeneration. The mean silicate ($SiOH_4$) concentration over the study period was 42.2 ± 30.2 sd μmol Si L^{-1} and ranged from 1.66 to 161 (Figure 7). Like PO_4, silicate concentrations were highest under low river flow conditions in Trinity Bay. The low concentrations may reflect utilization by diatoms when DIN concentrations were high. Alternatively, higher concentrations of silicate may be due to benthic regeneration during the warm summer months. Silicate concentrations peaked after the passage of tropical storm Allison and may indicate the input of silicate (as sediment) in stormwater runoff or possibly sediment turnover following the storm.

The DIN: P molar ratio averaged 3.55 ± 7.51 sd and ranged from 0.01 to 80.5. A plot of the paired DIN and P values shows that this ratio was rarely above the Redfield ratio of 16: 1 and suggests that phytoplankton in Galveston Bay were usually N-limited (Figure 8). In February and June 2001, the ratio exceeded 16: 1 due to the high concentrations of nitrate associated with riverine N loading. During these brief events, phytoplankton in the bay may have been P-limited.

Discussion

Freshwater discharge from the Trinity River into Galveston Bay plays a major role in the fluctuations of salinity, nutrients, and phytoplankton biomass. During periods of high river flow, nitrate is the major form of inorganic N added to the estuary and is a source of *new* N for the system. When river flows are low, especially during dry summer months, the bay receives little N input from the river and the major source of N must be through regeneration, either from the benthos or the water column. During these periods, benthic remineralization and release is a likely source for DIN (Zimmerman and Benner 1994; Joye and An 1999; Warnken *et al.* 2000). However, phytoplankton uptake of DIN may prevent measurable accumulations of DIN in the water column. Spring inputs of ammonium from the Trinity river are likely due to agricultural runoff, since this is the time of year when farmers fertilize their fields with ammonium or urea-based fertilizers. Warnken *et al.* (2000) reported benthic ammonium fluxes as high as 5.1 mmol N m^{-2} d^{-1} while An and Joye (2001) found rates of 0.82 mmol N m^{-2} d^{-1} in Galveston Bay during the summer months. Therefore, ammonium regeneration from the benthos is likely a consistent source of N for the estuary. During dry periods, benthic release of ammonium is probably the primary source of N for phytoplankton in the bay.

The primary source of phosphate for the bay seemed to be benthic regeneration. The highest phosphate concentrations occurred during low river flow in the summer months. Additionally, phosphate concentrations were lowest during high river discharge

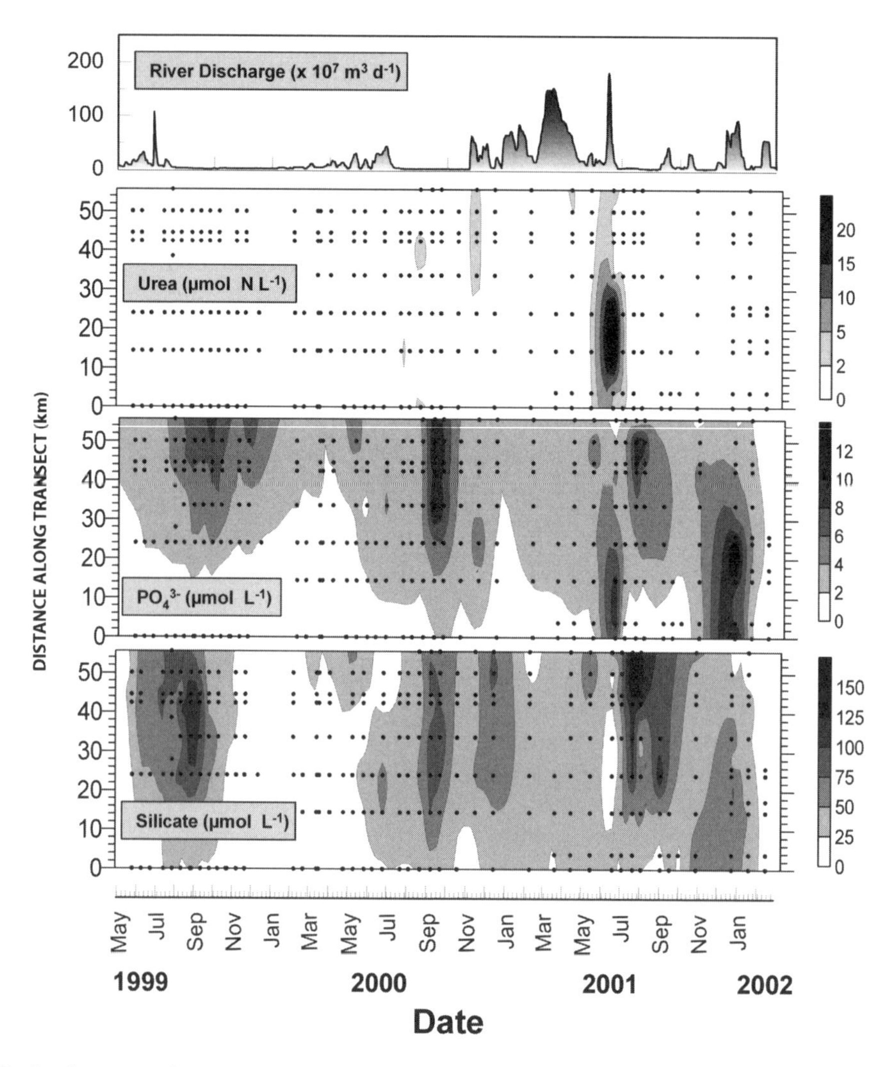

Figure 7. Spatio-temporal contour plots of urea, phosphate, and silicate for Galveston Bay (1999-2002). The distance along transect refers to the distances given in Figure 1. The dots depict sampling dates and locations.

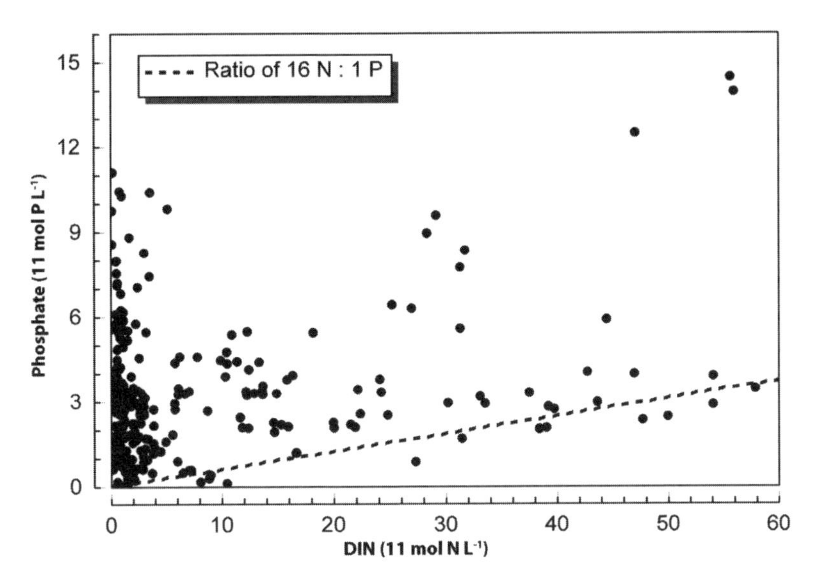

Figure 8. Plot of phosphate vs. total dissolved inorganic nitrogen for individual samples taken from Galveston Bay (1999-2002). The dashed line represents the Redfield ratio of 16 N : 1 P. Points above the line indicate N-limitation and values below the line suggest P-limitation for phytoplankton.

conditions. Upon examination of a 9-year data set, Santschi (1995) reported that phosphate concentrations regularly reach a maximum in September. The data in the present study reflect this observation exactly. Zimmerman and Benner (1994) report small phosphate fluxes of < 4 μmol P h^{-1}, but their study was conducted from March through July, missing the period when P release was usually highest. Warnken *et al.* (2000) found P flux rates as high as 0.53 mmol m^{-2} d^{-1} in Galveston Bay. The inverse relationship between river discharge and phosphate concentration could be due simply to dilution. However, during high flow conditions with high nitrate concentrations, the phytoplankton demand for P would be higher than under low N conditions. Thus the fluctuations in phosphate concentrations were likely a dynamic balance between benthic regeneration and phytoplankton uptake. Silicate concentrations closely tracked phosphate and similar regulating mechanisms may apply to this nutrient as well.

Urea and nitrite were usually minor nutrients Galveston Bay, except following the passage of tropical storm Allison. High concentrations were measured in the lower bay along the deep ship channel. Tropical storm Allison was an important event in the biogeochemistry of Galveston Bay. Following the passage of the storm, ammonium, urea, and phosphate concentrations briefly increased by an order of magnitude in lower Galveston Bay (Figs. 6, 7). These high nutrient concentrations suggest there was significant remineralization of organic matter, possibly particulate organic matter washed downstream following the storm (Paerl *et al.* 1998). With further flushing of the system by freshwater input, DIN concentrations were reduced to low levels (total

DIN < 5 µmol L^{-1}), probably due to dilution by high rainfall and rapid runoff associated with the tropical storm. Thus the tropical storm 'flushed' the system of dissolved nutrients. The bay seems to have recovered within 3-4 months after storm passage, with moderate concentrations of DIN resulting from late winter rains and river discharge.

Phytoplankton blooms closely tracked nitrate concentrations in Galveston Bay, although peaks in biomass were displaced downstream from peaks in nitrate concentration. The lower salinity (< 10 psu) in the upper bay during these periods may have prevented the formation of large blooms. Blooms likely formed as the DIN was mixed with higher salinity waters in the lower bay. Twelve nutrient (NO_3^-, PO_4^{3-}) addition bioassays, conducted from May 1999 to July 2000, indicated that the phytoplankton community in Galveston Bay was consistently N-limited (Örnòlfsdòttir 2002). All major algal groups of the community showed significant increases in biomass in response to the addition of NO_3^- (10 µM final concentration), sometimes resulting in a doubling in biomass (relative to controls) over a 24 hour incubation period. Evidence of phosphate or silicate limitation was not detected in these bioassays. In terms of size distributions, the picoplankton (< 2 µm), nanoplankton (2-20 µm), and microplankton (20-200 µm) fractions constituted 18%, 50%, and 32% of the total community biomass, respectively. All size ranges had representatives from most of the major algal groups and the nanoplankton had the highest diversity. The DIN: P ratio for Galveston Bay further supports the finding of N-limitation for phytoplankton. Over much of the 3-year study period, the DIN: P ratio was <16, the critical value according to the Redfield ratio. On the few occasions when the ratio was >16, the P limitation was likely the result of high DIN loading.

The ability of phytoplankton to take advantage of high nutrient conditions is also influenced by turbidity, which results in light limitation. Moderate winds that blow over bay waters resuspend sediments in the shallow bay waters. The average diffuse attenuation coefficient (k_d) over the 3-year period was 2.28 m^{-1}, which translates into a light level of 1% of incoming irradiance at a depth of 2.02 m. Thus, on average, benthic microalgal productivity is likely very low in waters deeper than 2 m. With an average bay depth 2.4 m, much of the benthos (and benthic microalgae) was usually light-limited. The shallow water column in Galveston Bay is usually well-mixed and phytoplankton may be able to sustain photosynthesis due to vertical mixing processes. Therefore, one could speculate that most of the primary production in Galveston Bay can be attributed to phytoplankton rather than benthic microalgae, especially during turbid conditions. Furthermore, benthic microalgal biomass in the bay is less than 10 mg Chl *a* m^{-2} (Pinckney and Lee, *in prep.*), which roughly approximates the areal biomass of phytoplankton.

Summary

This review has focused on the spatial and temporal scales of nutrient fluctuations in Galveston Bay, a large shallow subtropical estuary typical of the Gulf of Mexico.

Decadal and annual estimates of total N loading highlight the importance of interannual variation and long-term trends, which are useful for developing nutrient input management strategies and assessing the effectiveness of mitigation programs. However, long-term averaging of large spatial areas dampens the frequency and magnitude of seasonal events and system perturbations. Monthly estimates of N loading capture some of this variability and illustrate that monthly variation is higher (in relative terms) than annual variations. The comparison of Brock's (2001) loading estimates based on annual averaging (38,350 x 10^3 kg N y^{-1}) and monthly averaging (22,559 x 10^3 kg N y^{-1}) for the same time period (1977 - 1990) shows that the temporal averaging scale makes a big difference when estimating loading. Ideally, the best temporal scale for estimating loading rates is the frequency at which loading events occur to the system. In the case of Galveston Bay, major N loading events are directly related to Trinity River discharge. Since discharge is a function of meteorological conditions, which are unpredictable at time scales > 2 weeks, water samples for the determination of nutrient concentrations (to calculate loading) should be collected daily to obtain accurate and reliable loading estimates. Unfortunately, this sampling frequency is unreasonable for most monitoring agencies. The further development and implementation of *in situ* nutrient analyzers may provide the resolution needed for real-time, cost-effective estimates of nitrate loading.

The dynamic nature of nutrient distributions in Galveston Bay is clearly illustrated in the spatio-temporal contour plots. Nutrient concentrations in this estuary reflected a balance between river discharge, benthic regeneration, and water temperature (season). Major perturbations, such as the passage of tropical storm Allison, upset this balance and profoundly changed the biogeochemistry of the estuary. After this 'flushing' event, the system seemed to slowly return to a 'normal' pattern of nutrient distributions. High-resolution contour plots provided useful insights into the spatio-temporal relationships between different nutrients and the causal mechanisms responsible for the observed distributions.

Biotic responses to nutrient inputs were rapid, on time scales of 1 day. Estuarine phytoplankton biomass quickly increased, taking advantage of higher nutrient concentrations. In Galveston Bay, the phytoplankton community was N-limited and showed nearly instantaneous increases in biomass when DIN (primarily NO_3) was loaded into the estuary. Thus nutrient fluctuations on daily time scales can have profound effects on primary productivity in this estuary. In addition, the timing of loading events may also play a role in determining phytoplankton community structure. For example, short-term pulses of high NO_3 loading may favor diatoms while long periods of low loading could promote growth of smaller phytoplankton and phytoflagellates (Huisman and Weissing 1995).

Although atmospheric inputs of N are small (1.4% of total N loading) for Galveston Bay, this input could be very important, especially during periods of low river discharge. Afternoon showers during the summer could provide a significant source of *new* N to the estuary. Benthic fluxes of nutrients provide a nearly constant (depending on season) source of nutrients for phytoplankton. Again, during periods of low river discharge, the benthic flux is responsible for sustaining phytoplankton productivity. Denitrification appears to be the largest N sink for this estuary. Although there is disagreement on the actual rates, denitrification is responsible for removing from 14% to 52% of the total N loading. Of course, this estimate will change depending on the actual loading for the estuary.

Documenting nutrient fluctuations in estuaries is relatively straightforward provided the funding, time, and resources are available. Understanding the relative importance of the different processes driving these fluctuations on a system scale is a much more formidable task. Mechanistic studies of benthic and planktonic biogeochemical processes, coupled in space in time, are a necessary prerequisite for developing reliable models to simulate, and possibly predict, estuarine nutrient fluctuations.

Acknowledgements

Funding for the Galveston Bay research was provided by Texas Institute of Oceanography and Texas Department of Agriculture. T.L. Richardson, E.B. Örnòlfsdòttir, S.E. Lumsden, A.R. Lee, and A. Salazar assisted with field sampling and sample analyses. Special thanks to D. Brock for providing monthly N loading estimates.

References

AHNER, B.A., MOREL, F.M.M., and MOFFETT, J.W. 1997. Trace metal control of phytoplankton production in coastal waters. Limnol. Oceanogr. **42**: 601-608.

ALONGI, D.M. 1998. Coastal ecosystem processes. CRC Press.

AMON, R.M.W., and R. BENNER. 1994. Rapid cycling of high-molecular-weight dissolved organic matter in the ocean. Nature **369**: 549-552.

AN, S., and S.B. JOYE. 2001. Enhancement of coupled nitrification-denitrification by benthic photosynthesis in shallow estuarine sediments. Limnol. Oceanogr. **46**: 62-74.

ATLAS, E.L., S.W. HAGER, L.I. GORDON. and P.K. PARK. 1971. A practical manual for use of the Technicon AutoAnalyzer in seawater nutrient analyses; revised. Oregon State University, Corvallis, Technical Report 215.

AZAM, F., T. FENCHEL, G.J. FIELD, J.S. GRAY, L.A. MEYER-REIL, and F. THINGSTAD. 1983. The ecological role of water-column microbes in the sea. Mar. Ecol. Prog. Ser. **10**: 257-263.

AZAM, F., D.C. SMITH, G.F. STEWARD, and A. HAGSTROM. 1993. Bacteria-organic matter coupling and its significance for oceanic carbon cycling. Microb. Ecol. **28**: 167-179.

BENNETT, M.E., and J.E. HOBBIE. 1972. The uptake of glucose by *Chlamydomonas* sp. J. Phycol. **8**: 392-398.

BRICKER, S.B., C.G. CLEMENT, D.E. PIRHALLA, S.P. ORLANDO, D.R.G. FARROW. 1999. National estuarine eutrophication assessment: Effects of nutrient enrichment on the Nation's estuaries. National Oceanic and Atmospheric Administration, Washington, D.C.

Brock, D.A. 2001. Uncertainties in individual estuary N-loading assessments. p. 171-185. *In* R.A. Valigura *et al.* [eds.], Nitrogen loading in coastal waterbodies: An atmospheric perspective. Coastal and Estuarine Studies 57. American Geophysical Union, Washington, DC.

BROCK, D.A., R.S. SOLIS, and W.L. LONGLEY. 1996. Guidelines for water resources permitting: Nutrient requirements for maintenance of Galveston Bay Productivity. Final Report to Near Coastal Waters Program, EPA. by Texas Water Development Board, Austin, TX. 131 pp.

BURTON, J. D., and P.S. LISS. 1976. Estuarine chemistry, Academic Press.

CAPONE, D.G. 2000. The marine microbial nitrogen cycle, p. 455-493. *In* D.L. Kirchman [ed.], Microbial ecology of the oceans. Wiley-Liss, Inc.

CARLSSON, P., H. EDLING, and C. BECHEMIN. 1998. Interactions between a marine dinoflagellate (*Alexandrium catanella*) and a bacterial community utilizing riverine humic substances. Aquat. Microb. Ecol. **16**: 65-80.

CRINER, O., and M.D. JOHNICAN. 2002. Update 2000: Current status and historical trends of the environmental health of Galveston Bay. Galveston Bay Estuary Program. Webster, Texas.

CROCKER, P.A., and P.C. KOSKA. 1996. Trends in water and sediment quality for the Houston Ship Channel. Texas J. Sci. **48**: 267-282.
CROOT, P.L., J.W. MOFFETT, and L.E. BRAND. 2000. Production of extracellular Cu complexing ligands by eukaryotic phytoplankton in response to Cu stress. Limnol. Oceanogr. **45**: 619-627.
DAME, R., J. SPURRIER, and T. WOLAVER. 1989. Carbon, nitrogen, and phosphorus processing by an oyster reef. Mar. Ecol. Prog. Ser. **54**: 249-256.
DORTCH, Q. 1990. The interaction between ammonium and nitrate uptake by phytoplankton. Mar. Ecol. Progr. Ser. **61**: 183-201.
DROOP, M.R. 1974. Heterotrophy of carbon, p. 530-559. *In* W.D.P. Stewart [ed.]. Algal Physiology and biochemistry. Blackwell Sci.
FARRINGTON J.W. 1992. Marine organic geochemistry: review and challenges for the future. Mar. Chem. **39**: 1-244.
FISHER, T., E. PEELE, J. AMMERMAN, and L. HARDING. 1992. Nutrient limitation of phytoplankton in Chesapeake Bay. Mar. Ecol. Prog. Ser. **82**: 51-63.
GAINES, G., and M. ELBRÄCHTER. 1987. Heterotrophic nutrition. p. 224-268. *In* F.J.R. Taylor [ed.], The biology of dinoflagellates. Blackwell Sci.
GLIBERT, P., D. CONLEY, T. FISHER, L. HARDING, and T. MALONE. 1995. Dynamics of the 1990 winter/spring bloom in Chesapeake Bay. Mar. Ecol. Prog. Ser. **122**: 27-43.
GUO, L., P.H. SANTSCHI, and T.S. BIANCHI. 1999. Dissolved organic matter in estuaries of the Gulf of Mexico. p. 269-299 *In*: T.S. Bianchi, J.R. Pennock, and R.R. Twilley [eds.]. Biogeochemistry of Gulf of Mexico estuaries. John Wiley and Sons.
HARWOOD, J.E., and A.L. KUHN. 1970. A colorimetric method for ammonia in natural waters. Water Res. **4**: 805-811.
HELLEBUST, J.A., and J. LEWIN. 1977. Heterotrophic nutrition, p. 169-197. *In* W. Werner [ed.]. The biology of diatoms. Blackwell Sci.
HOBBIE, J.E. [ed]. 2000. Estuarine science: a synthetic approach to research and practice. Island Press. Pp. 593
HUISMAN, J. and F. WEISSING. 1995. Competition for nutrients and light in a mixed water column: a theoretical approach. Am. Nat. **146**: 536-564.
HUNG, C.C., D. TANG, K. WARNKEN, and P.H. SANTSCHI. 2001. Distributions of carbohydrates, including uronic acids, in estuarine waters of Galveston Bay. Mar. Chem. 73: 305-318.
JEFFREY, S., R. MANTOURA, and S. WRIGHT. [eds.]. 1997. Phytoplankton pigments in oceanography: Guidelines to modern methods. UNESCO.
JOYE, S.B., and S. AN. 1999. Denitrification in Galveston Bay Final Report. Submitted to Texas Water Development Board, Austin, Texas.
KEIL, R.G., and D.L. KIRCHMAN. 1994. Abiotic transformation of labile protein to refractory protein in sea water. Mar. Chem. **45**: 187-196.
KIEBER, D., J. MCDANIEL, and K. MOPPER. 1989. Photochemical source of biological substrates in seawater: Implications for carbon cycling. Nature **341**: 637-639.
KIRCHMAN, D.L., Y. SUZUKI, C. GARSIDE, and H.W. DUCKLOW. 1991. High turnover rates of dissolved organic carbon during a spring phytoplankton bloom. Nature **352**: 612-614.
LOWERY, T.A. 1998. Difference equation-based estuarine flushing model application to U.S. Gulf of Mexico estuaries. J. Coastal Res. **14**: 185-195.
MILLIMAN, J.D. 1974. Marine carbonates. Springer-Verlag. Pp. 374
MEYERS, T., J. SICKLES, R. DENNIS, K. RUSSELL, J. GALLOWAY, and T. CHURCH. 2001. Atmospheric nitrogen deposition to coastal estuaries and their watershed. p. 53-76. *In* R.A. Valigura *et al.* [eds.]. Nitrogen loading in coastal waterbodies: An Atmospheric perspective. Coastal and Estuarine Studies 57. American Geophysical Union.
MOFFETT, J.W., and L.E. BRAND. 1996. Production of strong, extracellular Cu chelators by marine cyanobacteria in response to Cu stress. Limnol. Oceanogr. **41**: 388-395.
MORSE, J.W., B.J. PRESLEY, R.J. TAYLOR, G. BENOIT, and P. SANTSCHI. 1993. Trace metal chemistry of Galveston Bay: water, sediments and biota. Mar. Environ. Res. **36**: 1-37.
NEILSON, A.H. and R.A. LEWIN. 1974. The uptake and utilization of organic carbon by algae: an assay in comparative biochemistry. Phycologia **13**: 227-264.
NIXON, S. 1995. Coastal marine eutrophication: a definition, social causes, and future concerns. Ophelia **41**: 199-219.
NRC (National Research Council). 2000. Clean coastal waters: understanding and reducing the effects of nutrient pollution. National Academy Press.
NOAA (National Oceanic and Atmospheric Administration). 1989. Galveston Bay: Issues, resources, status, and management. NOAA Estuary of the Month Seminar Series No. 13. US Department of Commerce.

OGAWA, H., Y. AMAGAI, I. KOIKE, K. KAISER, and R. BENNER. 2001. Production of refractory dissolved organic matter by bacteria. Science **292**: 917-920.
ÖRNÓLFSDÓTTIR, E.B. 2002. The ecological role of small phytoplankton in phytoplankton production and community composition in Galveston Bay, Texas. Ph.D. Dissertation. Texas A&M University.
PAERL, H.W. 1990. Physiological ecology and regulation of N_2 fixation in natural waters. Adv. Microb. Ecol. **11**: 305-344.
PAERL, H.W. 1995. Coastal eutrophication in relation to atmospheric nitrogen deposition: current perspectives. Ophelia **41**: 237-259.
PAERL, H.W., J.L. PINCKNEY, J.M. FEAR, and B.L. PEIERLS. 1998. Ecosystem responses to internal and watershed organic matter loading: consequences for hypoxia in the eutrophying Neuse River estuary, North Carolina, USA. Mar. Ecol. Prog. Ser. **166**: 17-25.
PILSON, M.E.Q. 1998. An introduction to the chemistry of the sea. Prentice-Hall.
PINCKNEY, J., D. MILLIE, K. HOWE, H. PAERL, and J. HURLEY. 1996. Flow scintillation counting of ^{14}C-labeled microalgal photosynthetic pigments. J. Plankton Res. **18**: 1867-1880.
PROSPERO, J., K. BARRETT, T. CHURCH, F. DENTENER, R. DUCE, J. GALLOWAY, H. LEVY, J. MOODY, and P. QUIN. 1996. Atmospheric deposition of nutrients to the North Atlantic basin. Biogeochemistry **35**: 27-73.
ROBINSON, L., P. CAMPBELL, and L. BUTLER. 2000. Trends in commercial fishery landings, 1972-1988. Texas Parks and Wildlife Department, Coastal Fisheries Division, Austin, Texas.
SANTSCHI, P.H. 1995. Seasonality of nutrient concentrations in Galveston Bay. Mar. Environ. Res. **40**: 337-362.
SHARP, J. 1991. Review of carbon, nitrogen, and phosphorus biogeochemistry. Rev. Geophys. 1991: 648-657.
SHERR, E.B. 1988. Direct use of high molecular weight polysaccharide by heterotrophic flagellates. Nature **335**: 348-351.
STANLEY, D. 2001. An annotated summary of nitrogen loading to US estuaries. p. 227-252. *In* R.A. Valigura *et al.* [eds.]. Nitrogen loading in coastal waterbodies: An Atmospheric perspective. Coastal and Estuarine Studies 57. American Geophysical Union.
TANG, D., K.W. WARNKEN, and P.H. SANTSCHI. 2001. Organic complexation of copper in surface waters of Galveston Bay. Limnol. Oceanogr. **46**: 321-330.
TOBIAS, C.R., J.W. HARVEY, and I.C. ANDERSON. 2001. Quantifying groundwater discharge through fringing wetlands to estuaries: seasonal variability, methods comparison, and implications for wetland-estuary echange. Limnol. Oceanogr. **46**: 604-615.
WARD, B.B. and D.A. BRONK. 2001. Net nitrogen uptake and DON release in surface waters: importance of trophic interactions implied from size fractionation experiments. Mar. Ecol. Prog. Ser. **219**: 11-24.
WARNKEN, K.W., G.A. GILL, P.H. SANTSCHI, and L.L. GRIFFIN. 2000. Benthic exchange of nutrients in Galveston Bay, Texas. Estuaries **23**: 647-661.
ZIMMERMAN, A.R. and R. BENNER. 1994. Denitrification, nutrient regeneration and carbon mineralization in sediments of Galveston Bay, Texas, USA. Mar. Ecol. Prog. Ser. **114**: 275-288.

VII. Mudflat ecosystem models

Laurent Seuront and Céline Leterme

Microscale patchiness in microphytobenthos distributions: evidence for a critical state

Abstract

The two-dimensional microscale (for scales ranging from 5 cm to 1 m) distribution of microphytobenthic biomass is investigated from superficial sediment samples taken on two intertidal flats characterized by sharp differences in terms of hydrodynamic exposure, sediment nature and biotic properties. Microphytobenthos biomass exhibited a very intermittent behaviour at both study sites, with the occurrence of sharp local fluctuations. The exposed sandy and the muddy flats are nevertheless respectively dominated by high-density and low-density patches, leading to significantly positively and negatively skewed distributions. It is also shown that the patch patterns exhibit specific power-law behaviours, involving the appearance of a self-organized critical state. The implications of critical versus subcritical states in microphytobenthos distributions are theoretically investigated on the basis of very simple numerical models, and a mechanistic explanation for the emergence of criticality in microphytobenthic populations is introduced.

Introduction

A central issue in ecology is the spatio-temporal organization of community structure and dynamics (Wiens 1989; Levin *et al.* 1997). In particular, the goal of spatial ecology is to determine how space and spatial scales influence population and community dynamics (Tilman and Kareiva 1997). Theoretical studies have suggested that biotic properties of individuals and populations interact to produce spatio-temporal complexity in homogeneous environments (e.g. Deutschman *et al.* 1993; Bascompte and Solé 1995). Potentially, environmental complexity interacts with biotic processes and influences spatial patterns (Roughgarden 1974; Pascual and Caswell 1997). In addition, theoretical and empirical studies show that the analysis of large (i.e. regional) scale patterns must integrate processes occurring at the small (local) scales (Levin 1992). Biomass and species are thus rarely dispersed uniformly (e.g. Kolasa and Pickett 1991). Instead patchiness (also referred to as 'spatial heterogeneity'; Seuront & Lagadeuc 2001) is the norm, and ecological field studies and environmental monitoring programs must be designed accordingly (Green 1979; Hurlbert 1984; Andrew and Mapstone 1987; Eberhardt and Thomas 1991).

In intertidal ecology, many studies focused on the interplay between abiotic processes and biotic community structure at different spatial scales (Archambault and Bourget 1996; Cusson and Bourget 1997; Guichard and Bourget 1998, 2001; Blanchard and Bourget 1999). More specifically, microphytobenthic communities are at the core of benthic primary production and the matter fluxes between benthic and pelagic domains. However, to our knowledge, only a few studies (e.g. Blanchard 1990; Pinckney and Sanduli 1990) have been devoted to investigate the distribution of microphytobenthos biomass on scales smaller than 1 m², i.e. usually the finest grain considered in landscape ecology (He *et al.* 1994) and intertidal ecology (MacIntyre *et al.* 1996; Blanchard and Bourget 1999). Alternatively, none have been confronted with the crucial question related to the phenomenology of the organization of microphytobenthic biomass at these scales where the most ecologically relevant processes of infection, nutrient uptake, cell division and behavior occur.

In this framework, the objective of the present work is (i) to demonstrate the heterogeneous nature of microphytobenthos biomass for scales smaller than 1 m², (ii) to quantify this heterogeneity in terms of critical behavior, (iii) to infer the nature of the observed behavior on the basis of a simple modeling approach, and (iv) to introduce a general phenomenological background likely to lead to critical dynamics in microphytobenthos communities.

Materials and methods

Sampling sites

The two study sites, located on the French coast of the Eastern English Channel, were chosen because of their intrinsic sharp differences in terms of hydrodynamic exposure, sediment nature and biotic properties. (Figure 1)

The first study site, an intertidal flat of sand in Wimereux (50°45'896 N, 1°36'364 E) is typical of the hydrodynamically sandy beach habitats that dominate the littoral zone along the French coast of the Eastern English Channel. Measurements were performed on a flat area located in the upper intertidal zone, without sharp topographical features such as ripple marks, high pinnacles or deep surge channels. The substrate was homogeneous medium size sand (200-250 µm, modal size), typical of the surrounding sandy habitat. Because of the substrate homogeneity and the weak biomass, productivity and production of both phyto- and zoobenthic organisms, the microphytobenthos biomass distribution is *a priori* expected to be rather homogeneous (Seuront and Spilmont 2002). In addition, due to the highly dynamic environment, microphytobenthos is expected to be resuspended and surface concentrations at low tide are low.

The second study site is located in the Bay of Somme, at Le Crotoy (50°13'524 N, 1°36'506 E) which is the second largest estuarine system, after the Seine estuary, and the largest sandy-muddy (72 km²) intertidal area on the Frech coasts of the Eastern English Channel. The sampling site was chosen in a topographically homogeneous

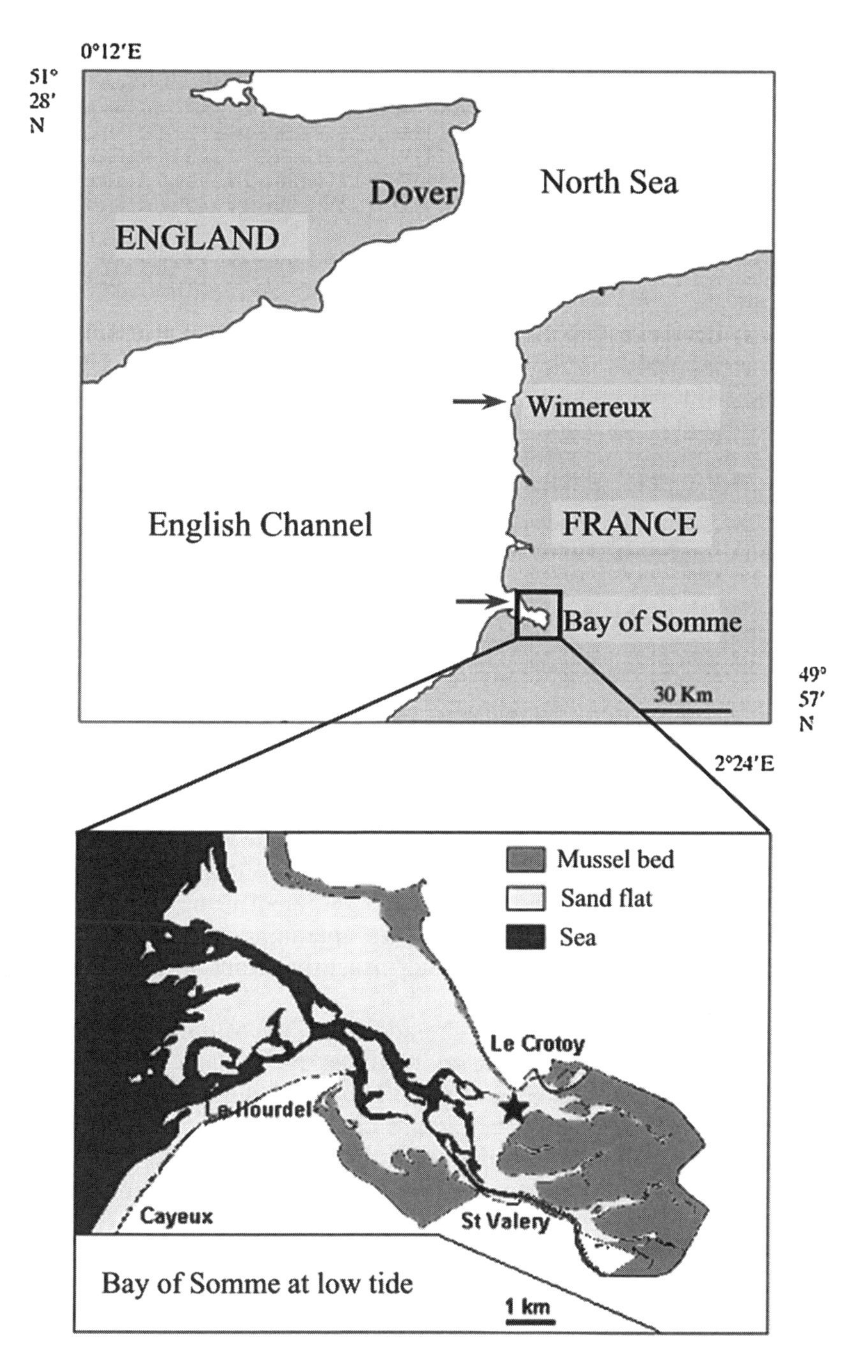

Figure 1. Location of the sampling stations in Wimereux and Le Crotoy on the east coast of the Eastern English Channel.

area, where the substrate grain size typically varied between 125 and 250 μm (modal size), and is characterized by higher phyto- and zoobenthos biomass and activity compared to the Wimereux site. Because of the weak hydrodynamic conditions, the microphytobenthos biomass is only weakly influenced by resuspension processes and surface concentraitons at low tide are high.

Micro-scale sampling

All measurements were performed at low tide, at the middle of the emersion period, on October 9 and 10, 2003 at Wimereux and Bay of Somme study sites, respectively. Samples were collected with a rigid 1m^2 aluminium quadrat constructed from 225, 1.9 cm^2 plastic cores resulting in an intersample distance of 6.67 cm. The cores were pushed into the sediment down to a depth of 1 cm, where the majority of the active cells are concentrated (Cadée and Hegeman 1974; Baillie 1987; Admiraal *et al.* 1988; Delgado 1989; de Jonge and Colijn 1994). This ensures that the observed spatial structure is not biased by any change in the spatial organisation of the microphytobenthos during the sampling process. Samples were then carefully removed, mixed to 5 ml of methanol and stored in a cool box, returned to the laboratory and stored in the dark at -20 °C.

Chlorophyll content analysis

Standard lab techniques for chlorophyll *a* extraction from samples is time consuming and not easily compatible with large numbers of samples. A standard procedure (e.g. Brunet 1994; Seuront and Spilmont 2002) is to place a sediment section in 8 ml acetone and to extract pigments for 4 hours in the dark at 4 °. After extraction, samples are centrifuged at 4000 rpm for 15 min. Chlorophyll *a* concentrations (, mg) in the supernatant are determined by spectrophotometry following the equation given by Lorenzen (1967). However, processing 225 sediment samples using this procedure would require more than 14 hours for two operators, we proposed hereafter an improved, faster method for extracting and measuring microphytobenthos biomass from sediment samples.

The proposed procedure consists of the addition 5 ml of methanol directly to the sampled sediment sections, and then assaying the extractant in a Turner 450 fluorometer previously calibrated with a pure Chlorophyll *a* solution (*Anacystis nidulans* extract, Sigma Chemicals) after an extraction time as short as 1 hour. Chlorophyll *a* concentrations in the sediment sections were then converted into Chl.*a* per surface unit (Chl.*a*, mg.m^{-2}) taking into account the surface (1.9 cm^2) of the sampling units. Using a set of homogeneous sediment sections we thus showed (i) that the chlorophyll extraction was complete after 1 hour, and remains stable in time (up to 10 days) when stored in the dark for temperature below 0 °C, and (ii) that the chlorophyll concentrations were not significantly different from those estimated from the above standard procedure ($p > 0.01$). In addition, processing 225 samples now takes no more than 5 hours to a single operator.

Self-organized criticality

Defining criticality. The most widespread example of self-organized criticality (SOC) is a pile of sand to which grains are continually added (Figure 2; Bak *et al.* 1987, 1988). Initially when the pile is flat there is little interaction among the different regions of the pile and adding a single grain will only affect a few other grains nearby. The system is in a subcritical state (Figure 2). As the pile grows by adding grains of sand, avalanches of grains spill down the sides such that adding a single grain can initiate a cascade affecting many other grains. Eventually, the slope of the pile grows until the 'angle of repose' is reached. The pile reaches a critical state and essentially does not get any steeper (Figure 2). Now, if grains are added avalanches occur with a wide range of sizes. The critical state is defined by a stationary statistical distribution of avalanches which propagate across all spatial and temporal scales (only limited by the finite size of the pile). Alternatively, the pile could be started in a supercritical state by forming a vertical cylinder of sand. A supercritical pile is highly unstable and is expected to collapse down to a critical state as grains are added (Figure 2). Thus, one can think of the critical state as an attractor for the dynamics of the pile.

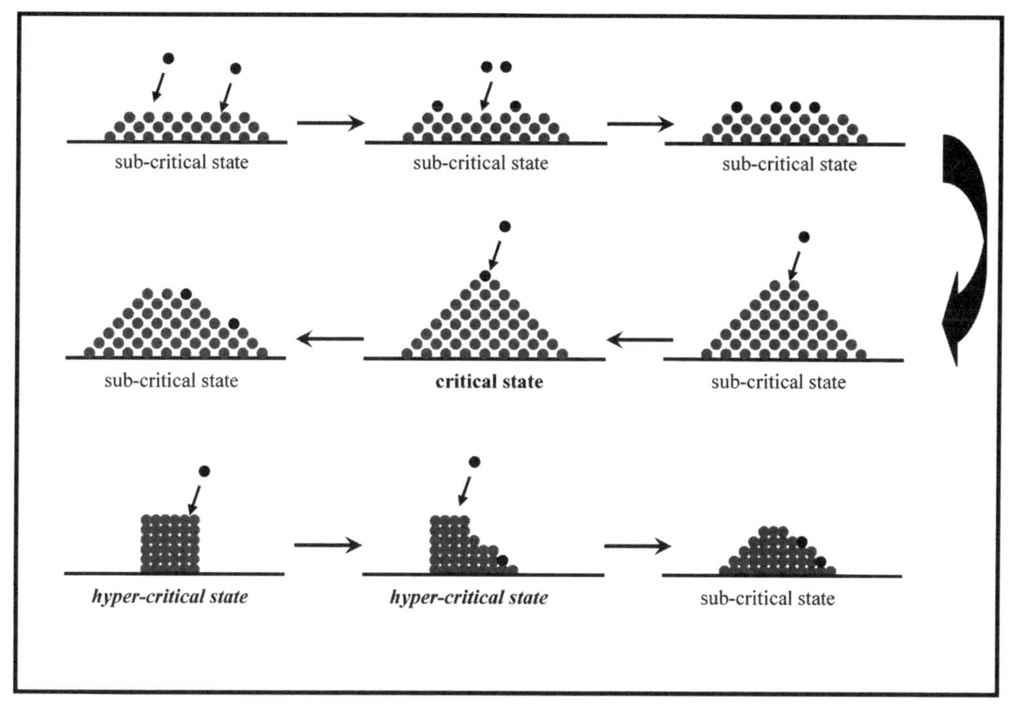

Figure 2. Schematic illustration of the dynamics of the sandpile as the most widespread example of self-organized criticality (SOC); see text for explanations.

Self-organized criticality signature. Because of the dynamical and structural properties of self-organized criticality it can be characterized through several scaling laws. In the case of the sand pile, the number of grains $n(l)$ falling a distance l at the same time step follows the power law form $n(l) \propto l^{-D}$, where D is the fractal dimension of the avalanches. More generally, for a critical system, the distribution of fluctuation sizes is describes as:

$$N(s) \propto s^{-D} \tag{1}$$

where s is the intensity of a given event (e.g. the size of an 'avalanche' in the sand pile example) and $N(s)$ the frequency of its occurrence. In practice, to estimate the fractal dimension D, the system under interest is observed over a period of time and the frequency of events of size s is recorded. In the sandpile case, the events are avalanches of sand grains, and the size of an event is the number of grains in a particular avalanche. A signature of self-organized criticality will be a straight line in a log-log plot of $N(s)$ vs. s. The slope of the straight line provides an estimate of the fractal dimension, D.

Self-organized criticality occurs in systems that build up stress and then release the stress in intermittent pulses. Indeed, the negative exponent in Eq. (1) leads to many small events (or fluctuations) punctuated by progressively rarer larger events. This intermittent behavior can thus be described by a power law stating that the probability of events with intensity X greater than a given threshold x as:

$$\Pr (X > x) \propto X^{-\phi} \tag{2}$$

where ϕ is the scaling exponent describing the distribution. As above, evidence for self-organized criticality will be given by the linear behavior of $\Pr (X > x)$ vs. X in a log-log plot, when X is a continuous or a discrete function (e.g. biomass and abundance measurements, respectively). However, because the computations of Eqs. (1) and (2) might not be as straightforward as it appears at first glance, we introduce in the next section a method that is strictly equivalent to Eq. (2) to identify self-organized criticality, but also and most importantly to differentiate random and non-random structures readily in any data sets.

Identifying self-organized criticality vs. randomness in ecological data. First, one must note that Eq. (2), also known as the Pareto's law (Pareto 1896), can be equivalently rewritten in terms of the probability density function (PDF) as:

$$P[X = x] \propto x^{-\gamma} \tag{3}$$

where γ ($\gamma = \phi + 1$) is the slope of a log-log plot of $P[X > x]$ vs. x. Now following the Harvard linguistic professor G.K. Zipf (1902-1950), consider the Zipf's law that states that the frequency f_r of the rth largest occurrence of an event is inversely proportional to its rank r as (Zipf 1949):

$$f_r \propto r^{-\alpha} \tag{4}$$

where α is the slope of the log-log plot of f_r vs. r. Eq. (4) can be generally written as:

$$X_r \propto r^{-\alpha} \tag{5}$$

where X_r is the '*weight*' of an occurrence of an event relative to its rank r. The concept of '*weight*' is very general and refers without distinction to frequency, length, surface, volume, mass or concentration. Discrete processes such as linguistic or genetic structures would nevertheless still require frequency computations, and thus refer to Eq. (4). Alternatively, Eq. (5) can be thought of as a more practical alternative that can be directly applied to continuous processes such as microphytobenthos distributions. From Eq. (5), it can be directly seen that there are kr variables X_r (where k is a constant) greater than or equal to $r^{-\alpha}$. This leads to rewrite Eq. (2) as:

$$P[X > kr^{-\alpha}] \propto r \tag{6}$$

and

$$P[X > X_r] \propto X_r^{1/\alpha} \tag{7}$$

From Eqs. (3), (5) & (7), the relationship between the exponents α, ϕ and γ is expressed as:

$$\begin{cases} \alpha = \frac{1}{\phi} \\ \gamma = 1 + \frac{1}{\alpha} \end{cases} \tag{8}$$

As a consequence, the Zipf and Pareto distributions can be regarded as being strictly equivalent. More specifically, the x-axis of the Zipf distribution is conceptually identical to the y-axis of the Pareto distribution (Eqs. 6 & 7). As a consequence, the use of one or the other distribution is simply a matter of convenience, although we stress that Zipf's law can be more easily and directly estimated than the Pareto's one.

Because of the one-to-one correspondance between Zipf and Pareto distributions, for the sake of simplicity, in the next section we will only use the Zipf's law, i.e. Eq. (5), to infer the presence of a self-organized critical state in microphytobenthos distributions. From the above statement, self-organized criticality will be identified by a linear behavior of X_r vs. r in a log-log plot. Alternatively, a random behavior steming e.g. from a white noise or a Normal distributions will manisfest itself as a continuous roll-off from a horizontal line (i.e. $\alpha \to 0$) to a vertical line (i.e. $\alpha \to \infty$). This is representative of the fact that no value is more likely to be more common than any other value.

To estimate the scaling exponents α, linear regression on the log-transformed data is preferred to nonlinear regression on the raw data because in the analysis of the log-transformed data the residual error will be distributed as a quadratic and thus minimum error is guaranteed. This is not the case with nonlinear regression (Seuront and Spilmont 2002). Finally, because an objective criterion is needed for deciding upon the appropriate range of ranks to include in the regression, we used the ranks which maximized the coefficient of determination and minimized the total sum of the residuals in the regression (Seuront and Spilmont 2002).

Results

Descriptive statistics of microphytobenthos distributions

Microphytobenthos chlorophyll *a* biomass ranged from 10.1 and 29.0 mg m^{-2}, i.e. 21.78 ± 4.04 mg m^{-2} ($\bar{x} \pm SD$) in Wimereux and between 42.54 and 113.98 in Le Crotoy, 77.83 ± 10.17 mg m^{-2} ($\bar{x} \pm SD$). The biomass estimates in Le Crotoy were in the range of microphytobenthos biomass taken from biologically rich and active muddy flats, i.e. bounded between 50 and 200 mg m^{-2} (Seuront and Spilmont 2002; Carrère *et al.* 2004). In contrast, the values observed on the sandy flat in Wimereux are (i) rather high for a hydrodynamically exposed sandy flat and (ii) significantly higher than microphytobenthos biomass estimated at the same location, one and two years earlier, i.e. 10.79 ± 4.15 mg m^{-2} (25 September 2001, $p < 0.05$; Seuront and Spilmont 2002) and 2.75 ± 0.88 mg m^{-2} (25 September 2000, $p < 0.01$; Seuront and Spilmont 2002). This may be a consequence of the relative intensity of the autumn phytoplankton blooms that occur in the coastal waters of the Eastern English Channel as well as the dominance of diatom species at that period of the year. Indeed, over the same period, phytoplankton chlorophyll *a* concentrations were estimated as 13.94 ± 2.52 μg l^{-1} in 2003 (Seuront, unpubl. data), 10.72 ± 3.29 μg l^{-1} in 2002 (Leterme and Seuront, unpubl. data) and 4.22 ± 1.12 μg l^{-1} in 2001 (Seuront, unpubl. data) in the shallow water moving onto or off the investigated sandy flat. In addition, considering that diatoms dominate the autumn phytoplankton assemblages, it is reasonable to think that most of the chlorophyll *a* measured in the sediment comes from phytoplankton deposition on the sediment at low tide.

More specifically, microphytobenthos biomass exhibited a very intermittent behavior, with the occurrence of sharp local fluctuations clearly visible at both study sites (Figure 3). The nature of the distributions differed however, with dominance of 'hotspots' in Wimereux (Figure 3a) samples and 'coldspots' in the Le Crotoy samples (Figure 3b). Results of descriptive analysis, including skewness and kurtosis estimates, specify the previous observations by showing that the 225 microphytobenthos biomass estimated from Wimereux and Le Crotoy sampling are not normally distributed (Kolmogorov-Smirnov test, $p < 0.01$). Their frequency distribution rather exhibits positively and negatively skewed behaviors, reflecting a distribution characterized by a few low density patches over a wide range of high density patches in Wimereux (g_1 = -0.73) and a few dense patches and a wide range of low density patches in Le Crotoy (g_1 = 0.48). Finally, the positive kurtosis (i.e. g_2 = 0.14 in Wimereux and g_2 = 1.39 Le Crotoy) show a distribution that is peakier than expected in the case of normality, especially in the Bay of Somme. The comparisons of the

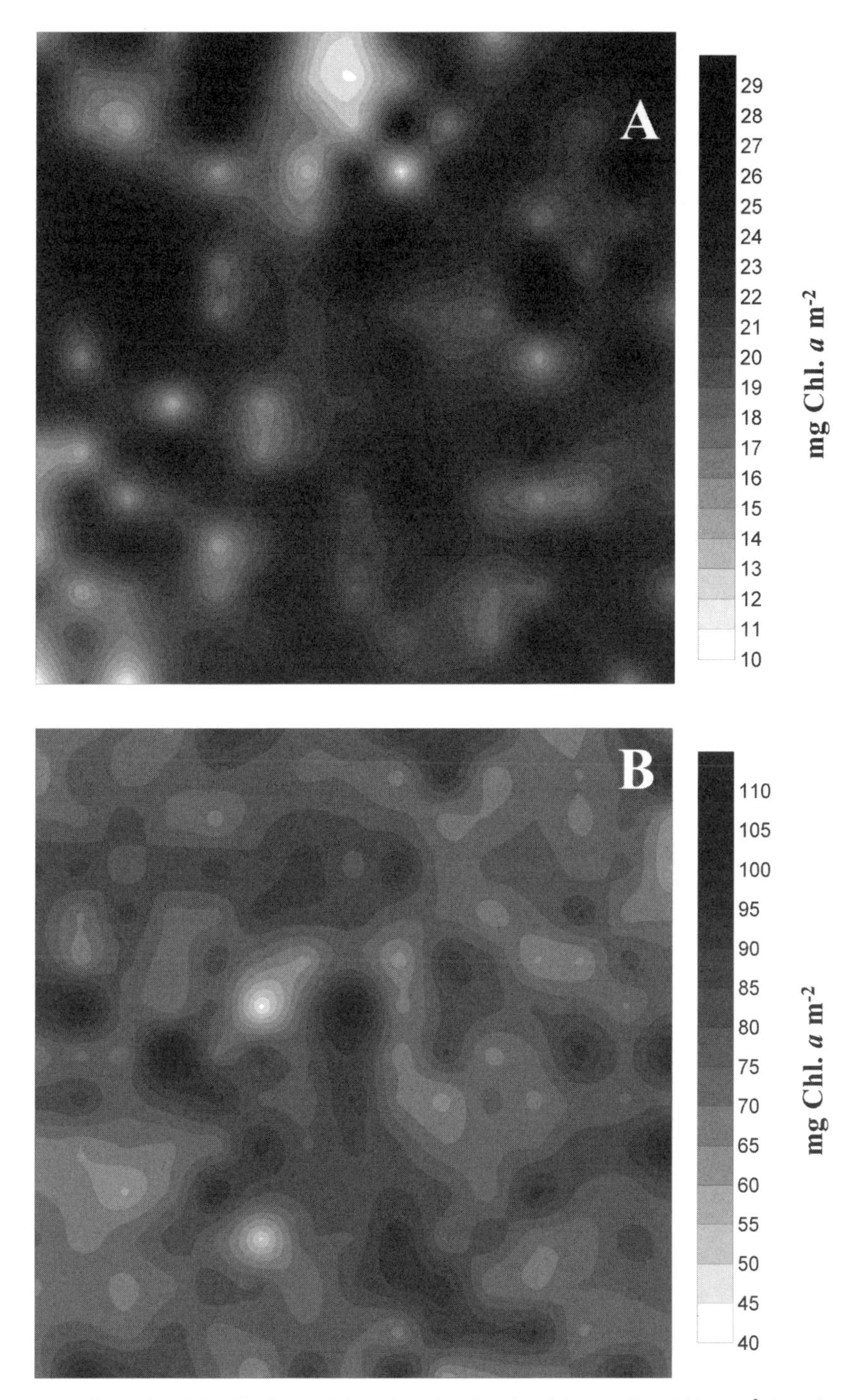

Figure 3. Two dimensional distributions of the microphytobenthos biomass (mg Chl. *a* m^{-2}) in Wimereux (A) and Le Crotoy (B).

observed distributions and simulated uniform distribution characterized by the same minimum and maximum values than the empirical data qualitatively confirms the non-random character of the microphytobenthos distribution in Wimereux (Figure 4) and Le Crotoy (Figure 5). In both cases, the differences between the field distributions (Figs. 4a & 5a) and uniform, homogeneous distributions, characterized by a regular alternance between high and low density areas (Figs. 4b & 5b) are clear.

Evidence for a self-organized critical state

The Zipf analysis of two-dimensional microphytobenthos patterns shows that microphytobenthos biomass was not randomly distributed (Figure 6). The Zipf plots show instead a very clear linear behavior with $\alpha = 0.071$ ($r^2 = 0.98$) for concentrations ranging from 24.15 to 28.18 mg m^{-2} in Wimereux (Figure 6a) and with $\alpha = 0.079$ ($r^2 = 0.98$) for concentrations ranging from 82.60 to 113.98 mg m^{-2} in Le Crotoy (Figure 6b). While the power law behavior expands to the maximum microphytobenthos concentration in Le Crotoy (Figure 6b), in Wimereux, the Zipf plot clearly diverges from a power law for concentrations higher than 28.18 mg m^{-2} (Figure 6a). This indicates that the probability of the occurrence of high density patches is lower than expected in the case of a power law. On the other hand, for concentrations lower than 24.15 mg m^{-2} in Wimereux and 82.60 mg m^{-2} in Le Crotoy, the Zipf plots progressively roll-off towards the behavior expected in the case of randomness (Figure 6). Now, this indicates that the probability of the occurrence of low density patches that is lower than expected in the case of a power law. However, the continuous roll-off towards the lowest concentrations is clearly different in Wimereux (Figure 6a) and Le Crotoy (Figure 6b). This could be indicative of differential driving processes competing with the pure power law behavior observed for higher concentrations. In the next section, we present several potential functional hypotheses likely to reproduce the shapes of the Zipf plot observed in Wimereux and Le Crotoy that may help to corroborate the conjecture that the microscale microphytobenthos distribution is a living system with critical dynamics.

Discussion

Critical versus subcritical states in microphytobenthos distribution

The fact that microphytobenthos concentrations greater (or smaller) than a given threshold do not follow the same law as other events indicates that there is something unique about these events. In particular, the differences observed between microphytobenthos distributions in Wimereux (Figure 6a) and Le Crotoy (Figure 6b) could be related to differences in grazing pressure likely to occur in these two study sites.

Consider a pure power law relationship of the form:

$$C_r \propto r^{-\alpha} \qquad (9)$$

where C_r is the microphytobenthos biomass and r its rank; see Eq. (5). Now consider a situation where the grazing pressure is assumed to be a random function of food

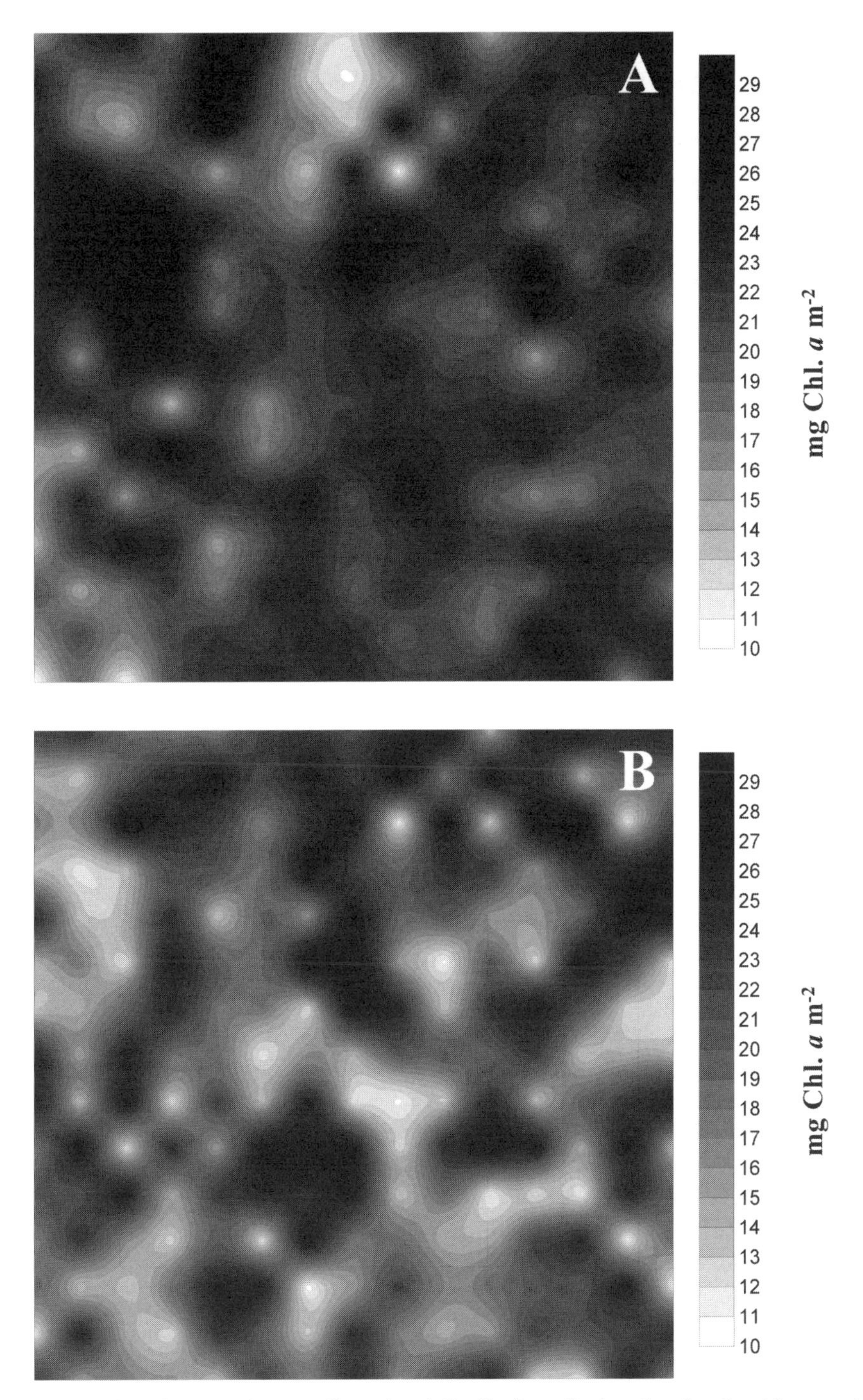

Figure 4. Comparison between the two-dimensional distribution of microphytobenthos biomass observed in Wimereux (A) and a simulated uniform distribution with the same minimum and maximum values (B).

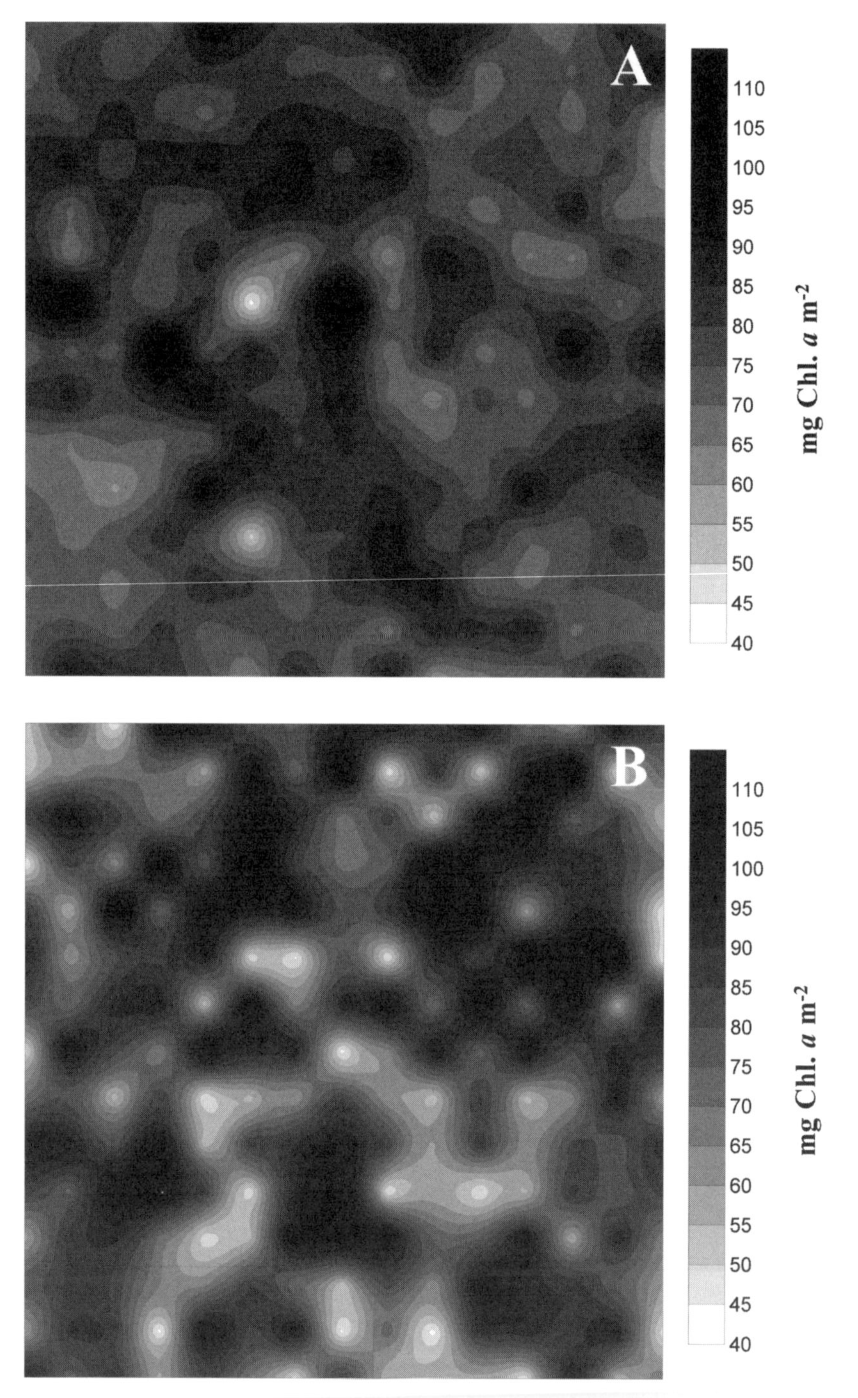

Figure 5. Comparison between the two-dimensional distribution of microphytobenthos biomass observed in Le Crotoy (A) and a simulated uniform distribution with the same minimum and maximum values (B).

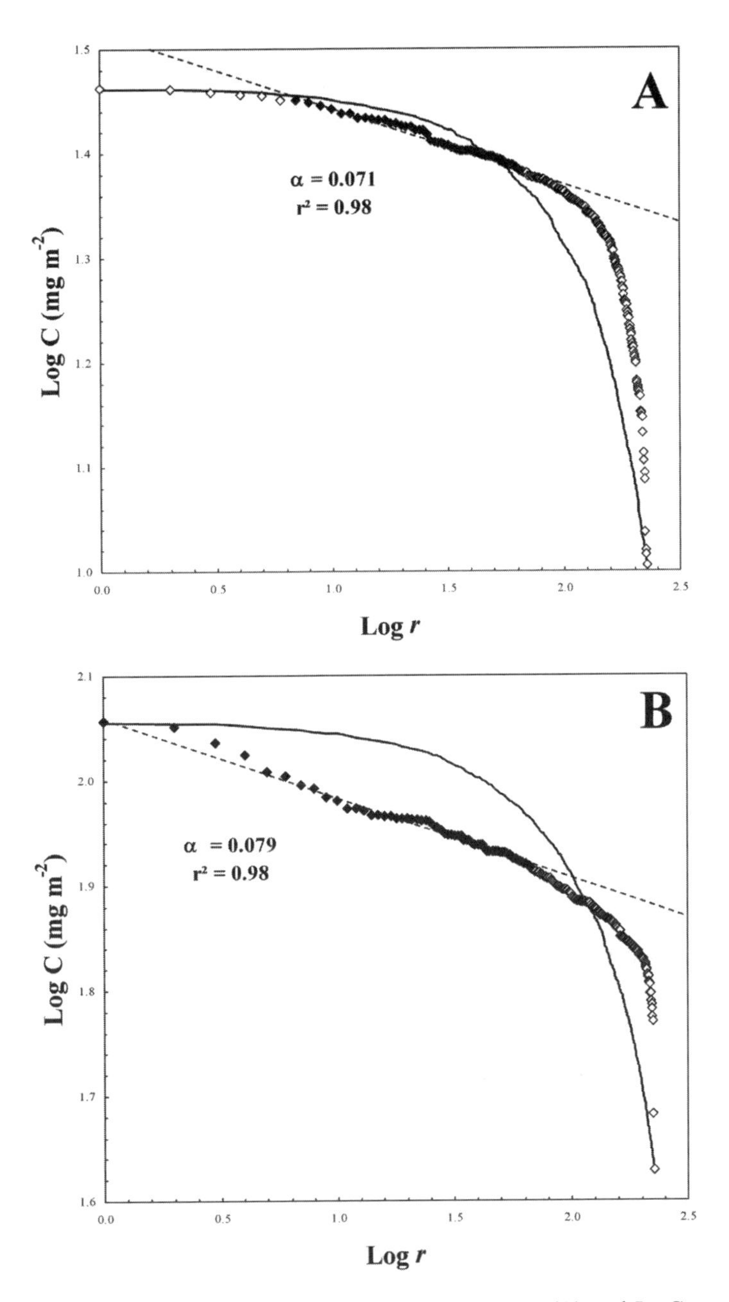

Figure 6. Zipf plots of the microphytobenthos biomass in Wimereux (A) and Le Crotoy (B). In both cases, the black diamonds correspond to the range of microphytobenthos concentrations exhibiting a power law behavior, i.e. a critical state, and used to estimate the exponent α as the slope of the linear fit maximizing the coefficient of determination and minimizing the total sum of the residuals in the regression (discontinuous lines). The continuous lines correspond to the Zipf plot obtained from 100 simulated uniform distributions with the same minimum and maximum values than the empirical ones.

availability. Considering the diversity of microphytobenthos grazers observed in Le Crotoy (McLusky *et al.* 1996), their differential grazing rates and abilities, as well as their differential spatial distribution, the grazing pressure on microphytobenthos biomass is likely to be of a purely stochastic nature. Eq. (9) can then be rewritten as:

$$C_{1r} = C_r - \varepsilon C_r \qquad (10)$$

where ε is a random noise process, i.e. $\varepsilon \in [0,1]$. For increasing amount of noise, the characteristic 'noise roll-off' occurring for low rank values is more violent and mimics the behavior observed in Wimereux and Le Crotoy (Figure 7). In addition, one must note here that the fluctuations generated by the noise contamination around a power law behavior are fully compatible with the irregularities observed over the scaling ranges identified in Wimereux and Le Crotoy Zipf plots (See Figure 6). The smoother roll-off observed in Wimereux (Figure 6a) when compared to Le Crotoy (Figure 6b) thus suggests a lower grazing impact in Wimereux where the meiobenthic biomass is negligible (Seuront and Spilmont 2002).

The previous approach, however, did not take into account the potential behavioral adaptation of grazers to varying food concentrations (Johnson *et al.* 1997). If one considers that remote sensing abilities can lead to aggregation of grazers and/or preferential grazing in areas of high microphytobenthos concentrations as investigated both empirically and numerically (e.g. Decho and Fleeger 1988; Montagna *et al.* 1995; Johnson *et al.* 1998), Eq. (9) can be modified as:

$$C_{2r} = C_r - 10^{(C_r/k)} \qquad (11)$$

where k is a constant and the ingestion function $I(C_r) = 10^{(C_r/k)}$ represents an increased predation impact on higher phytoplankton concentrations. The advantage of the function $I(C_r)$ is that it can be regarded as a representation of the aggregation of grazers with constant ingestion rates and/or evenly distributed grazers with increasing ingestions rates in high density microphytobenthos patches. Decreasing values of the constant k increases the grazing impact on high density patches. The grazed microphytobenthos population then diverges from a power law form for high values of C_{2r}, but asymptotically converges to the original power law for the smallest values of C_{2r}, i.e. $C_{2r} \propto r^{\alpha}$ for $r \rightarrow r_{\min}$ (Figure 8). The divergence from a power law observed in Wimereux (Figure 6b) could thus be explained by the low microphytobenthos biomass. Indeed, according to the optimal foraging theory (Pyke 1984) grazers living in food depleted and/or heterogeneous environments develop strategies to exploit high density patches and then to optimize the energy required to capture a given amount of food. Grazers living in a low microphytobenthos concentration envrionment (i.e. Wimereux) are thus more likely to graze preferentially on high density patches than grazers living in a high microphytobenthos concentration environment (i.e. Le Crotoy). As a consequence, the microphytobenthos communities investigated here can be regarded as being in a critical state when they follow a power law relationship (Figure 6). Alternatively, below and above a critical biomass in Wimereux and below a critical biomass Le Crotoy they are in a subcritical state.

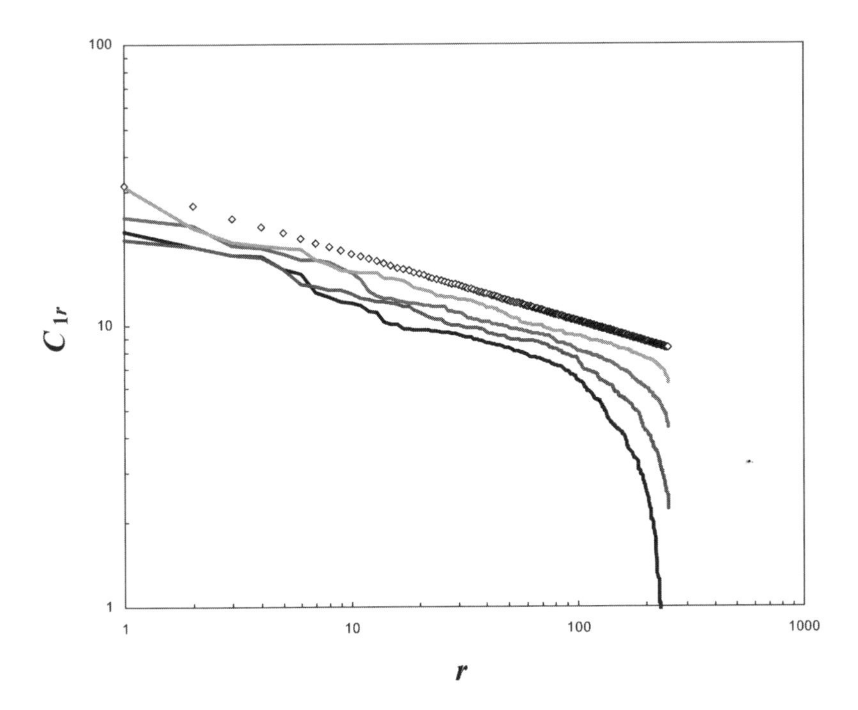

Figure 7. Log-log plot signature of the Zipf behavior expected in case of a power law C_r (open diamonds) competing with a random mortality component ($C_{1r} = C_r - \varepsilon C_r$), where $\varepsilon = 0.05, 0.25, 0.50$ and 0.75 (from top to bottom).

Toward a mechanistic explanation of SOC in microphytobenthos communities

The decrease in the number of patches above a critical biomass observed in Figure 6 suggests that the development of patches is structured by conflicting constraints. In the case of the sandpile model, the constraints are gravity which acts to lower the height of the pile and the addition of sand grains which raises the height of the pile. The structure of the pile thus emerges from the interaction of these forces. While we previously mainly addressed the constraints likely to lead to a divergence from self-organized criticality in microphytobenthos distribution, it is crucial to understand that it is actually the constraints, and their potential effects, acting on the structure and dynamics of a microphytobenthos assemblage that are responsible for the emergence of a critical behavior. The microscale distribution of microphytobenthos biomass is thus a function of exogenous (e.g. tides, hydrodynamism, sediment microtopography, porosity and cohesivity, competition for nutrient and light, predation) and endogenous processes (e.g. nutrient uptake, growth, migration, death) that can act to increase and/or decrease microphytobenthos biomass. A crucial issue is that most of these constrainsts do not act uniformly over the whole spatial domain. For instance, biomass gains related to cell division necessarily occur in the vicinity of a microphytbenthos individual. Alternatively, biomass losses related to cell death and grazing can occur anywhere and are

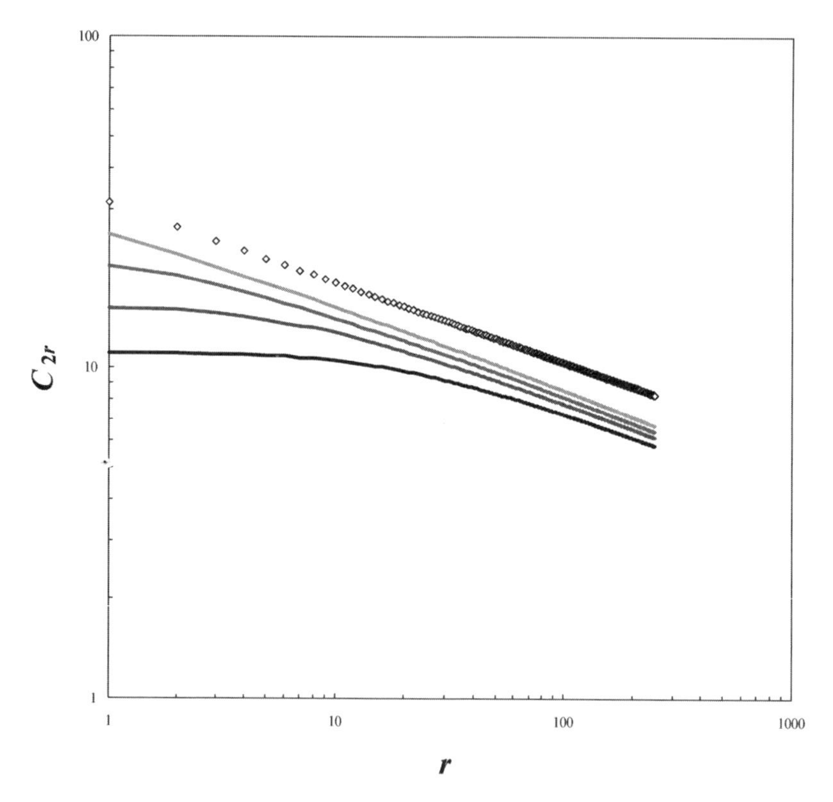

Figure 8. Log-log plot signature of the Zipf behavior expected in case of a power law C_r (open diamonds) competing with a preferential grazing component for high phytoplankton concentrations ($C_{2r} = C_r - 10^{(C_r/k)}$). The grazed microphytobenthos population diverges from a power law form for high concentrations, but asymptotically converges to the original power law for the smallest values. The extent of the observed divergence is controlled by increasing grazing pressure k (from top to bottom).

dependent on both the spatial distribution and foraging abilities of predators, respectively. Growth and death are dependent on nutrient and light availability that, in turn, is a function of the burying depth of microphytobnthos cells, the density, cohesivity and the spatial distribution of the sediment and the duration of the emersion. The microphytobenthos population may also be disturbed by turbulence and shear stress generated by tidal currents or wind-waves leading to the resuspension of microphytobenthos in the water column. One must also note that some contraints do not act uniformly in time. For instance, resuspension processes occur during the whole immersion period and lead to microphytobenthos biomass losses. Conversely, the resettling of cells occurs at the beginning of the emersion and can be regarded as a crucial event in the observed patch pattern. These constraints, acting to increase and/or decrease microphytobenthos biomass differentially in time and/or in space, result in a dynamic balance, or critical state, as in the sand pile model. Although we have shown on the basis of simple hypotheses related to predator-prey relationships how a disturbance can create a divergence from criticality it still remains unclear how a critical behavior can emerge. A simple mechanism is nevertheless suggested in the next section.

Interspecific competition is likely to be a driving force in structuring microphytobenthos community (e.g. Egge and Aksnes 1992; Sommer 1996; Stal 2001). The most relevant dynamics would then be observed in the niche space occupied by different species (MacArthur 1960; Hutchinson 1961; Odum 1971). Competitive pressure should be high in regions of niche space where species are densely packed, as would happen when a number of phytoplankton species share the same food source (Siegel 1998; Huisman and Weissing 2000). As in the steep region of the sandpile, avalanches are more probable, species occupying dense regions of niche space could be subject to higher extinction probabilities, thus reducing the probability of high density patches. The loss of species would change the distribution of species in niche space and, in turn, change the probability of extinction and patches. The system is in a critical state. In contrast, species occupying sparse regions of the niche space (i.e. areas depleted in microphytobenthos cells) are subject to weaker competition pressure and extinction probabilities. The system is in a more stable, or subcritical, state.

Conclusions

We have shown that microphytobenthos biomass sampled at scales ranging from 6.67 cm to 1 m in two structurally different intertidal ecosystems was far from homogeneous, but instead was strongly structured and exhibited some fingerprints for self-organized criticality. While the emergence of a self-organized critical state can be generally thought as being the result of conflicting constraints acting on microphytobenthos community, it is still difficult to identify precisely the processes responsible for the observed patterns. We have nevertheless proposed simple mechanisms likely (i) to generate a critical state and (ii) to induce a divergence from a critical state.

Finally, we stress that the above demonstrated structure in microphytobenthos distribution may have salient consequences on microphytobenthos biomass and production estimates. As illustrated from a recent literature survey (Seuront and Spilmont 2002) the variability observed from a localized high frequency microscale sampling strategy can be of the same order of magnitude as the seasonal and annual variability observed in the sediment in a wide variety of intertidal environments. This suggests that the sampling error might account for much of the variation in biomass at the seasonal and annual scales, especially when the microphytobenthos is sampled with relatively few cores as is usually the case, e.g. 6 in Sagan and Thouseau (1998), and 5 in Barranguet and Kromkamp (2000). In such a framework it is indeed doubtful that a small number of samples can be representative of a microphytobenthos population. While to our knowledge no alternative solution exists to unbiased regular biomass estimates, we believe that a proper parameterization of microphytobenthos microscale variability, as illustrated here in the framework of self-organized criticality, could form the basis of routine procedures devoted to infer variability from a limited number of samples.

Acknowledgements

We thank the organizing committee of the Royal Netherlands Academy of Arts and Sciences colloquium on Functioning of Microphytobenthos in Estuaries for their kind invitation, and the Royal Academy of Arts and Sciences for travel financial support. T. Meziane and N. Spilmont are acknowledged for their help and enjoyable company during the sampling experiment. This work is a contribution of the Ecosystem Complexity Research Group, and was supported by the ACI 'Jeunes Chercheurs' 3058, the CPER '*Phaeocystis*' and the Chantier '*Manche orientale*' of the Programme National d'Environnement Côtier (PNEC). R. Waters is greatly acknowledged for her valuable comments on the manuscript and for enjoyable and stimulating discussions.

References

Admiraal W., M.A. Arkel, J.W. van Baretta, F. Colijn, W. Ebenhöh, V.N. de Jonge, A. de Kop, P. Ruardij, and H.G.J. Schröder. 1988. The construction of benthic submodel. In: Baretta J., and P. Ruardij (eds.), Tidal flat estuaries. Simulation and analysis of the Ems estuary. Ecological Studies 71. Springer-Verlag, Heidelberg, 105-152.

Andrew, N.L., and B.D. Mapstone. 1987. Sampling and the description of spatial pattern in marine ecology. Oceanogr. Mar. Biol. Ann. Rev. **25**: 39-90.

Archambault, P., and E. Bourget. 1996. Scales of coastal heterogeneity and benthic intertidal species richness, diversity and abundance. Mar. Ecol. Prog. Ser. **136**: 111-121.

Baillie, P.W. 1987. Diatom size distribution and community stratification in estuarine intertidal sediments. Estuar. Coast. Shelf Sci. **25**: 193-209.

Bak, P., C. Tang, and K. Wiesenfield. 1987. Self-organized criticality: an explanation of 1/f noise. Phys. Rev. Lett. **59**: 381-384.

Bak, P., C. Tang, and K. Wiesenfield. 1988. Selforganized criticality. Phys. Rev. A **38**: 364-374.

Barranguet, C., and J. Kromkamp. 2000. Estimating primary production rates from photosynthetic electron transport in estuarine microphytobenthos. Mar. Ecol. Prog. Ser. **204**: 39-54.

Bascompte, J., and R.V. Solé. 1995. Rethinking complexity: modelling spatiotemporal dynamics in ecology. Trends Ecol. Evol. **10**: 361-366.

Blanchard, D., and E. Bourget. 1999. Scales of coastal heterogeneity: influence on intertidal community structure. Mar. Ecol. Prog. Ser. **179**: 163-173.

Blanchard, G.F. 1990. Overlapping microscale dispersion patterns of miofauna and microphytobenthos. Mar. Ecol. Prog. Ser. **68**: 101-111.

Brunet, C. 1994. Analyse des pigments photosynthétiques par HPLC: communautés phytoplanctoniques et productivité primaire en Manche orientale. Ph.D. Thesis. Université Pierre & Marie Curie, Paris.

Cadée, G.C., and J. Hegeman. 1974. Primary production of the benthic microflora living on tidal flats in the Dutch Wadden Sea. Neth. J. Sea Res. **8**: 260-291.

Carrère, V., N. Spilmont, and D. Davoult. 2004. Comparison of simple techniques for estimating Chlorophyll a concentration in the intertidal zone using high spectral resolution field spectrometer data, in press.

Cusson, M., and E. Bourget. 1997. Influence of topographic heterogeneity and spatial scales on the structure of the neighbouring intertidal endobenthic macrofaunal community. Mar. Ecol. Prog. Ser. **150**: 181-193.

Decho, A.W., and J.W. Fleeger. 1988. Microscale dispersion of meiobenthic copepods in response to food-resource patchiness. J. Exp. Mar. Biol. Ecol. **118**: 229-244.

de Jonge, V.N., and F. Colijn. 1994. Dynamics of microphytobenthos biomass in the Ems estuary. Mar. Ecol. Prog. Ser. **104**: 185-196.

Delgado, M. 1989. Abundance and distribution of microphytobenthos in the bays of Ebro Delta (Spain). Estuar. Coast. Shelf Sci. **29**: 183-194.

Deutschmann, D.H., G.A. Bradshaw, W.M. Childress, K. Daly, D. Grunbaum, M. Pascual, N.H. Schumaker, and J. Wu. 1993. Mechanisms of patch formation. In: Levin, S., T. Powell, and J. Steele (eds.), Patch Dynamics, Springer, Berlin, 184-209.

Eberhardt, L.L., and J.M. Thomas. 1991. Designing environmental field studies. Ecol. Monogr. **61**: 53-73.

Egge, J.K., and D.L. Aksnes. 1992. Silicate as regulating nutrient in phytoplankton competition. Mar. Ecol. Prog. Ser. **83**: 281-289.

GREEN, R.H. 1979. Design and statistical methods for environmental biologists. Wiley, New York.
GUICHARD, F., and E. BOURGET. 1998. Topographic heterogeneity, hydrodynamics, and benthic community structure: a scale-dependent cascade. Mar. Ecol. Prog. Ser. **171**: 59-70.
GUICHARD, F., and E. BOURGET. 2001. Scaling the influence of topographic heterogeneity on intertidal benthic communities: alternate trajectories mediated by hydrodynalmics and shading. Mar. Ecol. Prog. Ser. **217**: 27-41.
HE, F., P. LEGENDRE, and C. BELLEHUMEUR. 1994. Diversity pattern and spatial scale: a study of a tropical rain forest in Malaysia. Environ. Ecol. Stat. **1**: 265-286.
HUISMAN, J., and F.J. WEISSING. 2000. Biodiversity of plankton by species oscillations and chaos. Nature **402**: 407-410.
HURLBERT, S.T. 1984. Pseudoreplication and the design of ecological field experiments. Ecol. Monogr. **54**: 187-211.
HUTCHINSON, G.E. 1961. The paradox of the plankton. Am. Nat. **95**: 137-145.
JOHNSON, M.P., M.T. BURROWS, R.G. HARTNOLL, AND S.J. HAWKINS. 1997. Spatial structure on moderately exposed rocky shores: patch scales, and the interactions between limpets and algae. Mar. Ecol. Prog. Ser. **160**: 209-215.
JOHNSON, M.P., M.T. BURROWS, and S.J. HAWKINS. 1998. Individual based simulations of the direct and indirect effects of limpets on a rocky shore *Fucus* mosaic. Mar. Ecol. Prog. Ser. **169**: 179-188.
KOLASA, J., and S.T.A. PICKETT. 1991. Ecological heterogeneity. Springer-Verlag, New-York.
LEVIN, S.A. 1992. The problem of patterns and scale in ecology. Ecology **73**: 1943-1967.
LEVIN, S.A., B. GRENFELL, A. HASTINGS, and A.S. PERELSON. 1997. Mathematical and computational challenges in population biology and ecosystems science. Science **275**: 334-343.
LORENZEN, C.J. 1967. Determination of chlorophyll and phaeopigments: spectrometric equations. Limnol. Oceanogr. **12**: 343-346.
MACARTHUR, R.H. 1960. On the relative abundance of species. Am. Nat. **94**: 25-36.
MACINTYRE, H.L., R.J. GEIDER, and D.C. MILLER. 1996. Microphytobenthos: the important role of the 'secret garden' of unvegetated, shallow-water marine habitats. I. Distribution, abundance and primary production. Estuaries **19**: 186-201.
MCLUSKY, D.S., N. BRICHE, M. DESPREZ, S. DUHAMEL, H. RYBARCZYK, and B. ELKAÏM. 1996. The benthic production of the Baie of Somme, France. Biol. Ecol. Shallow Coast. Wat. **28**: 225-231.
MONTAGNA, P.A., G.F. BLANCHARD, and A. DINET. 1995. Effect of production and biomass of intertidal microphytobenthos on meiofaunal grazing rates. J. Exp. Mar. Biol. Ecol. **185**: 149-165.
ODUM, E.P. 1971. Fundamentals in ecology. Saunders, Philadelphia.
PARETO, V. (1896) Oeuvres complètes. Droz, Geneva.
PASCUAL, M., and H. CASWELL .1997. Environmental heterogeneity and biological pattern in a chaotic predator-prey system. J. Theor. Biol. **185**: 1-13.
PINCKNEY, J., and R. SANDULLI. 1990. Spatial autocorrelation analysis of meiofaunal and microalgal populations on an intertidal sandflat: scale linkage between consumers and resources. Est. Coast. Shelf Sci. **30**: 341-353.
PYKE, J.H. 1984. Optimal foraging theory: a critical review. Ann. Rev. Ecol. Syst. **15**: 523-575.
ROUGHGARDEN, J. 1974. Population dynamics in a spatially varying environment: how population size tracks spatial variation in carying capacity. Am. Nat. **108**: 649-664.
SAGAN, G., and G. THOUZEAU. 1998. Variabilité spatio-temporelle de la biomasse microphytobenthique en rade de Brest et en manche occidentale. Oceanol. Acta **21**: 677-693.
SEURONT, L., and Y. LAGADEUC. 2001. Towards a terminological consensus in ecology: variability, inhomogeneity and heterogeneity. J. Biol. Syst. 9: 81-87.
SEURONT, L., and N. SPILMONT. 2002. Self-organized criticality in intertidal microphytobenthos patch patterns. Physica A **313**: 513-539.
SIEGEL, D.A. 1998. Resource competition in a discrete environment: why are plankton distribution paradoxical? Limnol. Oceanogr. **43**: 1133-1146.
SOMMER, U. 1996. Nutrient competition experiments with periphyton from the Baltic Sea. Mar. Ecol. Prog. Ser. **140**: 161-167.
STAL, L.J. 1996. Coastal microbial mats: the physiology of a small-scale ecosystem. South African J. Bot., **67**: 399-410.
TILMAN, D., and P. KAREIVA. 1997. Spatial ecology: the role of space in population dynamics and interspecific interactions. Monographs in Population Biology, Vol. 30, Princeton University Press, Princeton.
WIENS, J. 1989. Spatial scaling in ecology. Funct. Ecol. **3**: 385-387.
ZIPF, G.K. (1949) Human behavior and the principle of least effort. Hafner, New York.

Jean-Marc Guarini, Gérard Blanchard and Pierre Richard

Modelling the dynamics of the microphytobenthic biomass and primary production in European intertidal mudflats

Abstract

A mathematical model was formulated in order to simulate the local dynamics of the microphytobenthic biomass of an intertidal mudflat community. The model was based on three sets of processes that drive the dynamics of the system. First, the intertidal benthic microalgae are able to migrate quickly to the mud surface to constitute a dense biofilm at the beginning of the daytime emersion period. At the end of the daytime emersion period, the microphytobenthos burrows into the aphotic layer of the mud and the biofilm disappears completely. Secondly, photosynthetic primary production occurs in the biofilm at the mud surface during the daytime emersion period, and production mainly depends on light and temperature forcing variables. And finally, the third set represents any losses of biomass that affect the dynamics slowly; this set of processes encompasses stress-induced mortality, grazing and resuspension of microphytobenthos into the water column during submersion periods. At the ecosystem level, primary production is controlled mainly by physical synchronizers (tidal and light-dark cycles).

The mathematical properties of the initial model were characterized by a stable steady state and a short resilience time. The underlying concepts of the formulation were tested with comparisons between simulations and observations and the model description was consistent with the observed patterns of biomass dynamics. Nevertheless, two groups of processes were not sufficiently well-described: vertical migration of the microphytobenthos and biomass loss. The losses were too poorly constrained; the grazing losses could not be distinguished from resuspension. The stimuli for vertical migration were also not fully understood and only an integrated vertical migration was taken into account. In particular, the trigger for the downward vertical migration, at the end of a daytime emersion period, is unknown and remains controversial.

Therefore, the model was reformulated with some new hypotheses linking the rhythms of vertical migrations to resuspension into the water column. New experiments were designed, mainly to characterize the biofilm and its ability to produce new biomass. Additional studies of the mathematical properties of the model provided new perspectives for validation and future work. This approach is the first step in proposing an integrated theory for the dynamics of the pelagic and benthic microalgal biomass and primary production in European-type semi-enclosed littoral ecosystems. Testing the validity of this integrated theory will require accurate quantification of biomass exchanges between the benthic and pelagic compartments.

Introduction

One of the main challenges in ecology is making accurate estimates of the productivity of marine ecosystems. This is particularly critical when ecosystems are exploited for their renewable resources. Semi-Enclosed Littoral Ecosystems (SELE including bays, lagoons, estuaries and deltas) are highly productive (Schelske and Odum 1962; Walker and Mossa 1982). Several types of SELE which generate large amounts of organic matter can be defined and in particular, McLusky (1989) has ranked SELE between two endpoints based on their primary producers. At one end of the McLusky classification scheme were American-Type SELE. These ATSELE are characterized by salt marshes with halophytes that provide a large amount of detritic organic matter through decomposition of the plant material. A part of the detritic biomass is used by the benthic and pelagic foodwebs, and the remaining fraction is exported to the open coastal zone (Long and Mason 1983; Wetzel and Sin 1998). At the opposite end, McLusky placed European-Type SELE, which are characterized by wide, bare intertidal mudflats without any macrophytic primary production. The high primary productivity in these ETSELE is supported mainly by microalgal communities that colonize both the water column and the sediments (Admiraal 1984; McLusky 1989; Underwood and Kromkamp 2000). This type of ecosystem is widely distributed along the European coastline, and building a quantitative description of the dynamics of these ecosystems is of fundamental importance to the understanding and management of the coastal resources of Europe (Figure 1).

The quantification of phytoplankton production in shallow water ecosystems (see Cloern 1996 for a review) is more advanced than the quantification of the dynamics of microphytobenthos living on intertidal mudflats. However, there is no existing general theory which describes the dynamics of phytoplankton in shallow-water ecosystems (Lucas *et al.* 1998). Theories developed for open oceanic areas cannot be applied because of the short-term variability and the absence of persistent stratification in shallow water tidal ecosystems (Pritchard 1967; Legendre 1981). In addition, both the benthic and pelagic microalgal compartments are tightly linked in ETSELE (Shaffer and Sullivan 1988). This linkage between the microphytobenthos and phytoplankton suggests that a part of the biomass observed in the water column is due to the resuspension of the benthic microalgae from the mudflat (Demers *et al.* 1987; de Jonge and van Beusekom 1992; 1995; Lucas *et al.* 2001). The resuspended fraction is important but has been difficult to determine accurately because it fluctuates as a function of both the hydrodynamic conditions and the microphytobenthic biomass in the surficial layer of the sediment. Thus the resuspension dynamics of the microphytobenthic biomass cannot be quantified as long as the dynamics of the microphytobenthic biomass itself remains unknown at the scale of the ecosystem.

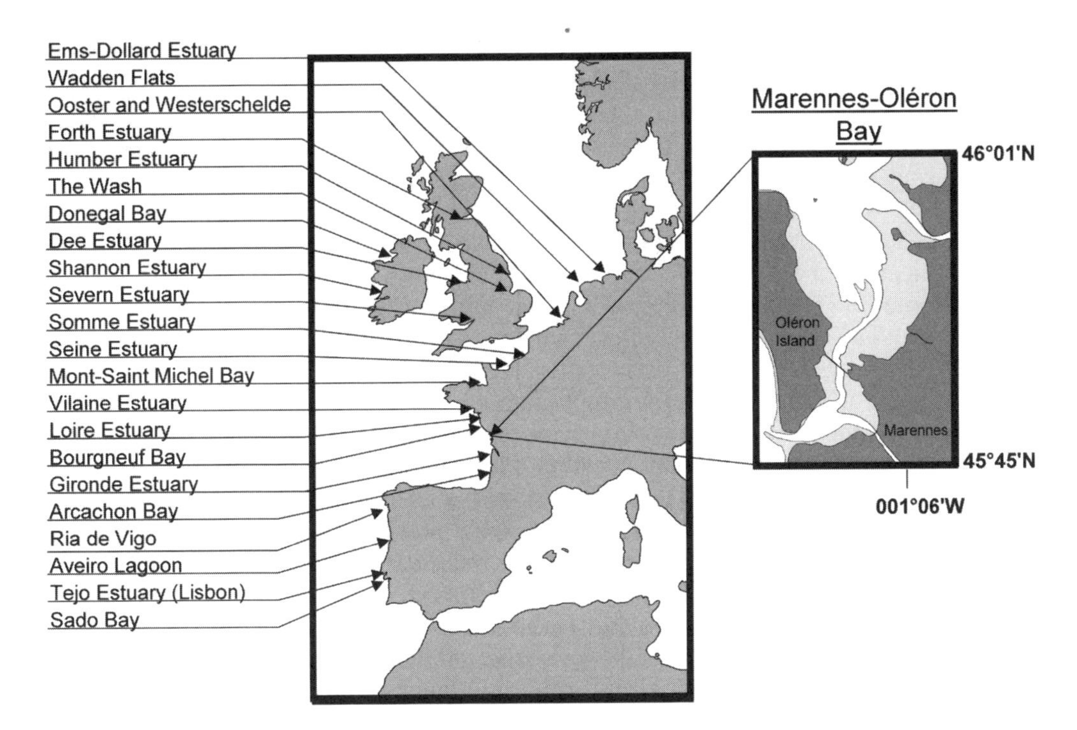

Figure 1. European-Type Semi-Enclosed Littoral Ecosystems (ETSELE) along the European coast. They are characterized by a high tidal range and wide, bare intertidal flats sustaining an important microphytobenthic production. The Marennes-Oléron Bay belongs to this category of ecosystems (60% of the total area are intertidal flats), and is one of the major sites for shellfish cultures in France.

Much conceptual progress has been made in characterizing intertidal microphytobenthos and its dynamics, mostly with respect to three areas: active vertical migration, the formation of the microphytobenthic biofilm during daytime emersion periods (Paterson *et al.* 1998; Serodio et Catarino 2000; Guarini *et al.* 2000a), and benthic primary production (Underwood and Kromkamp 2000). Important differences exist between the pelagic and the benthic systems of primary production. In particular, the way in which the microphytobenthos concentrates in the surficial layers of sediment and their active movement to reach the surface of the mud (Hay *et al.* 1993) contrast with the conditions in a well-mixed water column, in which the phytoplankton is transported passively and is exposed periodically to a variable light energy as a function of the vertical mixing intensity (Marra 1978; Legendre 1981). These differences require a new scheme to represent the dynamics of the microphytobenthos, and new techniques to study the processes over appropriate time intervals. Among them are separation techniques that exploit the active vertical migration of pennate diatoms (Couch 1989), and micro-scale techniques to measure the photosynthetic activity of the biofilm such as O_2 microelectrodes and Pulse Amplitude Modulated fluorimetry (Serôdio and Catarino 2000).

Quantification of the microphytobenthos resuspension remains difficult to estimate. The resuspension of the microphytobenthos is generally attributed to erosion processes induced by currents – mostly those generated by tidal oscillations – and waves (Demers *et al.* 1987; de Jonge and van Beusekom 1995). Therefore, in order to quantify the resuspension of the microalgal biomass, biological processes of production must be coupled with sedimentary processes of mud erosion. This coupled approach has been difficult to achieve because the results of hydro-sedimentary models are very sensitive to the critical shear-stress of the mud (McDonald and Cheng 1997) and the vertical distribution of the microphytobenthic biomass in the surficial layers of the mud is not well-known. This vertical distribution changes over time and the uncertainty in the thickness of the eroded layer of particles has the same order of magnitude as the thickness of the layer that contains the photosynthetically active biomass. Therefore, the entire photosynthetically active biomass can be resuspended all at once when this type of model is used. The fundamental question of population survival must be considered if the entire superficial layer is constantly eroded at each flood tide.

The objectives of this chapter are to review the processes that govern the biomass dynamics of the microphytobenthos inhabiting intertidal mudflats and to discuss how these processes can be evaluated and tested within a common conceptual framework by both experimental and mathematical (modelling) studies. A mathematical model specifically designed to quantify the dynamics of the biomass at the level of an ETSELE is proposed. This model is minimal; it only takes into account the main processes that govern the biomass dynamics, and the processes that are not well-known are represented by the simplest formulation possible (i.e. a linear function of the microphytobenthic biomass is used). The processes included in the proposed model are: the vertical migration, the primary production, the losses by grazing and stress-induced mortality, and the resuspension and transport of the biomass. Finally, the validation of the model is discussed using the first elements of a comparison between simulations and observations of a quantified benthic-pelagic coupling for the microalgal biomass.

Scales of variations and the conceptual scheme

Quantification of the intertidal microphytobenthos dynamics over an entire ecosystem requires a description that can include four spatial scales (microscale, mesoscale, macroscale and regional scale) each of which is associated with a deterministic component of the system. The microscale (between 1 µm and 1 cm) is the scale of the microalgae and is associated with the structure of the system. Electron microscope observations have shown that the formation of the microalgal biofilm at the mud surface during the daytime emersion period (Paterson 1989) and this biofilm (Figure 2) is a thin layer of contiguous cells which covers the surface of the mud. The mesoscale (between 1 cm and 1 meter) is an intermediary scale. It is the scale of the microphytobenthos patchiness, which varies as a function of the production and biomass (Seuront and Spilmont 2002; Saburova *et al.* 1995). The macroscale (between 1 m and 1 km) is associated with the geomorphological support of the system, and changes seasonally as a function of sedimentary processes (sedimentation, erosion or compaction events, Le Hir *et al.*

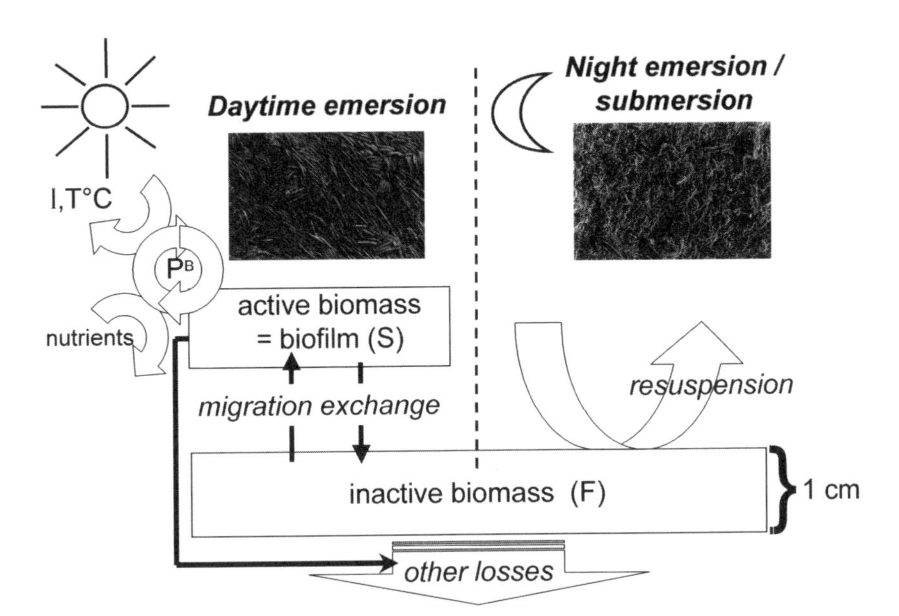

Figure 2. Conceptual scheme to describe the dynamics of the microphytobenthic biomass intertidal mudflat. The classical scheme of a unique compartment with a biomass distributed in a vertical gradient of light is replaced by a scheme of two discrete compartments, introducing a microalgal biofilm at the mud surface where the biomass (S) is concentrated and a reservoir where the biomass (F) is diluted into the first centimetre of the mud. The biofilm is a temporary structure that exchanges biomass with the reservoir compartment during daytime emersion only. The biofilm disappears during night emersion and submersion periods. Resuspension occurs only during immersion while losses by grazing or natural mortality affect the overall biomass at all times.

2000). At macroscale, the presence of the biofilm can be observed by a color change at the surface of consolidated muds; areas of unconsolidated mud, such as in runnels, do not develop a microalgal biofilm. Thus at the macroscale, the distribution of available mud surface for biofilm development must be considered. The regional scale (between 1 km and 100 km) is the scale of the tidal ecosystem. It is associated with the concept of habitats and intertidal subtidal areas must be delimited. At this scale, long-term changes affect the coastline and the bathymetry of the ecosystem.

The model design integrated the variability at mesoscale and represented the averaged dynamics of the microalgal biomass in a standard square meter of the mud surface. In the chapter devoted to the observation of the local dynamics of the biomass (Blanchard *et al.* this book), the sampling procedure used in the field studies was designed to be consistent with the modelling approach, and the size of the sampling unit was chosen to minimize mesoscale variability.

The mathematical model simulates the dynamics of the intertidal mudflat microphytobenthos at macroscale, based on the concepts and assumptions that come from the observations and experiments realized at microscale. The mathematical behaviour of the model was studied in order to characterise the properties of the microphytobenthos dynamics at the regional scale. The persistence of the benthic biomass is characterised by studying the steady state conditions of the simulations. The convergence to steady state conditions, representing the resilience of the system, describes how the benthic biomass responds to disturbances (*e.g.* suspension and sedimentation). Global trends are addressed by examining long-term changes of the steady state reached in the simulations.

At macroscale, horizontal movements of the diatoms can be neglected, and a local model (built with ordinary differential equations) that does not take into account spatial variations can be formulated, representing the dynamics of the benthic biomass in one square meter of mudflat. The square-meter of consolidated mud is assumed to be homogeneous and emerged at least once per neap-to-spring tide cycle. Inside this ideal square-meter takes place – potentially – the vertical migration and the formation of the biofilm during a daytime emersion period.

Only the biomass in the first centimetre of the mud is considered potentially active for photosynthesis and it is a part of this biomass that migrates up to the surface during a daytime emersion period to create the biofilm on the mud surface. The depth of light penetration is so shallow (not more than few hundred microns) that it can be neglected in the calculations. The light penetration depth is small compared to the micro-topography of the mud surface and has the same order of magnitude as the diatoms themselves. In addition, the layer of cells at the surface are so densely packed that they reduce the light penetration by self-shading, and therefore, it is assumed that only the microalgal biomass in the biofilm can produce and benefit from all the incident light energy available at the surface. This primary production and the vertical migration of the cells are controlled by the combination of the tidal and the light-dark cycles, which are considered to be deterministic.

This conceptual description (Figure 2) and approach differs considerably from previous modelling attempts (Pinckney and Zingmark 1993; Serôdio et Catarino 2000) since the representation of a partially active biomass as a continuous mass gradient that migrates in a light gradient (Pinckney and Zingmark 1991), is replaced with a biofilm which is a discrete, temporary and renewable structure, that ensures both production and growth of the community during daytime emersion periods. The biofilm is thus a functional structure created by the microalgal community that has migrated from the aphotic uppermost centimetre of the mud and also a structure that disappears near the beginning of the submersion period, or under a light threshold, when the microalgal cells either bury themselves into the sediment or become resuspended into the water column.

Modelling the local dynamics of the microphytobenthos

Microphytobenthic biofilm and active vertical migration

The modelling approach is based on the two main components of the conceptual description: the biofilm and the active vertical migration of the microphytobenthos.

The net vertical migration rhythm has been studied and described by many authors (Palmer and Round 1965; 1967, Hay *et al.* 1993; Serôdio *et al.* 1997) and a formal quantification is proposed here. Two state variables were defined: S, which represents the concentration of chlorophyll *a* per square meter in the biofilm on the mud surface (S in (mg Chl *a*) m^{-2}), and F which represents the concentration of chlorophyll *a* per square meter in a layer of mud one centimetre thick (F in the uppermost centimetre, in (mg Chl *a*) m^{-2}).

At the start of a daytime emersion period, the net vertical migration is oriented upwards and a dense layer of cells progressively covers the mud surface until all available space is occupied (Figure 2 and 3). Thus, the net upward migration process can be considered a density-dependant process that stops when space becomes limited, even if individual cells are renewed, without being detected. The surface space remains "saturated" with respect to diatom cells as long as the biofilm can be renewed, and within the limits of the daytime emersion period.

The following three sets of differential equations (Equations 1a, 1b, 1c) describe the dynamics of the exchanges between the aphotic layer of sediment (the uppermost

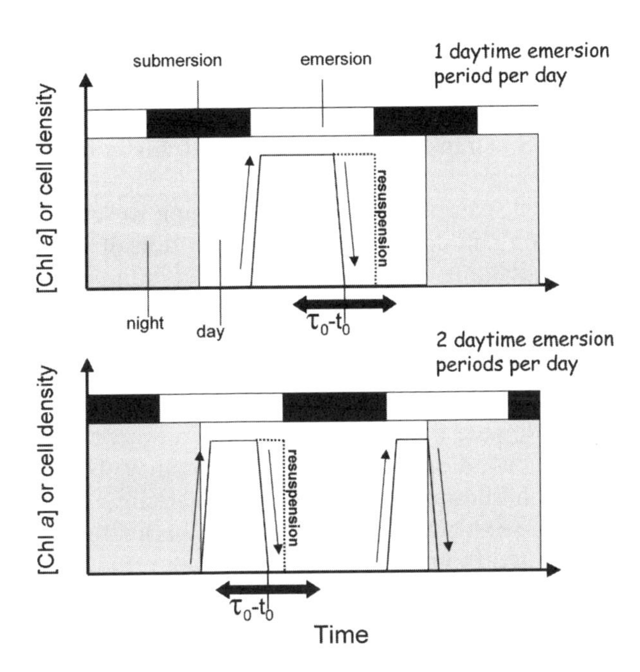

Figure 3. Conceptual scheme to describe the vertical migration in the context of the dynamics of the microphytobenthic biomass. The active vertical migration allows exchanges between S and F to occur. At the beginning of each daytime emersion period, the chlorophyll a concentration (the number of cells) increases rapidly at the surface to constitute the biofilm. The net vertical migration stopped when all the available space is occupied at the mud surface. The biofilm disappears completely at the end of the daytime emersion period, when the microalgae burrow into the sediment or if the biofilm is resuspended into the water column.

centimetre of the mud containing the biomass F) and the biofilm which contains the biomass, S. S and F have initial values equal to S_0 and F_0 respectively. A third variable, τ, is introduced to represent the time spent by the biofilm at the mud surface during a daytime emersion period. This time is set at the beginning of the daytime emersion by the initial condition, τ_0.

During daytime emersion periods:

If $\tau > 0$ If $\tau \leqslant 0$ [1a]

$$\begin{cases} \frac{dS}{dt} = + r_F F (1 - S/S_{max}) \\ \frac{dF}{dt} = - r_F F (1 - S/S_{max}) \\ \frac{d\tau}{dt} = -1 \end{cases} \qquad \begin{cases} \frac{dS}{dt} = - r_S S \\ \frac{dF}{dt} = + r_S S \\ \frac{d\tau}{dt} = -1 \end{cases}$$

The time, τ, controls net exchanges. As long as τ is positive, a part of F is transferred into S until the saturation value of the biomass in the biofilm (S_{max}) is reached. The transfer rate is: r_F (T^{-1}). When S is equal to S_{max}, net exchanges stop, and when τ becomes null or negative, even if the daytime emersion period is not finished, net exchanges are reversed; S is transferred into F, with a transfer rate of r_S (T^{-1}), until the biofilm disappears completely.

The initial condition, τ_0, can be determined according to several different hypotheses. The initial condition can coincide with the duration of the daytime emersion, assuming that pennate diatoms can anticipate (under internal or external factors) a submersion period and bury themselves into the mud before the flooding tide arrives. It could also be regulated by an internal rhythm that takes into account the combination of the tidal and the light cycles, including the phase difference between the synchronizers. This second hypothesis is slightly different from the first one, because the diatoms anticipate the flood, unless environmental conditions (changes in atmospheric pressure, wind speed or direction) modify its timing. For the third hypothesis, the initial condition is determined by the community itself, and depends on the total biomass in the first centimetre of the mud. In this case, τ_0 can be described by a function, f, of the total biomass (F_0+S_0) and S_{max}:

$$\tau_0 = f\left(\frac{F_0 + S_0}{S_{max}}\right)$$

The net migration behaviour during night emersion periods can be described as:

$$\begin{cases} \frac{dS}{dt} = - r_S S \\ \frac{dF}{dt} = + r_S S \\ \tau = \tau_0 \end{cases} \qquad [1b]$$

Under a light threshold, cells migrate downward into the sediment, and the chlorophyll *a* concentration S is transferred to F with the rate r_S (T^{-1}). Then, τ is set to its initial value, τ_0, for the next daytime emersion period.

During submersion periods, no vertical migration occurs and the system is described by:

$$\begin{cases} S = 0 \\ \frac{dF}{dt} = 0 \\ \tau = \tau_0 \end{cases} \qquad [1c]$$

No exchange occurs between S and F during a submersion period. The transition period between emersion and submersion is considered to be instantaneous at the scale of this dynamic. A fraction of the microphytobenthic biomass, S, is swept away by the flooding tide waters if the cells have not migrated down into the sediment before the end of the daytime emersion period. Afterwards, as in the night emersion period, τ is set to its initial value, τ_0, for the next daytime emersion period.

Simulations of this first model of vertical migration are performed by the commutation between the three systems [Equations 1a, 1b, 1c], and using the three hypotheses formulated about the variable τ (Figure 3). Figure 3A shows the results using the hypothesis of a complete synchronization between the vertical migration rhythm and the daytime emersion period. At the beginning, the biofilm is formed at the mud surface; S increases from 0 to S_{max}, and F decreases from F_0 to (F_0-S_{max})). Just before the end, the biofilm disappears, S decreases from S_{max} to 0, and F increases from (F_0-S_{max}) to F_0. No resuspension occurred for any initial conditions. Figure 3B corresponds to the second hypothesis. In this case, the vertical migration rhythm is fixed and cannot change. Therefore, if the flood is anticipated, S (which is equal to S_{max}) is resuspended in the water column, and the total biomass (F+S) decreased from any initial conditions (F_0+S_0), to reach zero, eventually.

The results from simulations based on the third hypothesis are shown in Figure 3C. The duration of the biofilm at the mud surface is a function of the initial biomass at the beginning of the daytime emersion period. A critical biomass B_c, above which the biofilm is resuspended in the water column, was defined and depends on the saturation value of the biofilm S_{max}, on the average duration spent by a unit of biomass (corresponding to S_{max}) at the mud surface, τ_s, and on the duration of the day-time emersion period, T_E:

$$B_c = S_{max} \frac{T_E}{\tau_s}$$

Under these conditions, the biomass F tended to decrease below the critical value of B_c such that no resuspension occurred.

Resuspension of the microphytobenthos always affects the value S (the biofilm) and occurred most of the time when S was equal to S_{max}. These simulations described the dynamics of the microphytobenthic biomass in the absence of production, grazing, mortality, and sediment erosion events that could induce the resuspension of a fraction of F. The situation becomes more complex if a neap to spring tide cycle is superimposed on the tidal cycle. In this case, two daytime emersion periods can occur per day with shorter durations (Figure 4). The daytime emersion period can be followed by a night emersion during which no resuspension occurs because the biofilm disappears when the incident light is under the threshold level, in other words the diatoms have buried themselves into the sediment.

Introducing local production and loss processes.

The dynamics of the biomass are governed by the production and loss processes and added to the system of differential equations that describes the net vertical migration behaviour of the microphytobenthic community. Loss processes include stress-induced mortality and benthic grazing by surface deposit feeders (thus decreasing S) and subsurface deposit feeders (decreasing F). For the compartment F, the loss terms can also represent any of several possible processes that bury cells in deeper layers (i.e. several decimetres), such as sedimentation events or bioturbation.

The system [1a] describing the dynamics during daytime emersion periods becomes:

if $\tau > 0$ If $\tau \leq 0$ [2a]

$$\begin{cases} \dfrac{dS}{dt} = (r_F F = p^b S)(1 - S/S_{max}) - m_S S \\ \dfrac{dF}{dt} = -r_F F(1 - S/S_{max}) + p^b S(S/S_{max}) - m_F F \\ \dfrac{d\tau}{dt} = -1 \end{cases} \qquad \begin{cases} \dfrac{dS}{dt} = - r_S S - m_S S \\ \dfrac{dF}{dt} = + r_S S - m_F F \\ \dfrac{d\tau}{dt} = - 1 \end{cases}$$

The system [1b], describing the dynamics during night emersion periods, becomes:

$$\begin{cases} \dfrac{dS}{dt} = - r_S S - \mu_S S \\ \dfrac{dF}{dt} = + r_S S - \mu_F F \\ \tau = \tau_0 \end{cases} \qquad [2b]$$

And the system [1c], describing the dynamics during submersion periods, becomes:

$$\begin{cases} S = 0 \\ \dfrac{dF}{dt} = - \nu_F F \\ \tau = \tau_0 \end{cases} \qquad [2c]$$

Where p^b is the production rate (T^{-1}); during daytime emersion periods, the flux of biomass, from S to F, which exists even if the net vertical migration is oriented upward, is represented by the term: $p^b S(S/S_{max})$ and reaches a maximum when $S = S_{max}$. This

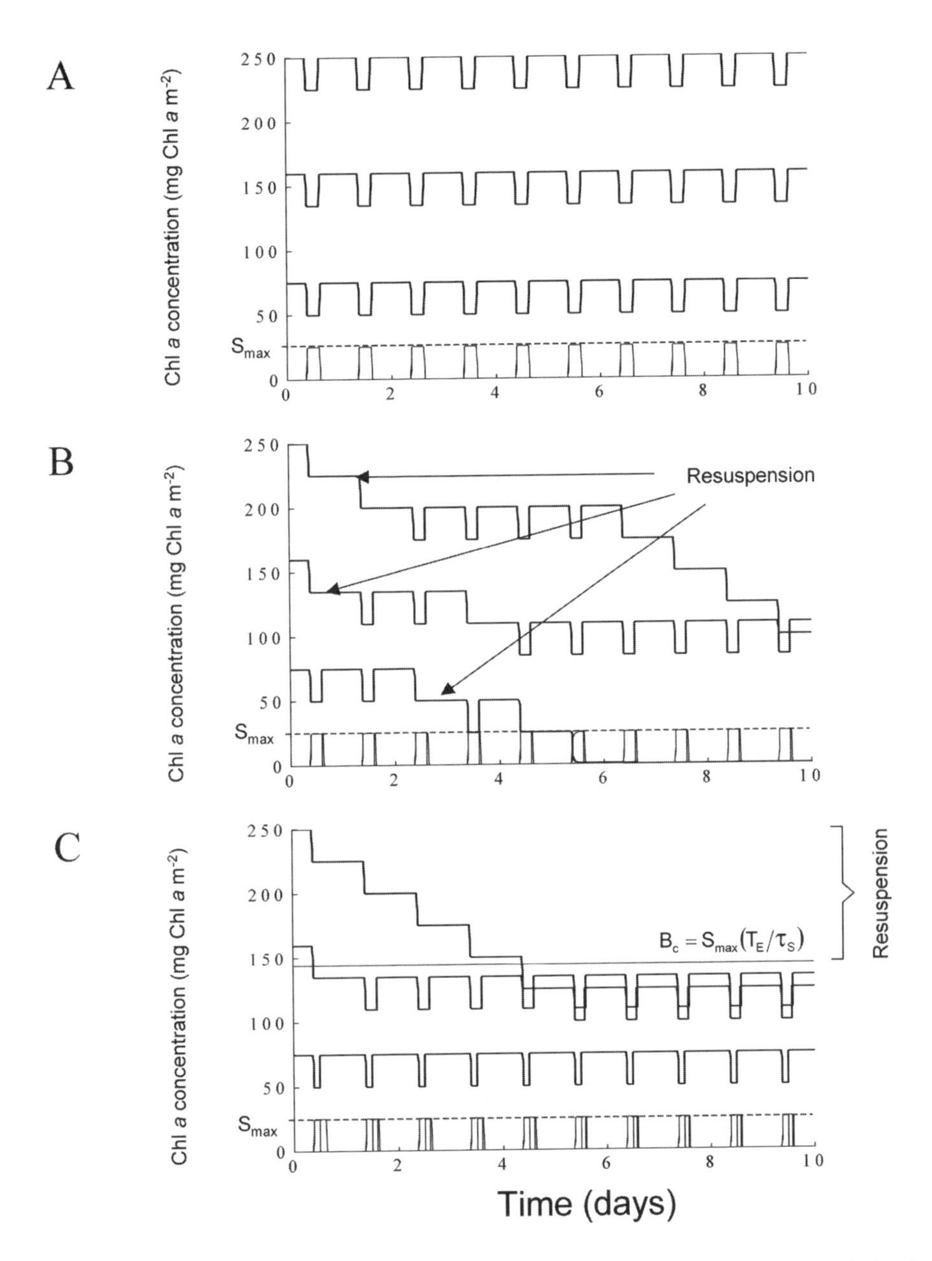

Figure 4. Simulations of vertical migration rhythm and net exchanges between S and F. At the beginning of each daytime emersion period, S (thin line) increased rapidly from 0 to S_{max}, and F (bold line) decreased from F_0 to (F_0-S_{max}). When each daytime emersion ends, S decreased rapidly from S_{max} to 0, when it is transferred into F or was resuspended into the overlying water at the beginning of a submersion. Three hypotheses were tested. A) no resuspension. B) random resuspension C) resuspension above a critical value for the overall biomass B_c. For each case, 3 initial values for F were chosen, 250, 160 and 75 mg Chl a m^{-2}. The 9 simulations were performed independently.

flux term allows production to happen even if the productive layer (S) is saturated, and furthermore, it implies that there is a constant renewal of cells at the mud surface during a daytime emersion period. m_S, μ_S, m_F, μ_F and ν_F are the loss rates (T^{-1}) for S and F and for each of the three different periods.

Modelling the microphytobenthos environment

The main environmental variables that affect the dynamics of the microphytobenthos in the model are hydrodynamic (water height and current velocities). The hydrodynamic variables control the alternating emersion and submersion periods, the light irradiance and the mud temperature, all of which synchronize and control the production in the biofilm at the mud surface during daytime emersion periods. Light irradiance changes the energy balance, and hence the mud surface temperature during daytime emersion periods. All these variables have been linked and represented in one single deterministic model formulation (Figure 5).

Hydrodynamics: The shallow-water hydrodynamic model is based on depth-averaged Navier-Stokes equations with hydrostatic and Boussinesq approximations (*i.e.* the fluid is considered to be incompressible). It simulates horizontal averaged velocities (u,v, in $m.s^{-1}$) and free-surface elevation ζ (m):

$$\begin{cases} \dfrac{\partial u}{\partial t} + u\dfrac{\partial u}{\partial x} + v\dfrac{\partial u}{\partial y} - fv = -g\dfrac{\partial \zeta}{\partial x} + \kappa_h\left(\dfrac{\partial^2 u}{\partial x^2} + \dfrac{\partial^2 u}{\partial y^2}\right) + \dfrac{\omega_x^w - \omega_x^f}{\rho(h+\zeta)} \\ \dfrac{\partial v}{\partial t} + u\dfrac{\partial v}{\partial x} + v\dfrac{\partial v}{\partial y} + fu = -g\dfrac{\partial \zeta}{\partial y} + \kappa_h\left(\dfrac{\partial^2 v}{\partial x^2} + \dfrac{\partial^2 v}{\partial y^2}\right) + \dfrac{\omega_y^w - \omega_y^f}{\rho(h+\zeta)} \\ \dfrac{\partial z}{\partial t} + \dfrac{\partial u(h+\zeta)}{\partial x} + \dfrac{\partial v(h+\zeta)}{\partial y} = 0 \end{cases}$$

where g represents the gravity ($m.s^{-2}$), f, the coriolis factor (s^{-1}), h the water height at the Lower Low Water (LLW water height for a tidal coefficient equal to 20), and ρ, the water density ($kg.m^{-3}$). κ_h is the diffusion coefficient ($m^2.s^{-1}$). ω^W and ω^f are the shear stress ($kg.m^{-1}.s^{-2}$) induced by basin bottom and wind, respectively and they are defined for directions x and y for the calculation of u and v respectively.

This model is minimal (*i.e.* it takes into account only the most essential processes to describe the dynamics) for any shallow water ecosystem where the vertical mixing is strong enough to prevent vertical gradients from developing. The tidal oscillation is induced by the open coastal boundary conditions that are only given in elevation, ζ, (Dirichlet boundary conditions). The two main first harmonic components of the tidal oscillations (M2 and S2) were included to describe the tide in the Marennes-Oléron Bay. They contribute respectively, and on average: 3.2 and 0.4 m to the tidal range, with a periodicity equal to 12H25 and 24H50 and a phase equal to 6.108 and 1.745 radians.

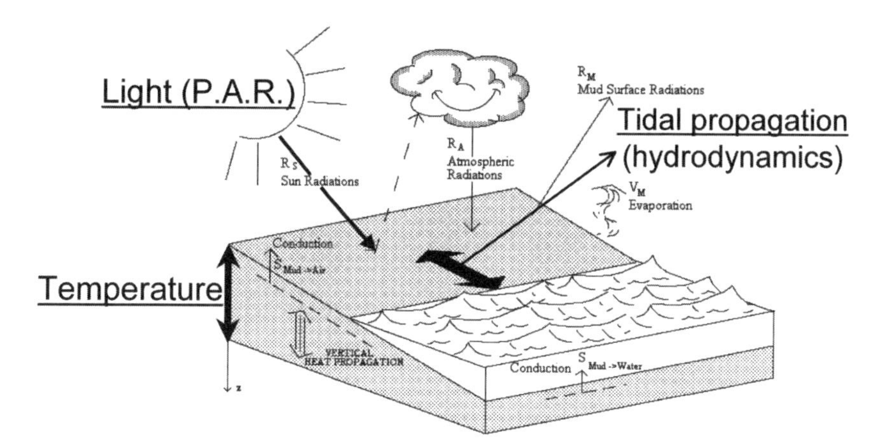

Figure 5. Conceptual scheme for the representation of the forcing synchronizers and factors influencing the production (tidal propagation, incident light irradiance and mud surface temperature). These three components are linked and were represented in the same deterministic model.

An Alternating Direction Implicit numerical method was used to solve the hydrodynamic equations on a staggered grid (ζ was solved at the middle of each mesh and u and v were solved on the right and upper sides, respectively). This modelling structure does not allow emersion to occur since the term $(h+\zeta)$ in the calculation of the stresses induced by the basin bottom and wind cannot be mathematically equal to zero. Therefore, a minimum water height was arbitrarily set to a small positive value ($(h+\zeta)min = 0.1$ m) which defined the layer of water remaining on the intertidal areas during emersion periods. When this minimum water height is reached, horizontal movements of water toward adjacent meshes are stopped, and consolidated mud (on the ridges) is considered as emerged.

The eulerian residual circulation in the Marennes-Oléron Bay, after a spring to neap tide cycle, is oriented from the north to the south of the ecosystem (Figure 6). Hence, as a general rule, when resuspended particles reach the central channel of the bay, they are quickly transported toward the southern end of the bay, and outside of this semi-enclosed ecosystem.

Incident light energy at the mud surface

The sun light intensity, E_S (W.m^{-2}) that reaches the mud surface during daytime emersion is described by the following equation:

$$E_S = E_{max} (\sin(d) \sin(\phi) + \cos(d) \cos(\phi) \cos(AH)) (1-A) (1-\zeta)$$

where E_{max} is the maximum irradiance value (W.m^{-2}), d is declination, ϕ is the latitude, AH the true hour angle, A the albedo, and ζ (dimensionless) the light attenuation

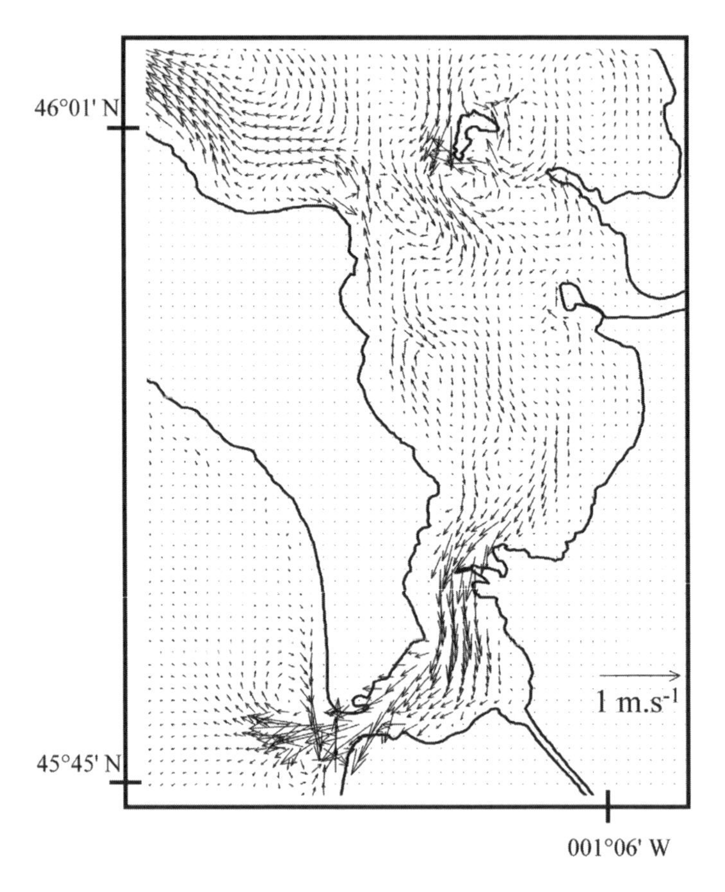

Figure 6. Eulerian Residual circulation in the Marennes-Oléron Bay, calculated from the hydrodynamic model and illustrating the north-south oriented residual flow. The residence time is, on average, 10 days in the bay. However, when particles reach the central channel, they are quickly transported toward the Maumusson Strait, at the southern oceanic boundary of the bay.

due to the cloud cover. According to Frouin (1990), E_{max} is equal to 1358.2 $W.m^{-2}$, between 250 and 4000 nm (total irradiance), and 584.9 $W.m^{-2}$ between 350 and 700 nm (Photosynthethic Active Radiations). E_S is equal to 0 $W.m^{-2}$ during both the night emersion and submersion periods. The light penetration into the mud is so small, that it was neglected in this model.

Temperature of the mud: The mud temperature $T_M(z,t)$ is calculated by the heat vertical propagation equation (Van Boxel 1986):

$$\rho_M C_{P_M} \frac{\partial T_M(z,t)}{\partial t} = \frac{\partial}{\partial z} \eta \frac{\partial T_M(z,t)}{\partial z}$$

where ρ_M is the mud density ($kg.m^{-3}$), C_{P_M}, the mud heat capacity at constant pressure ($J.kg^{-1}.K^{-1}$), η the mud conductivity ($W.m^{-1}.K^{-1}$), t is the time (s) and z the mud depth (m). The boundary conditions at the surface are described by the Heat Energy Balance (HEB, in $W.m^{-2}$) at the mud-air interface during emersion periods (Figure 5):

$$HEB = E_S + R_A - R_M - S_{MA} - V_M$$

Where E_S is the sun light energy flux, R_A the atmospheric radiation energy flux, R_M the mud radiation energy flux, S_{MA} the mud conduction energy flux between mud and air and V_M the evaporation latent energy flux.

During submersion periods, the HEB is described by:

$$HEB = S_{MW} = \frac{\eta}{h_S}(T_M(z_0, t) - T_W(t))$$

where S_{MW} is the mud conduction energy flux between mud and water, η the mud conductivity ($W.m^{-1}.K^{-1}$), h_S the height of the overlying water mixing layer (m) $T_M(z_0, t)$ the mud surface temperature (K) and $T_W(t)$ the water temperature (K). A complete description of the process formulations are available in Guarini *et al.* (1997).

The dynamics of the mud surface temperature (the uppermost centimetre of the mud) is characterized by 3 temporal scales of variations: long term seasonal scale, medium-term scale due to the neap to spring tide cycle (14.6 days) and short-term scale (due to the succession of submersion-emersion periods). Figure 7 represents only the medium-term and short-term variations in summer, when the MST variations are at a maximum. The highest MST (around 40 °C) are reached in Summer during spring tide – when low tide corresponds to midday – in the highest part of the mudflat. However, meteorological variability can hide neap to spring tide medium-term variations (Guarini *et al.* 1997).

Primary production processes

Production vs. Photosynthetic Active Radiations

The production rate is estimated by the photosynthetic activity, which was studied experimentally as a function of the photosynthetic active radiations, PAR, (Blanchard and Cariou-Le Gall 1994) and temperature (Blanchard *et al.* 1997). Experiments realized under controlled conditions, reflect the potential photosynthetic activity of the microphytobenthos. Microphytobenthos were separated from sediment exploiting their active vertical migration, according to the method developed by Couch (1989), and were then studied as a suspension of phytoplankton cells, in seawater. Primary production was estimated in terms of carbon assimilation, using ^{14}C labelling techniques

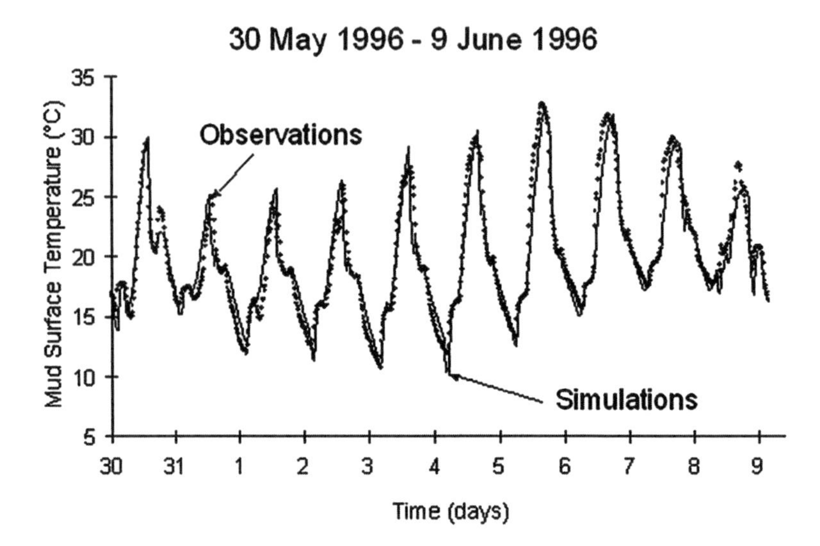

Figure 7. Mud surface Temperature (MST) variations during a neap-to-spring tide cycle. The MST was recorded during 11 days in early June 1996 and observations (dots) were compared to the simulations (line). The model predicts the MST with a precision of ca. 1 °C. During a tidal cycle, the MST dynamics is characterized by a sharp increase during daytime emersion, and a slower decrease during night emersion. During the submersion period, the MST reaches quickly an equilibrium with the water temperature.

(Blanchard *et al.* 1997). Since the incubation time used in the experiments was on the order of hours, any short-term changes that occur on the timescale of minutes were not taken into account. The experimental results showed an initial increase in the photosynthetic activity from 0 h^{-1} at 0 $W.m^{-2}$, with a slope α^b, which represents the photosynthetic efficiency. When the PAR increased, the curve converged to an asymptotic value, p^b_{max}, which represents the photosynthetic capacity of the microalgal community. Based on these results, the chl*a* normalized photosynthetic activity (h^{-1}) was described as a function of light energy by the following formulation:

$$p^b = p^b_{max} \tanh\left(\frac{E}{E_k}\right) = p^b_{max} \tanh\left(\frac{\alpha b E}{p^b_{max}}\right)$$

where p^b_{max} is the photosynthetic capacity (h^{-1}) and corresponds to the maximum asymptotic value of the P-E curve. E_k is the intersection between the slope at the origin (α^b, in J m^{-2} also called the photosynthetic efficiency) and the asymptotic value $p^b = p^b_{max}$ (h^{-1}). The seasonal evolution was characterized by a constant estimated value for E_k, which remained close to 100 $W.m^{-2}$ (Figure 8).

Maximum production vs. Photosynthetic Active Radiations

The asymptotic value, p^b_{max} or the photosynthetic capacity (h^{-1}), estimates the maximum production for a given environmental condition that varies with temperature.

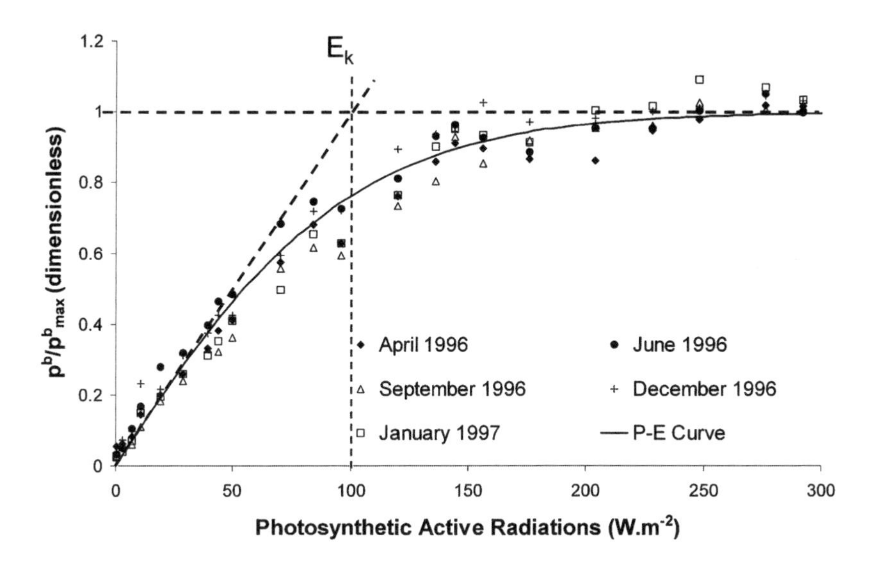

Figure 8. Standardized P-E curves. Photosynthetic activities were measured experimentally as a function of the Photosynthetic Active Radiations (PAR) using a ^{14}C assimilation method. The seasonal pattern is characterized by a constant value for E_k suggesting that the microphytobenthos did not acclimate to its light environment. E_k is estimated at around 100 $W.m^{-2}$, and most of the time, during daytime emersion periods, the microphytobenthos receives a saturating light for photosynthesis.

p^b_{max} increases when temperature increases up a maximum value which is the optimal value for photosynthesis, and then decreases down to zero until the maximum (or lethal) temperature for the photosynthesis is reached. The relationship between the photosynthetic capacity, p^b_{max} (h^{-1}), and the temperature T (°C) is described by the following formulation:

$$p^b_{max} = p^b_{max}(\varepsilon_T)^{\beta}\exp\{\beta(1-\varepsilon_T)\}$$

where β is the curvature coefficient of the curve, and ε_T is the relative deviation of the temperature T (°C) from the maximum value T_{max} (°C):

$$\varepsilon_T = \left(\frac{T_{max} - T}{T_{max} - T_{opt}}\right)$$

where, T_{opt} (°K) is the optimal temperature for the photosynthesis. The relationship between p^b_{max} and temperature was studied seasonally (Blanchard *et al.* 1997). The relationship was characterised by stable optimal and maximum temperatures for photosynthesis (Figure 9); only the maximum photosynthetic capacity varied showing a maximum in spring (of about 0.3 h^{-1}) and a minimum in winter (of about 0.1 h^{-1}).

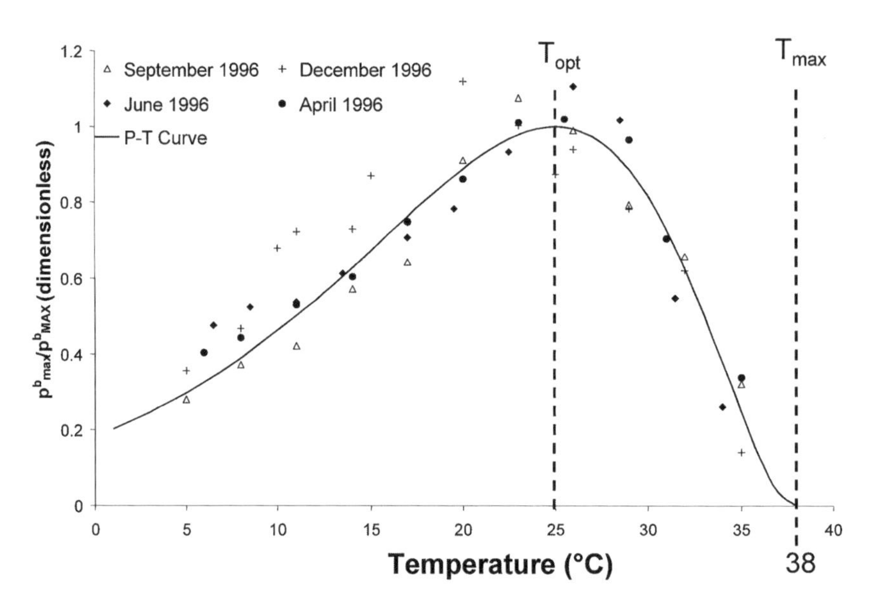

Figure 9. Standardized P-T curves. Photosynthetic capacity was measured experimentally as a function of the temperature (in the range of the field mud surface temperature variability) using a ^{14}C assimilation method. The seasonal pattern is characterized by a constant value for the optimal (T_{opt}=25 °C) and maximum (T_{max}=38 °C) temperature for the photosynthesis. The curvature coefficient did not differ significantly either. This suggested that the microphytobenthos did not present any acclimation to its thermal environment.

In order to quantify the influence of the mud surface temperature on the productivity of the microphytobenthos, the photosynthetic capacity was calculated as a function of the mud surface temperature, using the model that represents the photic, thermal, and hydrodynamic environment in the Marennes-Oléron Bay. Interestingly, during summer, the mud surface temperature reached a maximum of ca. 38 °C and is frequently above the optimal temperature for photosynthesis of 25 °C. Figure 10 shows that at the low tide of a spring tide in summer, the microphytobenthos is thermo-inhibited over most intertidal areas in Marennes-Oléron Bay: 75 % of the total intertidal area is affected by the phenomenon of thermo-inhibition at spring tide and 20 % at neap tide. In summer, microalgae undergo fast and large variations of temperature (from 18 °C at the beginning of the daytime emersion period, to 36 °C after 6 hours of emersion), which are unavoidable even by downward migration into the sediment, since the temperature increase propagates more than one centimetre deep into the mud during an emersion period (Harrison and Phizacklea 1987, Guarini *et al.* 1997). The microphytobenthos does not adjust its optimal photosynthetic capacity (p^b_{max}) to the seasonal changes in the maximum temperature reached at the surface of the intertidal mudflat. This lack of temperature acclimation may be due to the rapidity of the temperature increase during daytime emersion periods.

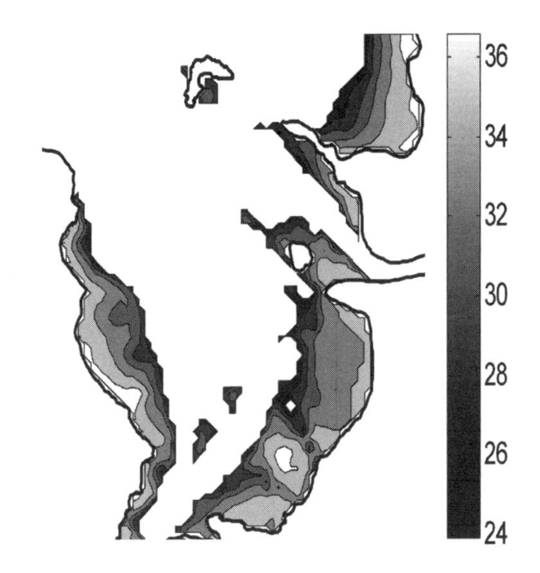

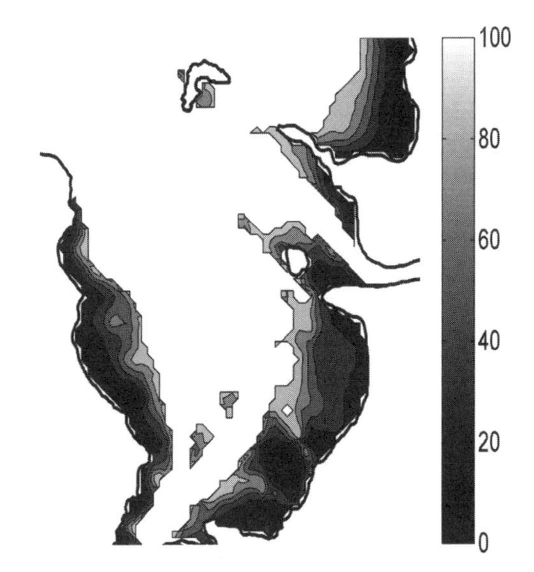

Figure 10. Simulation of the Mud Surface Temperature (MST °C) and the photosynthetic capacity (as a percentage of the maximum value) in the Marennes-Oléron Bay in summer (June, day 182). Results are for a low tide at spring tide. The MST temperature increased up to 36 °C and was above the optimal temperature for the photosynthesis (Topt = 25 °C) over 80% of the total area. The result is a general thermo-inhibition of the microphytobenthic productivity in the intertidal mudflats of the Marennes-Oléron Bay

Dynamic changes of the photosynthetic activity – photo-inhibition vs. photo-acclimatation

The photosynthetic activity varies not only as a function of light and temperature but also as a function of the ecophysiological state of the microalgal community. Two opposite reactions, photo-acclimation and photo-inhibition, occur when a long exposure to saturating light radiation affects the photo-systems of the microalgae. Photo-acclimation is a process that requires a physiological adjustment of the cells, which is difficult to achieve in a rapidly fluctuating environment. Photo-inhibition is a reaction to strong light irradiance, which provokes a decrease of photosynthesis, without changes in pigment concentration. Cells switch instantaneously from dark to saturating light for photosynthesis during daytime emersion and photo-inhibition is more likely to happen at the mud surface when microalgae are exposed to a saturating PAR, suddenly and for several hours.

The photo-inhibition process was studied experimentally (Blanchard *et al.* submitted) by exposing suspensions of benthic microalgae to a saturating light for the photosynthesis and at constant temperature; P-E curves were measured at regular time intervals (every 30 min). The Jassby and Platt (1976) equation was used to estimate the parameters: α^b (photosynthetic efficiency), E_k (light irradiance at saturation), and p^b_{max} (the photosynthetic activity).

The photo-inhibition process (Figure 11) was described by the following dynamic system:

$$\begin{cases} \dfrac{dp^b_{max}}{dt} = \gamma\left(p^b_{opt} - p^b_{max}\right) \\ \dfrac{dp^b_{opt}}{dt} = -p^b_{opt}\,\dfrac{(\delta + 1)}{t_{ph}}\left(\dfrac{t}{t_{ph}}\right)^d \\ \dfrac{dE_k}{dt} = \kappa\left(E_{max} - E_k\right) \end{cases}$$

which also determines the temporal evolution of $\alpha^b = \dfrac{p^b_{max}}{E_k}$

where p^b_{opt} (h^{-1}) is the optimal value of p^b_{max}, τ_{ph} (h) is the time threshold from which p^b_{opt} begins to decrease, γ (h^{-1}) the rate for p^b_{max} to converge towards its maximum value p^b_{opt}, δ is a dimensionless parameter representing the intensity of photo-inhibition, κ (h^{-1}) is the fitting rate to new saturating light conditions that leads to the maximum light irradiance at saturation, E_{max} ($W.m^{-2}$).

It has been suggested that photo-inhibition does not occur in intertidal mudflats, not only because the pattern which was characterised experimentally was not observed in the field, but also because the microalgae can migrate quickly down to aphotic conditions (in few hundred micrometers). This constant renewal of the biofilm, already suggested by Blanchard and Cariou-Le Gall (1994), is consistent with an increase of the overall biomass in the first centimetre of the mud during daytime emersion periods, even if the biofilm reaches its maximum cell density at the mud

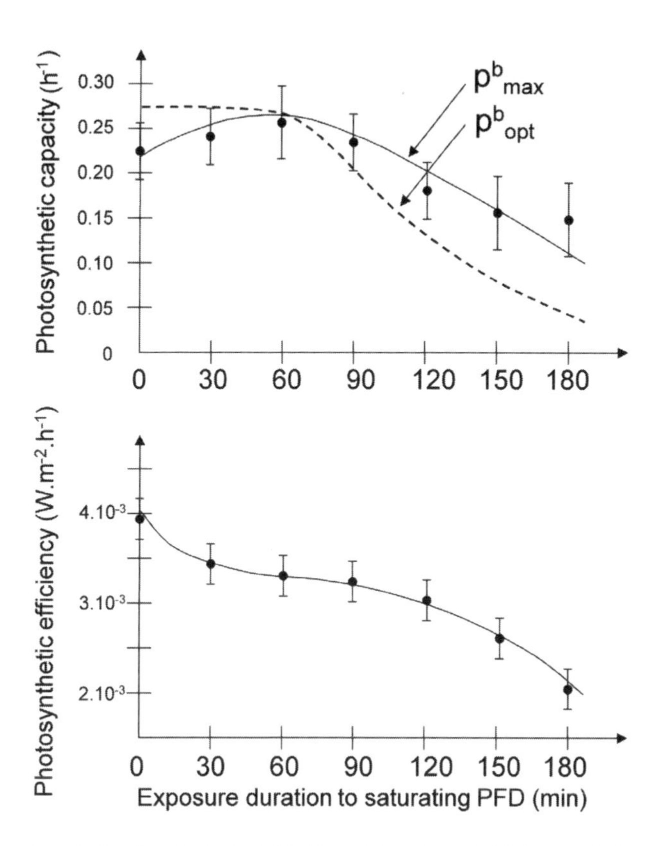

Figure 11. Measured and simulated photo-inhibition. The photo-inhibition of the microphytobenthic productivity was induced experimentally by maintaining a suspension of extracted cells in saturating light for the photosynthesis. The photosynthetic capacity and the photosynthetic efficiency were estimated every 30 min. during 180 min. The optimal value for the photosynthetic capacity is estimated from a dynamic model. Photo-inhibition (a decrease of the optimal photosynthetic capacity) started after ca. 1 hour of saturating light exposure.

surface. Thus, in the dynamic model, which describes the dynamic of the microphytobenthos biomass, the time spent by a group of cells (or by a unit of biomass) at the mud surface is considered as less than the time that is necessary to induce photo-inhibition when cells are exposed to saturating PAR The order of magnitude of this time, τ_s, is about 1 hour (Figure 11).

Mathematical properties of the dynamical model

Slow-Fast formulation

The first step in studying the mathematical properties of the system is to differentiate the dynamics into two temporal scales, qualified as 'slow' and 'fast'. The model parameters are S_{max}, r_F, r_S, p^b, m_S, m_F, μ_S and μ_F. S_{max} has the same unit as the state variables S and F (mg Chl *a* m^{-2}) and its estimated value is 25 (mg Chl *a*) m^{-2}. The other parameters are standardized rates (1/time unit) and can be ranked, according to their typical order of magnitude, from the highest value to the lowest:

$$r_S\,(100) \gg r_F\,(10) \gg p\,(1) \gg (m_S \approx m_F \approx \mu_S \approx \mu_F \approx v_F)\,(0.1)$$

In addition, a transformation can be applied to the model: B = S+F. This transformation is particularly valuable since B can be sampled easily on a mudflat, even while it remains difficult to observe the biomasses S and F separately. The model then becomes:

During daytime emersion periods:

if $\tau > 0$ If $\tau \leq 0$ [3a]

$$\begin{cases} \frac{dS}{dt} = (r_F(B - S) + p^b S)(1 - S/S_{max}) - m_S S \\ \frac{dB}{dt} = (p^b - m_S + m_F)S - m_F B \\ \frac{d\tau}{dt} = -1 \end{cases} \qquad \begin{cases} \frac{dS}{dt} = -r_S S - m_S S \\ \frac{dB}{dt} = (m_F - m_S)\,S - m_F B \\ \frac{d\tau}{dt} = -1 \end{cases}$$

During night emersion periods:

$$\begin{cases} \frac{dS}{dt} = -r_S S - \mu_S S \\ \frac{dB}{dt} = (\mu_F - \mu_S)S - \mu_F B \\ \tau = \tau_0 \end{cases} \quad [3b]$$

During submersion periods:

$$\begin{cases} S = 0 \\ \frac{dB}{dt} = -v_F B \\ \tau = \tau_0 \end{cases} \quad [3c]$$

These systems of equations describe slow-fast systems where S is the fast variable and B the slow one. Variations of S depend on fast processes (upward and downward migration) whereas variations of B only depend on slow processes (production and loss terms). The system can now be studied at these two scales of variation.

First, to examine variations at the fast-time scale the slow motions are “frozen” and the processes of production and loss terms become negligible relative to the vertical migrations. The model is then simplified by:

During daytime emersion periods:

if $\tau > 0$ If $\tau \leq 0$ [4a]

$$\begin{cases} \dfrac{dS}{dt} \approx r_F(B - S)(1 - S/S_{max}) \\ \dfrac{dB}{dt} \approx 0 \\ \dfrac{d\tau}{dt} = -1 \end{cases} \qquad \begin{cases} \dfrac{dS}{dt} \approx -r_S S \\ \dfrac{dB}{dt} \approx 0 \\ \dfrac{d\tau}{dt} = -1 \end{cases}$$

During night emersion periods:

$$\begin{cases} \dfrac{dS}{dt} \approx -r_S S \\ \dfrac{dB}{dt} \approx 0 \\ \tau = \tau_0 \end{cases} \qquad [4b]$$

During submersion periods:

$$\begin{cases} S = 0 \\ \dfrac{dB}{dt} \approx 0 \\ \tau = \tau_0 \end{cases} \qquad [4c]$$

For a short time, t, it is only possible to observe a variation in S; this happens during the daytime emersion periods (according to the condition that was set for τ) and at the beginning of the night emersion periods only. No variation in S is observed during the immersion periods. The sets of stationary points (L) of the fast equations are defined when dS/dt = 0, providing the “slow curves” of the systems. During the daytime emersion period, when $\tau > 0$, the slow curve is:

$$L_1 = \{(S,B) \in \mathbb{R}^+ \times [S_{max}, +\infty] \;/\; S = S_{max}\}$$

The condition $B \geq S_{max}$ ensures that the biomass content of the F-compartment is sufficient to fill up the biofilm at the mud surface. The domain of the definition is restricted to $[S_{max}, +\infty]$.

During the daytime emersion period, when $\tau \leq 0$, and during night emersion and submersion periods, the slow curve is described by:

$$L_2 = \{(S,B) \in \mathbb{R}^+ \times \mathbb{R}^+ \;/\; S = 0\}$$

Secondly, at the slow-time scale, fast motions appear instantaneous and at the beginning of each daytime emersion period, S 'jumps' from 0 to S_{max}, into the slow curves L_1. During daytime emersion period (when $\tau \leq 0$), night emersion or submersion period, S jumps from S_{max} to 0, if it is not already equal to 0, corresponding to the slow curve L_2. At this slow-time scale, and for limited period of time (*e.g.* a few neap-to-spring tide cycles), it can be assumed that the loss rates affecting S at the mud surface are not different locally from the loss rates affecting F at the sub-surface, and hence, $m_F \approx m_S \approx m$ during the daytime emersion period and $\mu_F \approx \mu_S \approx \mu$ during night emersion periods (m and μ affect B). The loss rates, ν_F, affecting F during submersion periods were also renamed ν, and represent the loss rates which affect B, consistent with the two previous simplifications. The simplified systems at a slow-time scale are:

During daytime emersion periods:

if $\tau > 0, S=S_{max}$ If $\tau \leq 0, S=0$ [5a]

$$\begin{cases} \frac{dS}{dt} \approx 0 \\ \frac{dB}{dt} \approx p^b S_{max} - mB \\ \frac{d\tau}{dt} = -1 \end{cases} \qquad \begin{cases} \frac{dS}{dt} \approx 0 \\ \frac{dB}{dt} \approx -mB \\ \frac{d\tau}{dt} = -1 \end{cases}$$

During night emersion periods:

$$\begin{cases} \frac{dS}{dt} \approx 0 \\ \frac{dB}{dt} \approx -\mu B \\ \tau = \tau_0 \end{cases} \qquad [5b]$$

During submersion periods:

$$\begin{cases} S = 0 \\ \frac{dB}{dt} \approx -\nu B \\ \tau = \tau_0 \end{cases} \qquad [5c]$$

During the daytime emersion periods, as long as $\tau > 0$, the production is described as a proportion of the constant biomass S_{max}. Then, the biomass B increases along the slow curve L_1 and converges asymptotically toward the equilibrium value $B^* \approx p^b S_{max} / m$ where m is the mortality rate during daytime emersion (h^{-1}); the smaller m becomes, the higher B* is. For $\tau \leq 0$ or during the night emersion and submersion periods, B decreases along the slow curve L_2 and converges asymptotically down to the equilibrium value $B^*\approx 0$, which represents the extinction of the microphytobenthos. The dynamics of the variable B are governed by the periodic succession of the four sub-systems of ordinary differential equations.

Mathematical behaviour of the system

Each sub-system (Equations 5a, 5b, 5c) was integrated analytically. The number of parameters is reduced by assessing that, locally, the loss rates are not significantly different where $m = (m \approx \mu \approx \nu)$. In addition, the two sub-systems that described the dynamics of the microphytobenthos during a daytime emersion period were combined in one integrated equation. Therefore, three integrated equations described the evolution of the system at the slow time-scale:

During daytime emersion periods:

$$B_E = B_{E_0} e^{-mT_E} + \frac{p^b S_{max}}{m} e^{-m(1-\alpha)(T_E - T_P)} - \frac{p^b S_{max}}{m} e^{-mT_E}$$

During night emersion periods:

$$B_N = B_{N_0} e^{-mT_N}$$

During submersion periods:

$$B_I = \left(B_{I_0} - \alpha S_{max}\right) e^{-mT_I}$$

where T_E, T_P, T_N and T_I are the durations of the daytime emersion period, the production period (corresponding to the presence of the biofilm at the mud surface), the night emersion period and the submersion period respectively, and with $\alpha=0$ if $T_P<T_E$ and $\alpha=1$ if $T_P \geq T_E$. If $\alpha=0$, no resuspension occurs, if $\alpha=1$ and the submersion period follows a daytime emersion period, S_{max} is resuspended in the water column.

The first and second hypotheses (Figure 12A,B) assume that the duration of the biofilm is synchronized with the duration of the daytime emersion period. In the first case, the synchronization is exact and the migration phases correspond to the beginning and the end of each daytime emersion period. In the second case, some fluctuations in the duration may induce resuspension randomly when the reflood occurs before the expected (deterministic) moment. In both cases $T_E \approx T_P$, and without any random fluctuations on the duration of the period T_E, $\alpha=0$. The solution for a typical succession (1-periodic case) of the following periods (Daytime E.+Submersion+Night E.+Submersion) is then:

$$B_I^{n+1} = (e^{-mT}) B_I^n + \frac{p^b S_{max}}{m}\left(1 - e^{-mT_E}\right) e^{-m(T-T_E)} - \alpha S_{max} e^{-m(T-T_E)}$$

$$B_E^{n+1} = (e^{-mt}) B_E^n + \frac{p^b S_{max}}{m}\left(1 - e^{-mT_E}\right) - \alpha S_{max} e^{-mT}$$

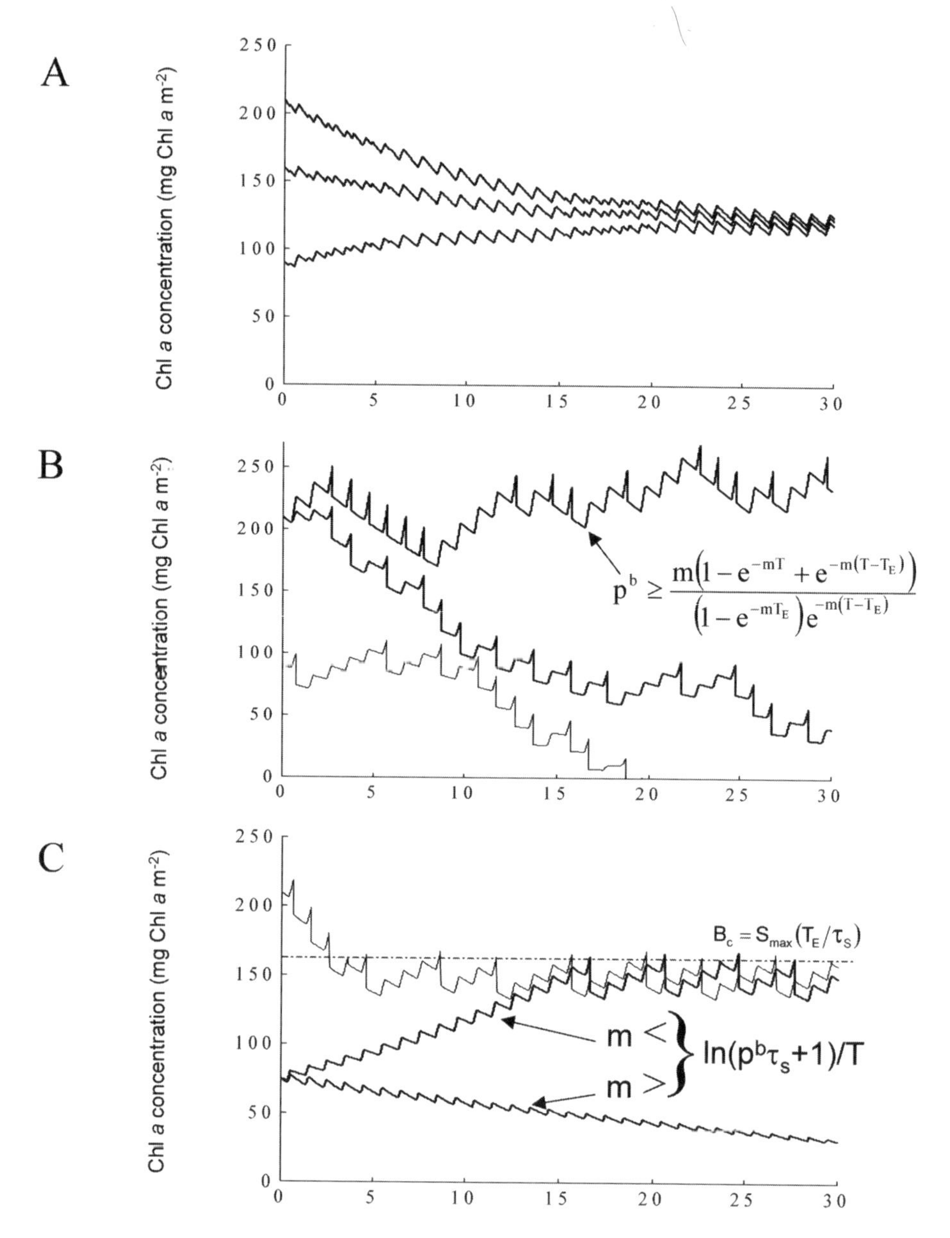

Figure 12. Simulations of the dynamics of the microphytobenthic biomass. The overall biomass (B=S+F) is represented in term of chlorophyll *a* concentration. The three figures A, B and C correspond to the three hypotheses that were tested for the vertical migration (see Figure 4). A) The chlorophyll *a* concentration converges to a unique positive, stable steady state condition. B) The chlorophyll *a* concentration converges to a positive, stable steady state according to the balance between the productivity and the global losses (grazing, natural mortality). C) The chlorophyll a concentration converges to the critical concentration $B_c=S_{max}(T_E/\tau_S)$, if the loss rates are lower than $(1/T)\ln(p^b\tau_s+1)$. Otherwise, it decreases down to zero. The steady state cycle is sensitive to the value of the initial condition.

with $T = T_I + T_N + T_I + T_E = 24h$. The discrete time series $B^{n+1} = aB^n + b$, where $a < 1$, converges to stable steady states ($B^* = b/(1-a)$), which are respectively:

$$B^*_I = \frac{\left((p^b/m)(1 - e^{-mT_E}) - \alpha\right) S_{max} e^{-m(T-T_E)}}{(1 - e^{-mT})}$$

and,

$$B^*_E = \frac{\left((p^b/m)(1 - e^{-mT_E}) - \alpha e^{-mT}\right) S_{max}}{(1 - e^{-mT})}$$

The steady state of the biomass dynamics is thus characterized by a series of oscillations, and the amplitude of these oscillations at steady state is equal to:

$$\Delta B^* = B^*_E - B^*_N = \frac{S_{max}}{(1 - e^{-mT})}\left(\frac{p^b}{m}(1 - e^{-mT_E})(1 - e^{-m(T-T_E)}) + \alpha e^{-mT}(e^{mT_E} - 1)\right)$$

The amplitude is always positive and is defined as long as m is greater than zero. The period T is always greater than T_E. If $\alpha=0$, no resuspension of the microalgal biomass occurs and the biomass converges to a series of positive steady state values. On the contrary, if α alternates randomly between 0 and 1, there is an episodic resuspension of the biofilm and the biomass deviates from the series of steady state values. If α is always equal to 1, there is a constant resuspension of the biofilm, and p must respect the following inequality to ensure that the biomass will maintain a minimum steady state value greater than S_{max}:

$$pb \geq \frac{m\left(1 - e^{-mT} + e^{-m(T-T_E)}\right)}{(1 - e^{-mT_E})e^{-m(T-T_E)}}$$

Similar calculations can be performed using a more realistic case with a changing phase difference between the tidal and light-dark cycles over a 15 day period (15-periodic case). The conclusions remain the same (Guarini *et al.* 2000b) since the dynamics of the biomass can by described by a similar series $B_{n+1} = aB_n + b$ where $a = \exp(-mT) < 1$. The biomass again converges to a series of steady state values $B^* = b/(1-a)$, where b can be calculated at the beginning of each daytime emersion, night emersion and submersion period. Finally, the model accounts for variations in p^b as a function of both incident light irradiance and mud surface temperature. In this case, p^b is replaced by $p^b(t) = p^b_{MAX}.f(t)$ with f(t), a periodic function of the time t, which limits the maximum production rate p^b_{MAX}. During daytime emersion period, the total biomass increases, from an initial value B_0, to reach the value, B_E:

$$B_E = e^{-mT_E}B_0 - e^{-mT_E} p^b_{MAX} S_{max} \int_{T_E} f(t)e^{mt}dt$$

For any sequence or combination of daytime emersion, night emersion and submersion periods, the dynamics can be described by a numerical series $B_{n+1} = aB_n + b$.

b includes the integral of the function of production ($f(t)\exp\{mt\}$), which can be integrated numerically. $a = \exp(-mT) < 1$, ensures that the numerical series converge to steady state series of values $B^* = b/(1-a)$. For any set of positive parameters, and according to the assumptions that we made, the model converges toward a series of steady state values whatever the frequency of the environmental synchronizers is.

The third hypothesis (Figure 12C) assumes that the duration of the biofilm at the mud surface during daytime emersion period is determined by the initial amount of biomass able to migrate. In the model, this initial biomass is B_0, which is also equal to F_0 since there is no biofilm at the mud surface, at the beginning of the daytime emersion. The potential duration of the biofilm is given by τ_0. It is calculated by dividing the initial biomass by S_{max}, and by multiplying this value by the average time spent by a unit of biomass (equal to S_{max}) at the surface of the mud: $\tau_0 = \tau_s \times (B_0/S_{max}) = \tau_s \times (F_0/S_{max})$. From these estimates, and using T_E, the duration of the daytime emersion period, the critical biomass for resuspension was calculated, $B_c = S_{max} \times (T_E/\tau_s)$. In the theoretical 1-day periodicity, for the case of the succession of the sequence (Daytime E.+Submersion+Night E.+Submersion) with a period T = 1 day, the discrete model is equal to:

$$B_{n+1} = B_n e^{-mT_E} + \frac{S_{max}p^b}{m} e^{-m(1-\alpha)(T_E-T_P)} e^{-m(T-T_E)} - \frac{S_{max}p^b}{m} e^{-mT} - \alpha S_{max} e^{-m(T-T_E)}$$

Two cases were investigated: when $\alpha = 1$, B is greater than B_c and the biofilm is resuspended in the water column each time the submersion period follows the daytime emersion period. When $\alpha = 0$, B is lower than B_c and there is no resuspension at all.

When $\alpha = 1$, $B_0 \geq B_c$, the series can be summarized by $B_{n+1} = aB_n + b$ where $a = \exp(-mT_E) < 1$. Therefore, the biomass converges to a series of steady state values $B^* = b/(1-a)$, which are given by:

$$B^* = S_{max} e^{mT_E} \left(\frac{p^b}{m} - \frac{p^b}{m} e^{-mT_E} - 1 \right) \Big/ \left(e^{mT} - 1 \right)$$

which decreases lower than B_c if :

$$p^b < \frac{m\left(T_E e^{mT} - T_E + \tau_S e^{mT_E}\right)}{\tau_S\left(e^{mT_E} - 1\right)}$$

When $\alpha = 0$, $B_0 < B_c$, the model has a more complex behaviour. The numerical series:

$$B_{n+1} = B_n e^{-mT_E} + \frac{S_{max}p^b}{m} e^{-mT} \left(e^{m\tau_S B_n} - 1\right)$$

increases if,

$$p^b > \frac{\left(e^{mT} - 1\right)}{\tau_S} \quad \text{or,} \quad m < \frac{1}{T}\ln(p^b\tau_S + 1)$$

and otherwise decreases to zero. According to these different conditions B_c will converge to a unique steady state value (greater than B_c), or to a steady state value that will depend on the initial condition (around B_c), or down to zero, if the grazing is too high to be compensated by the daytime production. As for the previous case, similar calculations can be made in more complicated cases for the 15-periodic case and with a fluctuating production. The complexity becomes higher and the series of coefficients α_j (j= 1, …, 15) are difficult to calculate, because they depends on an increasing number of conditions. Nevertheless, the numerical simulations have shown that the three types of behaviour exist in the model (convergence to a unique cyclic steady state greater than B_c, an initial condition dependant stead-state around B_c, or a local extinction of the microphytobenthos if the grazing is not compensated by the production).

Discussion and perspectives

Definition of the restrictions on the microphytobenthic model. Several conditions must be respected when studying this system. The first condition, $B_c > S_{max}$, ensures that the biomass (F) does not become negative. Two comprehensive surveys (Guarini *et al.* 1998) of the microphytobenthic biomass in the Marennes-Oléron Bay (France) in June 1995 and the other in January 1996 (Figure 13), have shown that the biomass per unit of area and in the first centimetre of the mud (in terms of the chlorophyll a concentration) did not decrease below S_{max}, which was estimated at 25 mg Chl a m^{-2} (Guarini 1998).

The second condition is that the duration of the daytime emersion period does not decrease below $1/r_F$ (where r_F is the biomass transfer rate from F to S) which corresponds to ca. 30 min. In addition, for a numerical purpose, this condition was combined with another condition that assumes that T_E (the duration of the daytime emersion period) has to be greater than τ_S. When this was the case, no vertical migration, no production and mostly no resuspension could occur in the model system. The third condition is that the duration of the daytime emersion period is not greater than the period of the semi-diurnal tidal cycle (12H25). This means that the mud surface is submerged at each tidal cycle. Numerically, the suppression of the submersion period is not problematic (the submersion period and the ‘resuspension’ coefficient, α, are set to zero), but it induces other problems linked to changes in sediment properties (drying and consolidation) and hence in vertical migration and production of the diatoms. The upper part of the mudflat, which is submerged only during spring tides, was not considered in these simulations.

Sensitivity of the simulations to parameter variability

Sensitivity analyses were performed at steady state, using the typical order of magnitude value of each of the parameters to calculate the nominal solution. The deviation ∂D from the nominal solution, D_0, was calculated at the beginning and at the end of each daytime emersion period, for any variations ∂P_i in each of the initial parameters,

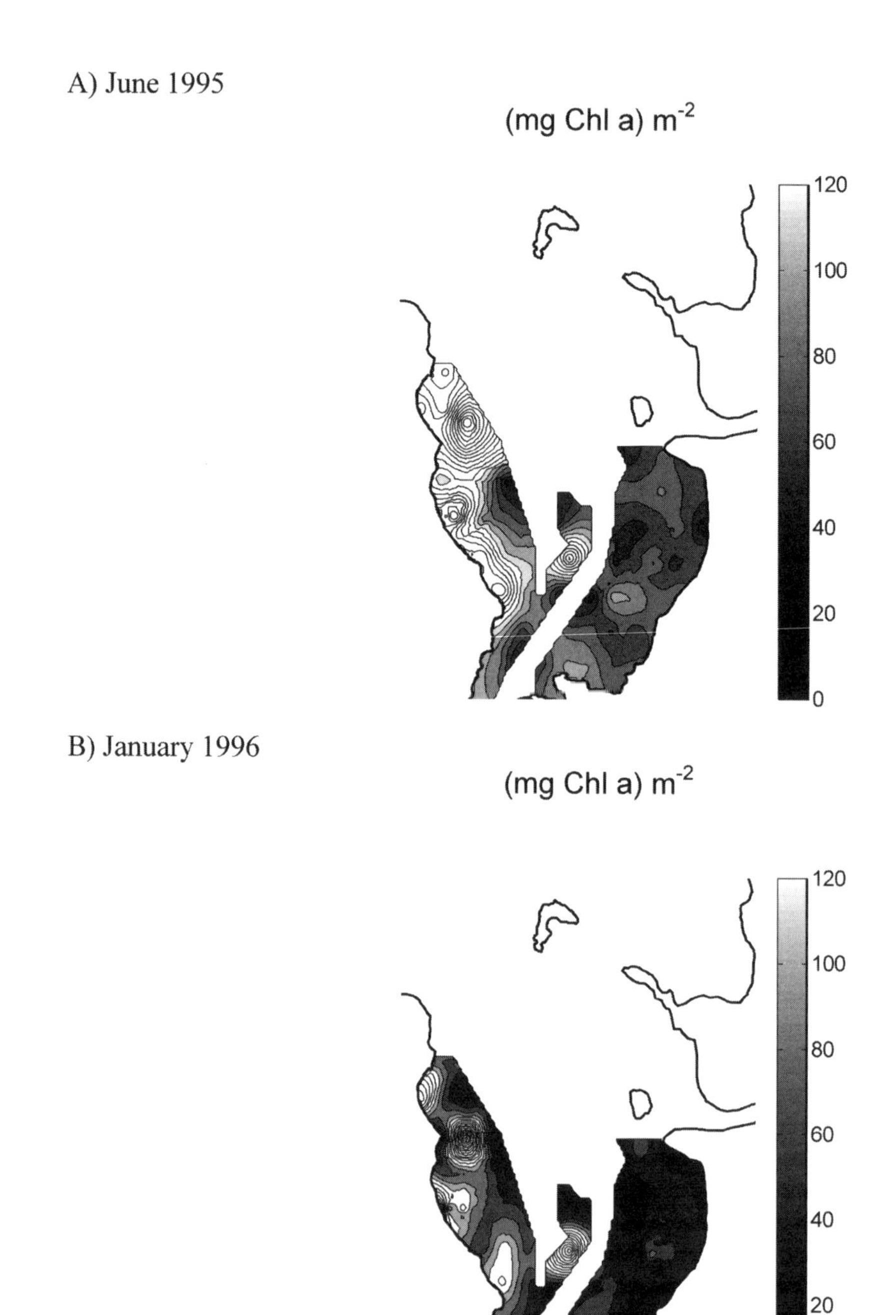

Figure 13. Spatial distribution of the microphytobenthic biomass in term of chlorophyll a concentration in the intertidal mudflats of the Marennes-Oléron Bay. A) Spatial distribution in June 1995, and B) spatial distribution in January 1996.

P_{i0} (Miller, 1974). The first order sensitivity coefficient, S_i, was calculated as a linear deviation from the nominal value: $S_i = \partial D / \partial P_i$. The relative sensitivity coefficients, R_i, were then calculated: $R_i = S_i P_{i0} / D_0$. The simulations were highly sensitive to the parameters p^b, S_{max}, and T_E. The average relative sensitivity coefficients for these parameters were equal to 1.0, meaning that any variations of these parameters have a proportional effect on the steady state estimate (a percentage of variation of these parameters induces the same percentage of variation in the steady state estimate). This justified the efforts made to estimate p^b and S_{max} as a function of the forcing variables and species composition (Guarini 1998); the average relative sensitivity coefficient for τ_s (in the third modelling hypothesis only) was equal to –1.0; T_E and τ_s have an antagonistic effect on the simulations and τ_s has to be estimated accurately.

On the contrary, the sensitivity of the mortality rate is not as high. The average relative sensitivity coefficient for this parameter is equal to -0.5. The main problem with this parameter is not with the sensitivity of the simulations with respect to its variability, but rather with the difficulty in estimating it spatially. Indeed, it depends on a variable, but high number of grazers (from the meio- and macrofauna), as well as on the physiological condition of the microphytobenthic community. It is then necessary to estimate the mortality rate by fitting simulations to observations and in this case, local estimates included a large part of the effect of the variability of the other parameters on the simulated biomass.

Properties of the system

The dynamics of the microphytobenthic biomass in the first centimetre of the mud (B), as it is described in the model, is characterized by a series of strong oscillations which can be observed at the beginning and at the end of each daytime emersion period. Thus, in the field, several samplings were carried out that have shown this particular pattern (see Guarini *et al.* 2000a,b and this chapter). This is an important result of this model that allowed us to define accurately the relevant temporal scale at which to observe the system. In return, the field observations confirmed the relevance of the conceptual basis of the model formulation.

Three modelling hypotheses were tested. The first hypothesis did not include any resuspension of the biofilm and the production period (and the presence of the biofilm at the mud surface) coincided with the daytime emersion period. The microphytobenthic community was synchronized with the tidal and the light dark cycles, and anticipated any refloodling in this scheme. However, the determinism of the downward migration is not well-known. Because of the variability in the tidal oscillation (due to the variability of the meteorological conditions of pressure and wind), the downward migration, in this scheme, has to be induced by an external forcing factor. The stimulus must come from the hydrodynamic processes.

The second modelling hypothesis included resuspension provoked by the variability in the tidal oscillation and assuming that the microphytobenthic community cannot anticipate an early refloodling. The production periods still coincided with the daytime emersion periods. Episodic resuspensions appeared like local disturbances that deviated the biomass from its trajectory. The dynamics for the first two different modelling hypotheses converged to a unique cyclic equilibrium that was unconditionally stable. The recovery-time (*i.e.* the time necessary to reach values close to the asymptotic steady state from an initial condition $B_0 \geq S_{max}$) can be calculated. The approximate solution of the model was described by the series:

$$B_{n+1} = aB_n + b$$

which is also:

$$B_n = aB_{n-1} + b$$

And which can be developed as:

$$B_n - b(a^0+a^1+\ldots+a^{n-1}) = a^n B_0$$

If B_n is considered as the value close to the approximate solution B^*, $B_n=yB^*=yb/(1-a)$, where $y \in]0, 1[$. Practically, y is close to 1. $B^*(y-1) + a^n B^* = a^n B_0$ and then:

$$a^n = \frac{B^*(1-y)}{B^* - B_0}$$

or, after transformation :

$$n = \frac{\ln(B^*(1-y)/(B^*-B_0))}{\ln(a)}$$

with $\ln(a) = -(mT)$. n is to the number of sequences necessary to reach the value yB^*. The formulation of the recovery time, t_R, is:

$$t_R = \frac{\ln(B^*(1-y)/(B^*-B_0))}{-m}$$

The recovery time mainly depends on the relative difference between the initial condition B_0 and the steady state value B^*. It will also increase when the value of y get closer to 1. The recovery time depends on the value of the loss rate m (and depends on its units): the smaller the loss rate, the longer the recovery time. The recovery time does not depend on either the production rate, or on the saturation value S_{max}. According to the order of magnitude of the parameters, the estimated recovery time to reach 95% of the steady state is about two weeks long, and decreases as B_0 approaches B^* (Figure 14).

The third modelling hypothesis was significantly different from the two previous ones. It assumed that the microalgal biomass is regulated by the community itself according to the following principle: when it reached a critical biomass, which depends locally on the saturation value at the mud surface and the duration of the daytime emersion period, the biofilm was resuspended into the water column. The

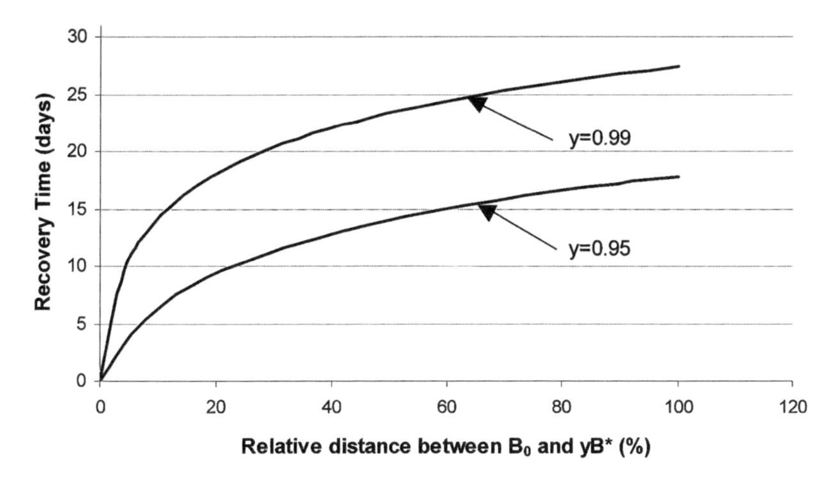

Figure 14. Recovery time as a function of the distance between the initial condition and the steady-state condition of the chlorophyll a concentration for case A (Figures 4 and 12) which assumed that no resuspension of the microphytobenthos occurred.

dynamic was allowed to converge to several steady states according to the relative value of the parameters describing the production and the losses (by grazing or stress-induced mortality). Steady state values around the critical biomass were not unique and depended strongly on the initial condition. They were nevertheless contained in the interval: $[B_c\text{-}S_{max}; B_c\text{+}S_{max}]$. If the mortality became too high and was not compensated by production, the biomass decreased down to zero. The system is conditionally persistent. Recovery times were estimated numerically using typical orders of magnitude of the parameters and were about as fast as the recovery times calculated for the two previous modelling hypotheses (about 15 days long).

Loss processes, grazing, stress-induced mortality and resuspension

The three modelling hypotheses that were tested were all based on the same concept of a biofilm that ensures production, and quantifying the variability of the primary production rate was crucial. The resuspension of the microphytobenthic biomass under standard conditions (without sediment erosion) depended on the presence of the biofilm at the moment of reflooding during daytime. Each of the three model hypotheses become increasingly complex in terms of the model structure and the resulting dynamics. First there was no resuspension (downward vertical migration prevents such an event to happen), secondly, resuspension could be induced by random variability in the synchronizers of the vertical migration, and finally that resuspension depended on the community itself through its deterministic property of biofilm renewal for a restricted number of times.

The loss rate included both stress-induced mortality and grazing. It was small and was represented by the simplest formulation: a linear function of the microphytobenthic biomass. Loss processes were nevertheless very important in the determination of steady state, recovery time and conditions of resuspension of the biomass. The lack of knowledge about these processes of loss remains an important limitation on our ability to simulate and predict accurately the dynamics of the microalgal biomass in littoral ecosystems. Methods for estimating quantitatively the grazing and mortality fluxes and rates have to be found that can be extrapolated over the entire ecosystem. Inverse and network analysis methods, which are able to estimate fluxes in foodwebs where information is lacking, could be used to estimate loss rates (Vezina and Platt 1988).

How to choose between the three model hypotheses: field work approach

One way to determine which modelling structure best represents the dynamics of the microphytobenthic biomass, is to compare simulations with observations of the system. The main differences concern the resuspension process and the duration of the biofilm during the daytime emersion period. From the benthic point of view, it is technically difficult at this time to observe simultaneously and continuously both the biomass and the biofilm at the scale of the intertidal mudflats. In addition, the model is strongly sensitive to the parameter variability which generates strong uncertainties in the biomass estimates.

In contrast, from a pelagic point of view, significant resuspension should be characterized at high tide by a gradient of high biomass from the shore to the subtidal area and by the transport and dilution of biomass in subtidal areas at low tide. Therefore, the phytoplankton biomass was sampled in the water column of the Marennes-Oléron Bay. A sampling cruise was carried out in anticyclonic conditions (atmospheric pressure greater than 1015 mbar), with low wind velocities (lower than *ca.* 2.0 $m.s^{-1}$). The cruise took place at the end of March 2003 when mortality rate was low and production rate was high (Blanchard *et al.* 1997). At this time of the year, the phytoplankton bloom did not yet start. The sampling was performed at neap tide when the daytime emersion period preceding the submersion was short and, hence, according to our third hypothesis, when the probability of the presence of the biofilm at the reflood was the highest. Finally, this sampling cruise was realized after a long period (greater than a month) of stable meteorological conditionsso as to be close to steady state conditions. The cruise was designed to sample a large area in a minimum amount of time; the chlorophyll *a* concentration and the turbidity were recorded along the route of the ship using a continuous multi-parameter station. Two consecutive sampling plans were realized, at high tide occurring at noon, and at low tide occurring *ca.* 6 hours later. Full details of this sampling are provided in Guarini *et al.* (submitted).

The results of the sampling (Figure 15) showed a gradient of biomass at high tide, with a maximum concentration greater than 12 (µg Chl *a*).l^{-1}. The suspended matter concentration values were lower than 100 $mg.l^{-1}$. Assuming that the microphytobenthic biomass is diluted in the first centimetre of the mud, and considering an average value of 100 $mg.m^{-2}$ of chlorophyll *a* in the first centimetre of the mud (Guarini *et al.* 1998), associated with an average dry sediment density equal to 16.10^{6} $g.m^{-2}.cm^{-1}$,

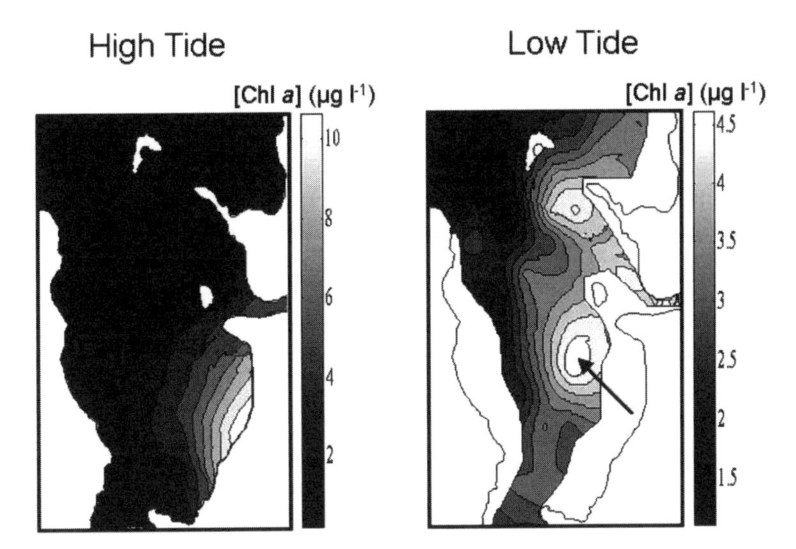

Figure 15. Spatial distribution of the phytoplankton biomass in terms of chlorophyll a concentration in the water of the Marennes-Oléron Bay, at the end of March 2003. The gradient of chlorophyll a concentration at high tide, followed by the advection-dispersion induced transport and the dilution at low tide of chlorophyll a concentration were evidence for the resuspension and export of the microphytobenthic biomass in the Marennes-Oléron Bay.

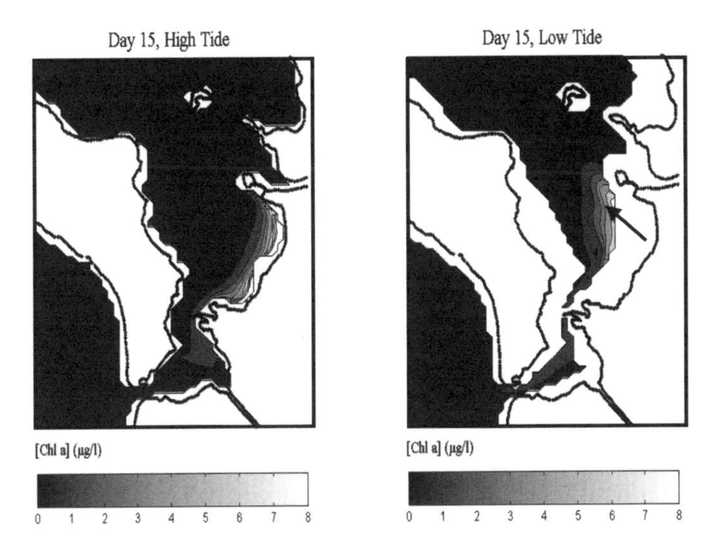

Figure 16. Simulation of the transport of the microphytobenthos in the Marennes-Oléron Bay. After 15 days, the results of the resuspension and transport of the microphytobenthos are consistent with the observed patterns (Figure 15).

the suspended matter concentration in *ca.* 1 m water column should have been equal to *ca.* 2.0 $g.l^{-1}$. This observation is strong evidence for the resuspension of the biomass without significant accompanying sediment erosion and the biofilm is the only structure that concentrates enough biomass to create a peak of resuspension at 12 µg Chl *a*. m^{-2}. At the following low tide, the resuspended biomass was observed to be transported in the central channel, at the northern part of the mudflat where it was diluted into deeper water. The chlorophyll *a* concentration decreased to 6 µg Chl $a.l^{-1}$. These patterns are consistent with the numerical modelling of the resuspension and transport of the biomass using the third hypothesis (summarized by the regulation of the population, Figure 16). Meteorological conditions were highly favourable to an increase of microphytobenthic biomass on the mudflat, but were not such that an early return of the tidal waters would be provoked. Therefore, the process of biofilm resuspension is self-regulated by the biomass and implied an absence of downward migration because the total biomass is greater than the critical biomass ($B_c = S_{max}(T_E/\tau_S)$). In addition, the resuspension of the biofilm may occurs without an associated sedimentary component. This hypothesis that biofilm resuspension occurs without sediment erosion, provides a reasonable and consistent explanation for the export of the microphytobenthic biomass from coastal semi-enclosed ecosystems that are usually considered as zones of net deposition.

Comparison between the dynamics of microphytobenthos and phytoplankton

The dynamics of the microphytobenthos can be characterized by a constant adjustment to steady state conditions that govern the seasonal trends. As a consequence, the within year variability is not very strong and depends more on meteorological conditions, which vary between years. Moreover, local conditions predominate since no spatial processes (passive or active transport) seem to influence the local dynamics. This is very different from the dynamics of the phytoplankton that is entirely governed by blooms, which can be seen as transitory phases, and stop when production nutrients become limited (Cloern 1996). Changes in limitation conditions often induce a succession of communities. Spatial processes of passive water mass transport control overall phytoplankton bloom dynamics (Lucas *et al.* 2000a,b).

In contrast, microphytobenthic production is mostly regulated by the space available at the surface of the mud. The mud surface is usually completely saturated during all or part of the daytime emersion period. Microphytobenthic production is also regulated by mud surface temperature that can vary quickly and widely during emersion periods. Light variations have a minimal impact, since most of the time light levels on the mudflat are saturating for photosynthetic activity (Guarini *et al.* 2002). For phytoplankton, light availability in fact, fluctuates strongly and completely controls production (specially in turbid waters), while temperature variations are small. Losses by stress-induced mortality and grazing, which are slow processes, influence the seasonal pattern of the microphytobenthos dynamics, while they do not have a strong impact of the phytoplankton dynamics during bloom events.

The net production of the microphytobenthos inhabiting intertidal mudflats was estimated in the Marennes-Oléron Bay by using the orders of magnitude which correspond to each parameter, and by considering an average value for the C: Chl *a* ratio equal to 40 (de Jonge 1980). The typical value of the net production is equal to 1 $gC.m^{-2}.day^{-1}$. This value is, in average, 4 times greater than the corresponding phytoplankton production in the bay (Raillard and Menesguen 1994). The microphytobenthic compartment is extremely productive, and has a strong potential for exportation to either the benthic foodwebs or to the water column through hydrodynamic resuspension.

Furthermore, our study pointed out that the resuspended microphytobenthos contributes to the biomass of the phytoplankton community. Resuspension may occurs on a daily basis; the microphytobenthic contribution varies in space and time, but is always significant at the scale of the ecosystem. Therefore, the comparison between the dynamics of the phytoplankton and microphytobenthos should be considered carefully, and should be replaced eventually by an integrated approach of the dynamics of benthic and pelagic microalgal biomasses in ETSELE.

Conclusions

The goal of this modelling study was to propose a consistent representation of the dynamics of the microphytobenthos inhabiting wide bare intertidal mudflats. This model was based on concepts that integrate the present knowledge. It nevertheless does not include all the concepts that were developed by many authors working on this topics, because some simplifications had to be made. For example, the species composition was not taking into account, because the species diversity is low, because no real pattern of succession was observed generally and the number of state variables would be too large, preventing any study of general mathematical properties. The model was designed to represent the dynamics of the microphytobenthos at the level of the ecosystem. Therefore, the small scale variability was filtered out. Finally a consistent representation has to be made, and controversial results could not be taken into account. For example, the presence of a biofilm, and a strong discontinuity at the mud surface could not be reconciled with the notion of a vertical profile of biomass and production.

The resulting model is one of the very few attempts to quantify specifically the dynamics of this compartment (see Blackford 2002 for subtidal areas). This contrasts with phytoplankton studies for which parts of theory already exist for many ecosystems: open ocean (Sverdrup, 1938), coastal ocean and shallow water ecosystems (Cloern 1996). This lack of coherent and consistent assessment strategies can be explained by a much shorter history of studying subtidal and intertidal mudflats, and by the enormous practical and technical difficulties in sampling mudflat systems. The most important problem currently is the lack of accurate measurements of the primary production under field conditions (Currin *et al.* 1996; Underwood and Kromkamp 2000). Estimates made on extracted microalgae need to be examined carefully, in order to decide if they were representative of *in situ* production, and if the method of microphytobenthos extraction does not select preferentially some species of the community. The model discussed here does not solve these problems, but it does provide an integrated theoretical framework that can be tested and be used to determine experimental strategies about the microphytobenthos.

The model presented and discussed here is based on concepts and methods that originated with studies of the microphytobenthos, and were not based on the pelagic system of microalgal primary production. Intertidal benthic microalgae undergo fast changes due to their periodic alternation between air and water exposures. The physical environment of the microphytobenthos is not comparable to the physical conditions present in a well-mixed water column, and secondly, the strategies of the respective communities for production are also completely different. For example, the microphytobenthos community growth is due to vertical migration cycles toward the mud surface and the hourly changes in light irradiance and temperature that control the photosynthetic activity of the cells present at the mud surface during a daytime emersion period. Losses of the microphytobenthic biomass are not only through benthic grazing, but also plausibly from the resuspension of the biomass when it is swept away from the mud surface during each submersion period.

The 3 main features of the model are as follows:

1. The cycle of the vertical migration of the microalgae. The process is fast and was density-dependant in the model, but it was only partially described and the cell behaviour must be separated from the integrated result observed at the level of the community.
2. The creation of the biofilm at the mud surface during daytime emersion periods. The biofilm as a functional structure ensures the photosynthesis (controlled by light and mostly temperature) and hence the production of new biomass for the community. The biofilm must be better characterized in terms of accurate statistical estimates for both the number of cells and the biomass per unit of area.
3. A discrete 2-compartments structure. The biomass is concentrated at the mud surface (in the biofilm), and the remaining biomass is diluted in the first centimetre of the mud, with exchanges (through vertical migration) between each compartment.

The model provides a basis for describing the dynamics of the microphytobenthos that inhabits intertidal mudflats. On one hand, the model can incorporate many other processes as they are shown to be relevant in the regulation of the community at the level of the ecosystem. For example, nutrient availability and control on the production were not taken into account, but could be important if both the use of nutrient pools by microalgae and the fluxes of nutrients in the sediment are described. On the other hand, the model suggests that in order to understand completely the dynamics of the microphytobenthos in European-type semi-enclosed littoral ecosystems, it will be necessary to develop an integrated approach for the total microalgal biomass (benthic plus pelagic), an approach which will need to take into account explicitly exchanges of biomass between the benthic and pelagic compartments. This is one of the major challenges in littoral ecology, especially to quantify the contribution of the microphytobenthos in the carbon cycle and budget of coastal ecosystems.

References

ADMIRAAL, W. 1984. The ecology of estuarine sediment-inhabiting diatoms. In: Progress in Physiological Research, Round F.E, Chapman D.J. (ed.), Vol. 3. Bristol: Biopress Ltd. 269-322.

BLACKFORD, J.C. 2002. The Influence of Microphytobenthos on the Northern Adriatic Ecosystem: a Modelling Study. Estuar. Coast. and Shelf Sci. **55**: 109-123.

BLANCHARD, G.F., J.M. GUARINI, P. RICHARD and PH. GROS. 1997. Seasonal effect on the relationship between the photosynthetic capacity of intertidal microphytobenthos and short-term temperature changes. J. Phycol. **33**: 723-728.

BLANCHARD, G.F., J.M. GUARINI, C. DANG and P. RICHARD. Characterizing and quantifying photoinhibition in intertidal microphytobenthos. Submitted to J. Phycol.

BLANCHARD, G.F and V. CARIOU-LE GALL. 1994. Photosynthetic characteristics of microphytobenthos in Marennes-Oleron Bay, France: Preliminary results. J. Exp. Mar. Biol. Ecol. **182**: 1-14.

CLOERN, J.E. 1996. Phytoplankton bloom dynamics in coastal ecosystems: a review with some general lessons from sustained investigation of San Francisco Bay, California. Rev. Geophys. **34**: 127-168.

COUCH, C.A. 1989. Carbon and nitrogen stable isotopes of meiobenthos and their food resources. Estuar. Coast. Shelf Sci. **28**: 433-41.

CURRIN, C.A., L. LEVIN and T. SINICROPE. 1996. Factors affecting intertidal benthic microalgal production in East and West Coast marsh habitats. Conference: 24th. Ann. Benthic Ecology Meeting, Columbia, SC (USA), 7-10 Mar 1996. page 94.

DEMERS, S., J.C. THERRIAULT, E. BOURGET and A. BAH. 1987. Resuspension in the shallow sublittoral zone of a macrotidal estuarine environment: wind influence. Limnol. Oceanogr. **32**: 327-339.

FROUIN, R., D.W. LINGNER, C. GAUTIER, K.S. BAKER and R.C. SMITH. 1989. A simple analytical formula to compute clear sky total and photosynthetically available solar irradiance at the ocean surface. J. Geophys. Res. **94**: 9731-9742.

GUARINI, J.M. 1998. Modélisation de la dynamique du microphytobenthos des vasières intertidales du bassin de Marennes-Oléron. Effet des synchroniseurs physiques sur la régulation de la production. Thèse de Doctorat – Océanologie - Université de Paris 6. 177pp.

GUARINI, J.M., G.F. BLANCHARD, C. BACHER, PH. GROS, P. RIERA, P. RICHARD, D. GOULEAU, R. GALOIS, J. PROU and P.G. SAURIAU. 1998. Dynamics of spatial patterns of microphytobenthic biomass: inferences from a geostatistical analysis of two comprehensive surveys in Marennes-Oléron Bay (France): results from a geostatistical approach. Mar. Ecol. Prog. Ser. **166**: 131-141.

GUARINI, J.M., G.F. BLANCHARD, PH. GROS and S.J. HARRISON. 1997. Modelling the mud surface temperature on intertidal flats to investigate the spatio-temporal dynamics of the benthic micro-algal photosynthetic capacity. Mar. Ecol. Prog. Ser. **153**: 25-36.

GUARINI, J.M., G.F. BLANCHARD, PH. GROS, D. GOULEAU and C. BACHER. 2000a. Dynamic model of the short-term variability of microphytobenthic biomass on temperate intertidal mudflat. Mar. Ecol. Prog. Ser. **195**: 291-303.

GUARINI, J.M., G.F. BLANCHARD and PH. GROS. 2000b. Quantification of the microphytobenthic primary production in European intertidal mudflats: a modelling approach. Cont. Shelf Res. **20**: 1771-1788.

HARRISON S.J. and A.P. PHIZACKLEA. 1987. Vertical temperature gradients in muddy intertidal sediments in the Forth estuary, Scotland. Limnol. Oceanogr. **32**: 954-63.

HAY S.I., T.C. MAITLAND and D.M. PATERSON. 1993. The speed of diatom migration through natural and artificial substrata. Diatom Res. **8**: 371-384.

JASSBY, A.D. and T. PLATT. 1976. Mathematical formulation of the relationship between photosynthesis and light for phytoplankton. Limnol. Oceanogr. **21**: 540-547.

JONGE, V.N. DE. 1980. Fluctuations in the organic carbon to chlorophyll a ratios for estuarine benthic diatoms populations. Mar. Ecol. Prog. Ser. **2**: 345-53.

JONGE, V.N. DE and J.E.E. VAN BEUSEKOM. 1992. Contribution of resuspended microphytobenthos to total phytoplankton in the Ems Estuary and its possible role for grazers. Neth. J. Sea. Res. **30**: 91-105.

JONGE, V.N. DE, and J.E.E. VAN BEUSEKOM. 1995. Wind- and tide-induced resuspension of sediment and microphytobenthos from tidal flats ins the Ems estuary. Limnol. Oceanogr. **40**: 766-778.

LE HIR, P., W. ROBERTS, O. CAZAILLET, M. CHRISTIE, P. BASSOULLET and C. BACHER. 2000. Characterisation of intertidal flat hydrodynamics. Cont. Shelf Res. **20**: 1433-1459.

LEGENDRE, L. 1981. Hydrodynamic control of marine phytoplankton production: the paradox of stability. In Ecohydrodynamics. J.C.J. Nihoul ed. Elsevier Oceanography Series, 32. Elsevier, Amsterdam, Oxford, New York., 191-208.

LONG, S.P. and C.F. MASON. 1983. Saltmarsh ecology. Blackie, Glasgow. 168 pp.

LUCAS, L., J.E. CLOERN, J.R. KOSEFF, S.G. MONISMITH and J.K. THOMPSON. 1998. Does the Sverdrup critical depth model explain bloom dynamics in estuaries. J. Mar. Res. **56**: 375-415.
LUCAS, C.H., C. BANHAM, and P.M. HOLLIGAN. 2001. Benthic-pelagic exchange of microalgae at a tidal flat. 2. Taxonomic analysis. Mar. Ecol. Prog. Ser. **212**: 39-52.
LUCAS, L., J.R. KOSEFF, J.E. CLOERN, S.G. MONISMITH and J.K. THOMPSON. 1999a. Processes governing phytoplankton blooms in estuaries. I: The local production-loss balance. Mar. Ecol. Prog. Ser. **187**: 1-15.
LUCAS, L., J.R. KOSEFF, S.G. MONISMITH, J.E. CLOERN and J.K. THOMPSON. 1999b. Processes governing phytoplankton blooms in estuaries. II: The role of horizontal transport. Mar. Ecol. Prog. Ser. **187**: 17-30.
MARRA, J. 1978. Phytoplankton Photosynthesis response to vertical movement in the mixed layer. Mar. Biol. **46**: 203-208.
MCDONALD, E.T. and R.T. CHENG. 1997. A numerical model of sediment transport applied to San Francisco Bay, California. J. Mar. Environ. Eng. **4**: 1-41.
MCLUSKY, D.S. 1989. The estuarine ecosystem. 2nd edition. New York: Chapman & Hall. 215 pp.
MILLER, D.R. 1974. Sensitivity analysis and validation of simulation models. J. theor. Biol. **48**: 345-360.
PALMER, J.D. and F.E. ROUND. 1965. Persistent, vertical migration rhytms in benthic microflora. I. The effect of light and temperature on the rhytmic behaviour of Euglena obtusata. J. Mar Biol. Ass. U. K. **45**: 567-582.
PALMER, J.D. and F.E. ROUND. 1967. Persistent, vertical migration rhytms in benthic microflora. VI. The The tidal and diurnal nature of the rhythm in the diatom Hantzschia virgata. Biol. Bull. **132**: 44-55.
PATERSON, D.M. 1989. Short-term changes in the erodibility of intertidal cohesive sediments related to the migratory behavior of epipelic diatoms. Limnol. Oceanogr. **34**: 223-234.
PATERSON, D.M., R. DOERFFER, J. KROMKAMP, G. MORGAN and W. GIESKE, 1998. Assessing the biological et physical dynamics of intertidal sediment ecosystems. A remote sensing approach (project BIOPTIS). Barthel, K.G. (ed.), Barth, H. (ed.), Bohle-Carbonell, M. (ed.), Fragakis, C. (ed.), Lipiatou, E. (ed.), Martin, P. (ed.), Ollier, G. (ed.). Third european MArine Science et Technology Conference,Lisbon (Portugal), 23-27 May 1998. pages 377-390.
PINCKNEY, J.L. and R.G. ZINGMARK. 1991. Effects of tidal stage and sun angles on intertidal benthic microalgal productivity. Mar. Ecol. Prog. Ser. **76**: 81-89.
PINCKNEY, J.L. and R.G. ZINGMARK. 1993. Modelling the annual production of intertidal benthic microalgae in estuarine ecosystems. J. Phycol. **29**: 3 96-407.
PRITCHARD, D.W. 1967. Observations of circulation in coastal plain estuaries. In Estuaries (G.H. Lauff, ed.) AAAS, Washington DC. **83**: 37-44.
RAILLARD, O. and A. MENESGUEN. 1994. An ecosystem box model for estimating the carrying capacity of a macrotidal shellfish system. Mar. Ecol. Prog. Ser. **115**: 117-130.
SABUROVA M.A., I.G. POLIKARPOV and I.V. BURKOVSKY. 1995. Spatial structure of an intertidal sandflat microphytobenthic community as related to different spatial scales. Mar. Ecol. Prog. Ser. **129**: 229-239.
SCHELSKE, C.L. and E.P. ODUM. 1962. Mechanisms maintaining high productivity in Geogia estuaries. Proc. Gulf Carrib. Fish Inst. **14**: 75-80.
SERÔDIO J., J.M. DA SILVA and F. CATARINO. 1997. Non destructive tracing of migratory rhythms of intertidal benthic microalgae using in vivo chlorophyll a fluorescence. J. Phycol. **33**: 542-553.
SERÔDIO, J. and F. CATARINO. 2000. Modelling the primary productivity of intertidal microphytobenthos: Time scales of variability et effects of migratory rhythms. Mar. Ecol. Prog. Ser. **192**: 13-30.
SEURONT, L. and N. SPILMONT. 2002. Self Organized criticality in intertidal microphytobenthos patch patterns. Physica A: statistical mechanics and its applications. **313**: 513-539.
SHAFFER, G.P. and M.J. SULLIVAN. 1988. Water column productivity attribuatable to displaced benthic diatoms in well-mixed shallow estuaries. J. Phycol. **24**: 132-140.
SVERDRUP, H.U. 1953. On conditions for the vernal blooming of phytoplankton. Journal du Conseil International pour l'Exploitation de la Mer. **18**: 287-295.
UNDERWOOD, G.J.C. and J. KROMKAMP. 2000. Primary production by phytoplankton and microphytoplankton in estuaries. In: Nedwell DB, Raffaelli DG (eds) Estuaries. Advances in Ecological Research, Academic Press. Pages 93-153.
VAN BOXEL, J.H. 1986. Heat balance investigations in tidal areas. PhD Thesis, University of Amsterdam, Netherland : 137 p.
VÉZINA, A.F. and T. PLATT. 1988. Food web dynamics in the ocean. I best estimates using inverse methods. Mar. Ecol. Prog. Ser. **42**: 269-287.
WALKER, H.J. and J. MOSSA. 1982. Effects of artificial structures on coastal lagoon processes and forms. Oceanol. Acta. **5**: 191-198.
WETZEL, R.L. and Y.-S. SIN. 1998. Ecosystem Process Models: Applications to Wetland Systems. Ocean. Res. **2**: 189-197.

VIII. Mudflats and socioeconomics

Colin J. Macgregor

Qualitative methods offer insights into socio-economic and landuse: Influences on coastal environments

Abstract

Nutrient concentrations (N, P and K) determine the extent to which surface water courses, and therefore coastal estuaries, are or may become eutrophic. The degree of eutrophication directly impacts on the biological health and even the geomorphic processes of such systems. Water quality testing easily identifies point sources of pollutants but when concerned with diffuse sources, such an approach yields little useful information.

Agriculture has been identified as a major diffuse polluter but the actual extent of its contribution and the reasons for its occurrence are not completely understood. If appropriate management strategies are to be developed to reduce its impact then we must understand the socio-economic circumstances that give rise to such pollution. Also crucial in management is the co-operation of farmers which implies we must understand the attitudes and values of farmers influencing the systems. Understanding present management regimes will also clarify the degree to which these practices are impacting on the environment as well as helping convince farmers of their roles and responsibilities. However, we must also acknowledge that change to more sustainable land management practices will bring impacts to the farmers themselves. These impacts may be minor or major depending upon individual circumstances. It follows therefore that it is important to identify those farmers who will be exposed to significant social and/or economic impacts as a result of conforming to current recommended practices (CRPs) and environmental legislation such as that associated with the EC's nitrate directive (Nitrate Vulnerable Zone).

This paper will discuss the value and use of interpretivist social science methods in natural resources management contexts generally but the Eden catchment in Scotland is a useful area to explore these in detail. Arable and livestock farming dominates landuse in the cathment and there is a history of diffuse pollution which is known to be impacting on the Eden estuary – an area regarded highly as a bird habitat. The catchment has also recently be designated a NVZ so farmers are already starting to think about what confirmation to more sustainable management practices will mean to their operations.

Introduction

Coastal pollution as a result of nutrient over-enrichment (eutrophication) associated with nitrogen and phosphorous in particular is now regarded as one of the most significant problems associated with coastal ecosystems. In the worst cases such pollution results in harmful and nuisance algal blooms but there are also many other subtler food chain effects. This paper explores some of the most important socio-economic issues associated with terrestrial coastal pollution.

Water pollution is typically categorised according to its source. Point source pollution originates from sources such as pipes or ditches, from wastewater treatment plants or industrial discharges, which drain directly into rivers or the sea. Such sources are easily identified and very often investment in treatment technologies is all that is required to resolve the problems. Often however such investment will only take place where legislative requirements demand but the direct socio-economic impacts are usually constrained to those factory owners and/or local authorities responsible although the economic costs are inevitably passed on to the consumers of their services and/or products.

Diffuse sources of pollution emanate from sources that have no defined point of entry but typically result from rainwater running off the land and picking up pollutants along the way. Identifying and controlling such pollution is a more complex problem because it tends to be associated with land uses rather than easily identifiable operations. Perhaps the most notable land use type known to be responsible for diffuse pollution is agriculture. Historically, coastal eutrophication increased dramatically in the developed world during the 1950s and 1960s when the agricultural benefits of super-phosphate and nitrogen fertilisers were first developed and used with enthusiastic abundance (Boesch 2001). Other agricultural chemicals also affect coastal environments. For example, herbicides can stress saltmarsh diatoms and higher plants, which in turn can increase erosion (Mason *et al.* 2003).

While there has been some progress in reducing point sources of pollution, the same cannot be said of diffuse causes, particularly those associated with agriculture (Skinner *et al.* 1997). As a result, there has to date been very little clearly demonstrated success in terms of improvements in coastal water quality. There is now worldwide consensus that eutrophication, and by implication, the excessive use of fertilisers, must be reduced to some less harmful level (Boesch 2001). Excess nitrogen losses to the environment in particular are acknowledged as having serious environmental impacts. Responding to this, the European Union (EU) has introduced its nitrate directive. Under this directive Nitrate Vulnerable Zones (NVZ) have been and are being established throughout Europe. In essence, NVZs identify pollution sensitive areas (often river catchments) where stricter management controls over the use of fertilisers and organic manure are enforced in an effort to prevent nitrogen pollution.

While legislative initiatives have an important role to play, improvements in agricultural pollution will only occur with the cooperation of farmers. It follows that it is necessary to understand the socio-economic conditions that give rise to agricultural pollution before strategies can be developed to address the causes. We must also understand the attitudes and values of farmers influencing coastal systems. Their present management practices must also be understood in order to determine the degree to which their specific practices are actually impacting on the environment. All such information should help 'convince' farmers of their roles and responsibilities and so gain the support necessary for change to take place.

We must also acknowledge that change to more sustainable farm management practices will have direct socio-economic impacts on the farmers themselves and unlike point source polluters who can pass on the economic impacts to their consumers, no such option is available to farmers. This is because prices for agricultural commodities are either pre-set by the European Union (EU) or the supermarket supply chains with little regard to the costs of production. Of course, the economic impacts on farmers will vary depending upon individual circumstances so it follows that it is also important to identify those farmers who will be exposed to significant social and/or economic impacts as a result of conforming to the current recommended practices (CRPs) and environmental legislation such as that associated with NVZ status.

The research behind this paper is concerned with two central research questions. First, what are the socio-economic and farm land use/management practices most likely to cause eutrophication in coastal areas; and second, what types of farming are the greatest concern and who will suffer the most as a result of enforced environmental management?

Methodology

The Eden catchment and estuary

The Eden catchment of Fife, Scotland makes an ideal case study site to investigate the socio-economic factors associated with adopting sustainable land management practices in an agricultural context. The catchment lies in eastern Scotland (Figure 1) and has two main rivers associated with it; the Eden River, which roughly dissects the catchment from west to east, and the Motray Water in the northern part of the catchment. Both these rivers drain into the Eden estuary at the village of Guardbridge.

The Eden catchment extends over an area of approximately 320 km^2 (TRPB 1994) the majority of which is drained by the River Eden (some 260 km^2). The Eden River itself rises in the Ochil Hills, not far from Loch Leven at around 220 meters above sea level but the majority of the catchment is low-lying and gently undulating.

The major land use in the catchment is arable farming and approximately 76 percent can be regarded as high quality agricultural land with very fertile soils or imperfectly drained brown forest and alluvial soil types (ECN 2002). The underlying geology comprises Devonian and a Carboniferous stratum, the former of which includes the most productive aquifer in Scotland, the Knox Pulpit formation (Robbins 1990). There is a small salmon run to the river and otters are also present (ECN 2002). Annual rainfall of the catchment varies from approximately 1,000 mm in the upper catchment to around 600 mm at the Eden estuary (TRPB 1994).

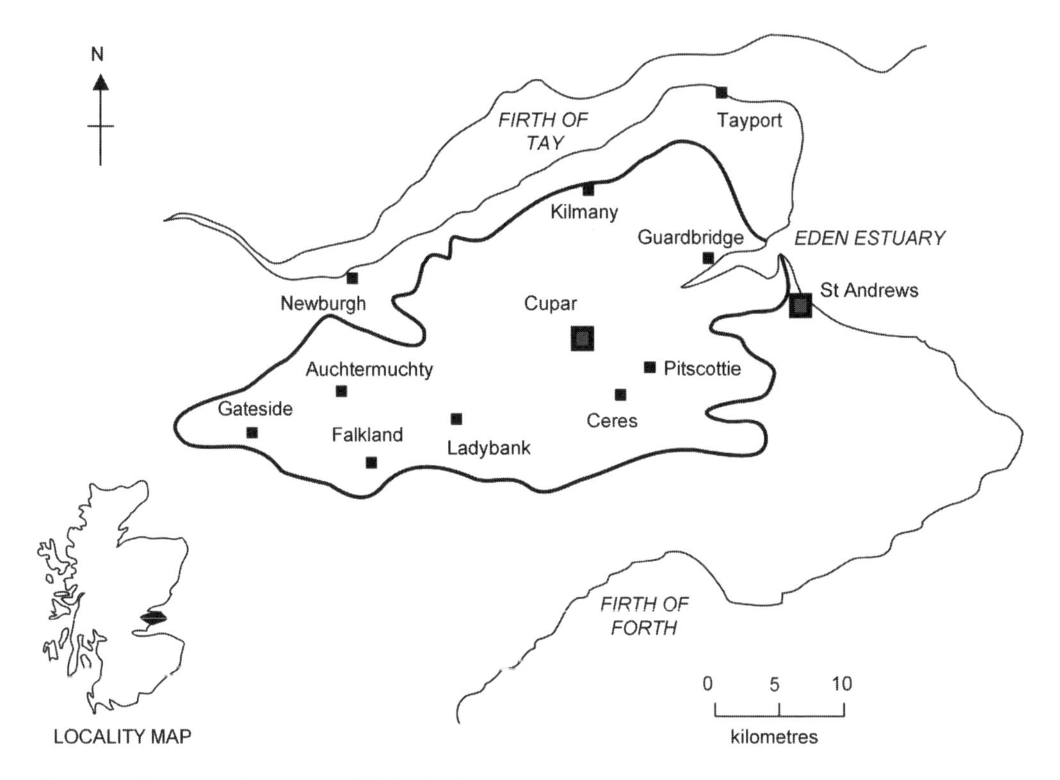

Figure 1. The Eden Catchment of Fife, Scotland

The Eden's estuary forms the Eden Estuary Local Nature Reserve, which is regarded as an important over-wintering site for wildfowl and waders (ECN 2002). Water is abstracted from groundwater, the river and its tributaries for irrigating crops (at present there is no licensing or restrictions on abstraction).

Although treated sewage is discharged to the rivers of the catchment from several small communities and from the town of Cupar (population 9,700), it is the effect of diffuse inputs from agriculture that is believed to be critical to river water quality and the ecological health of the estuary (SEPA 1996). Concern over the quality of drinking water from the Balmalcolm borehole led to a small area within the catchment (approx. 5 sq. kms.) just to the south of Ladybank, being designated a NVZ in 1996.

Despite some minor improvements in surface water quality in the catchment (the result of improvements to the various sewerage treatment plants in the 1980s), in January 2002 the Scottish Executive made it clear that the remainder of the Fife region, including the River Eden catchment, would also be designated a NVZ (Scottish Executive 2002). Later, in January 2003, the Action Programme measures under NVZ designation came into force (Scottish Executive 2003), making the timing of this research particularly relevant.

Qualitative methods in environmental social science contexts

Mention social science methods to people not familiar with the various techniques and their initial thoughts are of questionnaires or perhaps telephone interviews. It is certainly true that a considerable amount of social science research is conducted using such methods and they certainly have some notable assets. For example, they allow a sample of the population to present their views and where the sample is sufficiently large and random, they can relatively accurately represent the views of a community as a whole. However, such methods are usually associated with positivist approaches and these tend to be constrained and limited by *a-priori* assumptions about the research questions (hypothesis testing). As such, they afford limited dynamic interaction and can only usually represent a snapshot in time. The alternative to positive approaches are interpretivist approaches. These usually rely on qualitative data and so offer the opportunity to explore the full range of issues associated with the research question/s. Data acquisition is developed bottom-up rather than top-down and while this can generate enormous quantities of data, there is a risk that some of it may be regarded as superfluous or even irrelevant to the research question/s. In short, positive quantitative approaches imply smaller data quantities derived from a larger sample of the population. Interpretivist qualitative approaches imply large amounts of data from a small population sample.

The approach adopted in this type of research lies somewhere between these two. Commonly known as the General Inductive Method (GIM), the approach is typically used where there are specific research objectives as is the case in this study (this research had a number of specific research objectives). Such research objectives imply a deductive methodology however, as the name suggests, the inductive component is regarded very highly and by allowing the data to influence the outcomes the issues that are important to the informants are not lost.

Most researchers using the GIM collect data from interviews and since there are specific research objectives and questions, it is important to exercise some control over the interview process. This ensures the interviews stay on the subject of concern, otherwise considerable time can be spent discussing and ultimately examining issues that have little relation to the main aims of the research.

Following interviews, the inductive process essentially involves a series of steps that include close readings of transcribed interviews and coding sections of the text according to dominant themes or categories. These categories may also be linked with other categories in various ways such as a network, a hierarchy, or perhaps causal sequences.

Establishment of the Agricultural End Users (AEU) group

While this paper is restricted to reviewing the findings associated with the two research questions stated above there are other research questions which demand more continuous input from the informants. It was therefore decided that a group of farmers would have to be identified that would be prepared to remain in the project until the end of 2005 (Agricultural End Users (AEU) group).

There are approximately 127 individual farms in the Eden catchment[1]. Although true statistical representation is both difficult and unnecessary in interpretivist research of this kind, it was nevertheless acknowledged that it would be useful if the membership of the AEU group was generally representative of the whole farm population. It was considered that an AEU group of approximately 30 informants fell within realistic expectations for the project.

Identifying farmers to make up the AEU group involved a stratified sampling procedure, which considered the location of the farms, the type of farms and the size of farms. Most farmers identify themselves in British Telecom's *Yellow Pages* © and the various localities of the catchment can be mapped according to telephone prefix codes. Initial meetings with farmers enabled the stratification to take place.

Two interview times were agreed upon. The first of these explored attitudes and management practices. Questions first centred on developing a farm profile. Then environmental deterioration, river and estuary quality, general farm management, nutrient management, the Eden as a Nitrate Vulnerable Zone (NVZ), set-aside and buffer strips, governmental financial support, and a range of other 'progressive' farm management issues were explored. The second interview was used to gather more detailed farm data in order to allow an estimate of nutrient balance to be made for the farms (a nutrient budget).

Nutrient Budgeting

The Nutrient Budget (NB) is a management tool that can help land users, particularly farmers, account for the flow of nutrients (nitrogen (N), phosphorous (P), and potassium (K)) through the farm system, thereby identifying losses which can allow management decisions to be made that may reduce losses to a minimum (Sheaves 1999). Figure 2 summarises how nutrient inputs, outputs and losses typically occur on farms.

Figure 2 demonstrates that nutrient inputs enter the farm system in three ways; two are natural (deposition and mineralisation) and affect all land uses but the third, which is of most concern in terms of potential surpluses, is directly related to farming. Its sources are typically mineral fertilisers, concentrated feedstuffs, roughage (e.g. straw) and seeds or seedlings. Farmyard manure (FYM) and/or slurry can also be a major input when imported to the farm. Outputs take the form of agricultural produce of various kinds. However, the farm is generally inefficient at converting nutrients within the inputs to outputs as produce – this is particularly true of nitrogen, mainly because of its diffuse nature when in the environment. Phosphorus and potassium losses can also generate serious consequences when excessive amounts find their way to the wider environment but these two will not be discussed further in this paper because data analysis associated with them is still continuing.

Losses of nitrogen occur in many areas of the farming process but according to Bergius *et al.* (2002) there are four that are particularly noteworthy. The first two processes account for most of the nitrogen lost in the cycle. These are:

[1] The median farm size of farms involved in this study was 192 ha. There is approximately 24,320 ha of farmland within the catchment, therefore, 24320 / 192 = 126.66.

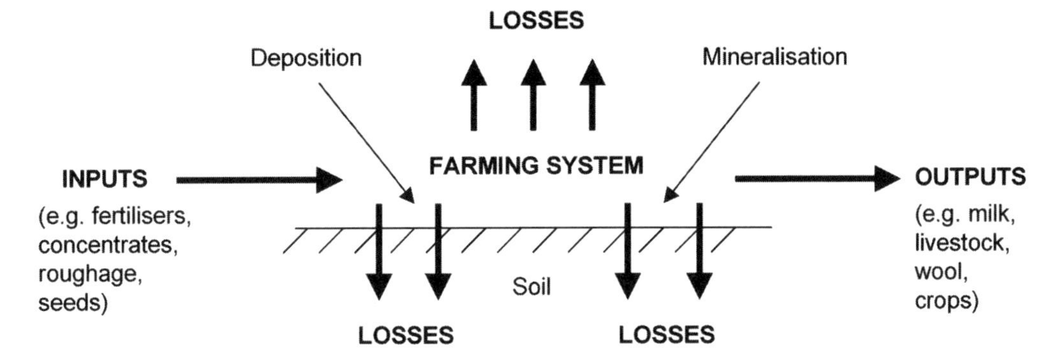

Figure 2. Nutrient flows within a typical farm enterprise (adapted from Sheaves 1999)

1. Denitrification – bacteria change nitrate in the soil to atmospheric nitrogen which joins the atmosphere.
2. Volatilisation – turns urea fertlilsers and manures on the soil surface into gases that also join the atmosphere.

Both of these are a serious concern; volatilisation produces ammonia which contributes to the greenhouse effect and acid rain, and while losses of this kind can vary considerably, they can still be significant. For example, Sheaves (1999: 5) notes that 'it is estimated that overall on an average dairy farm 100 kg of N/ha per year are lost this way from manure alone'. Fertliser use and the ageing and dying parts of plants can add considerably to this figure. As a general rule we can assume that the higher the nitrate content of the soil, the greater is the risk of evaporation.

All losses of nitrogen are environmentally undesirable but for this study it is terrestrial losses associated with the remaining two forms of loss that are of most concern because it is these that directly impact on the coastal system. These are:

3. Runoff – water carries nitrogen from fertlisers, manure and soil into rivers and streams.
4. Leaching – water carries nitrates deep into the soil profile where plants can no longer access them. This ultimately creates a groundwater quality problem.

In simple terms, nitrogen not lost to the atmosphere or incorporated in the outputs of the farm finds its way to streams and/or groundwater and ultimately to the Eden estuary. While some of this lost nitrogen enters the underlying aquifer much still comes to the surface or the estuary by way of springs and seeps. In terms of nutrient budgeting one must also note that much of the nitrogen lost to the atmosphere by way of (1) and (2) effectively returns to the land indirectly by way of deposition.

There are essentially two methods of conducting whole-farm nutrient budgets. The field-by-field method (e.g. AgResearch 2002) examines inputs, outputs and surpluses for all fields in the farm and then combines these to produce a whole-farm result. The advantage of this method is that it allows the farm manager to consider and target changes in management for individual fields. Clearly this method is particularly suited to individual farmers. However, a more immediate and less 'data-hungry' method is the farm-gate method (Lanyon and Beegle 1989). This method is useful when one is concerned with catchment-scale analyses involving a number of farming enterprises. Consequently, it is the second method that was adopted for this study.

Data variables required for farm-gate nutrient budget estimates include the following:

- The area of land farmed – this must be broken down into crops, vegetables, pastures and non-production areas.
- Livestock on the farm – this accounts for all animals on the farm for the year under consideration. Animal numbers are multiplied by their respective livestock units (LU) value to determine an overall carrying load. Manure and urea deposition can then be estimated from these. Stock weight gains over the year must also be accounted for as outputs.
- Milk production – total for the year.
- Wool production – total for the year.
- Livestock feeds – this takes account of all imported feed-stuffs of any kind. The N P and K values of the different feeds must also be noted.
- Roughage – any hay, silage or straw that is either imported or exported from the farm.
- Crops and seeds – seeds and/or plant seedlings brought onto the farm and their weight and N P K content must be identified. Likewise, in terms of outputs, all yields from crops must be noted along with their respective N P and K contents.
- Organic manure – for a farm-gate analysis the only details that necessarily must be recorded is FYM and/or slurry that is either imported or exported from or to the farm. Different types of organic manure have differing levels of N P and K (e.g. chicken manure compared with bovine) so this must be considered where necessary.
- Fertilisers – all mineral fertilisers used on the farm must be accounted for with clear statements about their respective N P and K concentrations.
- Legumes – all legume crops fix nitrogen into the soil from the atmosphere so the type of legume and the number of hectares of production must be considered.
- Deposition – precipitation deposits nitrogen directly to the farm. An estimate for the total over the year must be included.
- Mineralisation – again, an estimate of the generation of nitrogen as a result of mineralisation must be included.

Data associated with the above was collected during interviews with the farm managers. This usually took up to 3 hours each (sometimes involving more than one visit) although for farmers who maintained high quality records, the time necessary was considerably less.

At the time of compiling this paper, 33 farms were involved in the study and the median farm size for these was 192 hectares. As mentioned above, the sample was stratified according to farm size, farm type and locality within the Eden catchment. While the distribution of farm types within the sample cannot be considered statistically representative, the stratification is nevertheless approximately proportional to the whole farm population of the catchment. Since farming in the Fife area has above national average production levels and there is multigenerational participation, the selected farms are representative of the type of operation that will continue farming for the foreseeable future.

Results

Findings from initial interviews

None of the informants felt responsible for any negative environmental impacts either on or off the farm although twelve farmers did express some concern over nutrient exportation to rivers. About a third also said that soil erosion was a potential problem and there were other soil-related concerns mentioned including sub-soil compaction, structure decline and wind erosion.

Not surprisingly, the terrestrial environment was regarded as being slightly more important than the river or marine environment. Personal investments in the terrestrial environment (e.g. shooting interests) seem the likely cause of this slight bias. Productivity and marketing were the primary motivators for ensuring environmental integrity while lack of funds, together with government associated bureaucracy, were the main negative reasons that make continued environmental management difficult.

The health of the Eden estuary was not mentioned by any of the farmers in their responses about water quality despite the fact that this issue was specifically mentioned in a number of questions. More than a third of farmers thought that nitrates were a concern in rivers but further comments suggest this is perceived to be a diminishing problem. When asked about improving river and estuary quality, themes centred on present nutrient management regimes, denial (i.e. 'we're not responsible'), reconciliation (i.e. 'we'll adjust as we have to') and concern about possible future legislative or quality assurance scenarios.

Fences and hedgerows help prevent runoff which helps avoid river pollution and eutrophication (Unwin 2001). Even though most farmers were aware of this, more than half said that they had increased the size of their fields in recent years. While some suggested that this is more likely to occur on arable only farms because it is assumed they can benefit from the removal of fences and hedgerows, there was actually no evidence here to suggest that cereal producers were any more likely to take out fences than the livestock producers.

More than two-thirds of the farmers interviewed use a combination of animal waste and mineral fertiliser on their farms. When asked about management practices that are likely to cause nutrient exportation, the most important themes to emerge were applying excessive nutrients, the rain (including 'bad luck', which is associated with application followed by rain), applying manure to frozen ground, poor timing of applications and inappropriate manure and/or slurry storage. The implication here is that it is the livestock farms that are more likely to indulge in inappropriate practices.

Nearly all the farmers have their soils regularly tested for pH, phosphorous (P) and potassium (K), however, the frequency of testing varies considerably from almost monthly to periods greater than 5 years. Less than a third of the farmers that make up the AEU group test within recommended frequencies (i.e. every 2-3 years). Most farmers keep records of nutrient use and tests but only about half have records dating back more than 10 years, suggesting perhaps that it is only relatively recently that most have been monitoring consistently.

All farmers interviewed were aware that the Eden was being designated a NVZ in 2003 and just about all of the requirements associated with that legislation were mentioned collectively. However, individual's understanding of the requirements and implications varied widely. Most were aware of the closed seasons for nutrient applications and about a third new there would be restrictions on quantities. Storage requirements (silage and slurry) and record keeping were also mentioned by about a third of those interviewed. In terms of perceived impacts, nearly half the farmers thought NVZ status would affect their management practices in some way. Again, comments suggest it will be livestock producers more than cereals producers that will be hardest hit – storage was their biggest concern. Trading dung for straw with cereals producers may provide possible solutions to some livestock producer's storage capacity concerns.

Under the EU's Common Agricultural Policy (CAP) cereal farmers can receive annual payments (approximately £241.20/ha) for the area of land under cereal production so long as they 'set-aside' 10% of their cereal producing land upon which no production of any kind can take place (Moore, Allen and Innocent 2003). Not surprisingly, nearly all the farmers in the group had established some set-aside – usually the minimum required by the Agriculture Area Payments Scheme (AAPS) and all said they had established it to qualify for support under the scheme. Farmers clearly see more benefits associated with set-aside than there are costs. The only concern of any significance is the fact that these areas must be mown in accordance with the provisions of the scheme. Weeds and vermin may also be a minor issue.

The AAPS is by far the most common source of government funding support accessed by the farmers interviewed but this scheme effectively supports production of certain commodities. Other funding schemes of an environmental stewardship-type nature were mentioned but these were not nearly so popular e.g. four farmers were accessing the Countryside Premium Scheme (CPS), two the Rural Stewardship Scheme (RSS) and two the Woodland Grant Scheme (WGS). Views regarding accessing government funding varied. The AAPS was so widespread that most did not even acknowledge that it was funding support. But apart from the AAPS, many expressed apathy or even contempt for other funding schemes such as the RSS or CPS. For example, there were these comments: 'I don't want the commitment', 'I'm not interested', 'It's all too complicated' and 'It's too competitive'. Many also said they 'didn't know what was available' or they 'hadn't got round to it'.

When asked about possible changes in funding allocation under the CAP, nearly all were against such changes. Many also made comments to the effect, 'I'm a producer – not an environmental manager', suggesting perhaps that CAP reforms may bring cultural as well as economic and social impacts. About half the concerns were associated with the geographic distribution of funds and many felt they might be discriminated against on the basis of location. There was also concern about eligibility and the bureaucracy that they thought would inevitably be involved in applying for funds.

Buffer strips (non-production land along field margins) can enhance farm wildlife and help prevent run-off (Leeds-Harrison *et al.* 1996). Half the farmers interviewed said they had established buffer strips and four main reasons were given: to make the spraying of pesticides and herbicides easier, wildlife benefits, access to government funding and to help prevent erosion. Comments suggest there are probably more potential costs to the farmer associated with buffer strips than there are benefits; for example, loss of productive land, potential weeds and pests, fence establishment and maintenance (where there are stock). However, being able to avoid the spray regulations associated with Local Environment Risk Assessment for Pesticides (LERAPs) was regarded so highly that all felt these relatively minor costs were worth enduring.

Developing sustainable farming operations is clearly the way to ensure economic, social and environmental benefits. According to the NGO Linking Environment And Farming (LEAF) *Integrated Farm Management* (IFM) is a whole farm philosophy that provides 'the basis for efficient and profitable production which is economically viable and environmentally responsible…IFM integrates beneficial natural processes into modern farming practices using advanced technology. It aims to minimise environmental risks while conserving, enhancing and recreating that which is of environmental importance' (LEAF 2001). Less than half of the 30 farmers interviewed said they had heard of the integrated farming philosophy but it was clear from the descriptions of those that said they had that only a very few had anything but a very rudimentary understanding of what is implied; for example, 'sustainability' was never mentioned in their responses.

Precision farming involves utilising soil testing, computer and GPS technology to help improve the effectiveness of fertiliser distribution. The technology has the potential to reduce fertiliser costs and minimise wastage. In contrast to integrated farming, more than two-thirds of the farmers understood quite well what was implied by the technology and it turned out that two members of the group had already been experimenting with it.

The contrasting findings between integrated farming and precision farming perhaps demonstrate well how innovative methods will be adopted if it is perceived there are 'true' economic benefits; in this case, cost savings associated with reduced fertiliser use.

Farm Waste Management Plans (FWMP) can be developed for farmers with the intention of providing them with strategies to minimise the risk of diffuse pollution from their farms. When prepared properly, they should address at least four areas of concern: minimising dirty water around steadings, better nutrient use, risk assessment for manure and slurry and the management of water margins (SAC 2002). More than a third of the farmers in the AEU group said they had prepared a FWMP. However, further questions exploring these plans suggested that the vast majority were brief statements concerning the disposal of farm waste e.g. plastic containers, tyres etc. In fact, the responses here indicated that most farmers, in thinking of their FWMP, were actually referring to very minimal statements demanded by product quality assurance rather than comprehensive strategic waste management plans.

Further discussion associated with these interviews will follow after the section attending to nutrient budgeting.

Findings from the nutrient budgets

At the time of writing this paper, 30 farms had supplied full datasets for the NBs. These included 4 dairy or intensive livestock farms, 16 mixed farms and 10 arable only farms. The total area comprising these farms is 8,832 ha which is equal to approximately 34% of the farming area of the catchment. NBs were calculated for the year that started at the beginning of November 2001 and ended at the end of October 2002. While it is extremely difficult to estimate how much of the surplus was lost to the atmosphere, results suggest that farming in the area is making a very significant contribution to nitrogen pollution at the estuary. As Table 1 shows, all the farms in this sample were showing N surpluses to a greater or lesser degree. This finding is particularly significant in light of the fact that all farmers interviewed were unconvinced that they were responsible for any pollution to the system.

For management purposes, it is useful to know what the spatial distribution of farms having the greatest pollution impact on the catchment. With respect to nitrogen use, the NB delivers two notable findings. First, as discussed above, it estimates the nitrogen surplus/deficit for each farm. By expressing this in kgN/ha it offers a way of comparing surpluses between farms. However, it should be noted that comparing farms in this way can be misleading since every farm must be regarded as being unique (the true value of NBs lies in their capacity to monitor individual farms from one year to the next). Also useful is the level of efficiency i.e. what proportion of the N from inputs is converted to output products. Table 1 displays the findings of these

Table 1. Nitrogen Surpluses from Thirty Eden Catchment Farms

Farm Ref.	**Surplus Nitrogen (kgN/ha)**	*Nitrogen* **Efficiency (%)**	**Farm Ref.**	**Surplus Nitrogen (kgN/ha)**	*Nitrogen* **Efficiency (%)**
1	116.52	21	16	55.54	63
2	74.15	65	17	120.28	36
3	118.35	51	18	110.88	20
4	60.77	71	19	115.68	41
5	78.63	58	20	83.22	48
6	316.31	13	21	116.46	36
7	1460.94	28	22	50.54	73
8	370.05	17	23	120.28	36
9	76.66	52	24	159.31	28
10	50.78	68	25	54.36	72
11	97.44	27	26	48.3	69
12	138.31	37	27	263.01	30
13	111.67	44	28	82.65	49
14	105.91	62	29	45.5	82
15	58.26	56	30	83.61	66

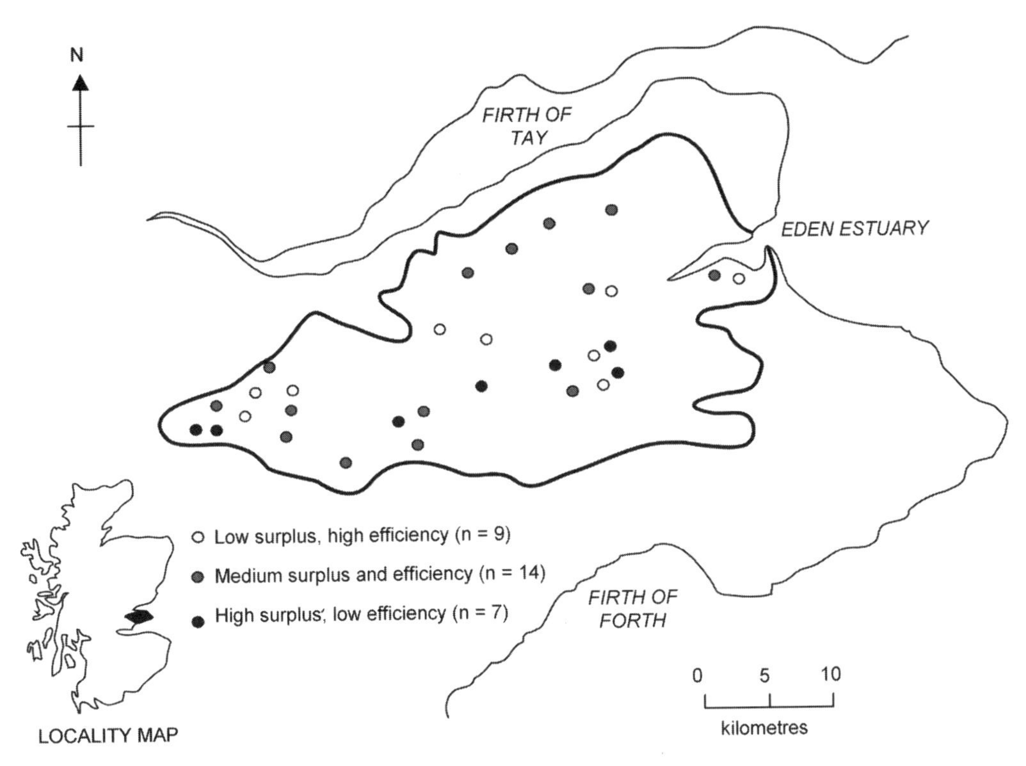

Figure 3. Spatial distribution of farms with estimates of their relative nitrogen pollution impacts

two results while Figure 3 displays the approximate location of the farms with completed NBs[2].

The circular symbols on the map (Figure 3) represent the nitrogen impact from each farm in the catchment. The symbology has been designed to take account of both efficiency and the amount of N per hectare when deriving estimates of impact. For the purpose of display, these have been summarised as: Low surplus, high efficiency (farms with < 100 kgN/ha and > 60% efficient); Medium surplus and efficiency (farms between 100 and 150 kgN/ha and 30 to 60% efficient); and, High surplus, low efficiency (farms > 150 kgN/ha and < 30% efficient).

A review of figure 3 suggests little if any catchment-wide spatial pattern in the distribution of the higher and/or lower agricultural polluters although it could be suggested that there would appear to more higher polluters along the course of the Eden, which runs east to west in the southern part of the catchment. The Motray Water also runs east to west but it is located in the northern part of the catchment, and it would seem that the four farms occupying this area are medium polluters only. There is some water quality evidence (TRPB 1994) to confirm that the Eden has more higher

[2] Confidentiality demands that the exact details of the farms' locations and references cannot be displayed.

polluters than the Motray but with limited cases in the sample, this comment must be treated with some caution. Of more interest perhaps is the relationship between farm type and pollution. Table 2 displays the median surpluses for the different farm types along with their respective median N efficiency levels.

Table 2: Median Nitrogen Surpluses and Efficiencies for 30 Eden Catchment Farms

Farm type	Surplus Nitrogen (kgN/ha)	Nitrogen Efficiency (%)
Arable (crops) only (n = 10)	75.4	61.5
Open mixed farming (n = 16)	113.7	42.5
Dairy / intensive livestock (n = 4)	343.2	18.5

What is apparent from Table 2 is that the arable farms generated the least N surplus while the dairy and intensive livestock producers produced the most. In fact, the four farms described here as dairy/intensive livestock producers contributed more than 21% of the total N from all thirty farms. In contrast, the arable producers contributed less than 14% despite the fact there were 10 of them in this category. There is a spatial component to this also; rainfall is heavier in the west of the catchment and it is here that most of the dairy farming in the catchment is carried out (higher rainfall produces the richer pastures necessary for milk production). The Tay River Purification Board (TRPB 1994) also observed higher nitrogen concentrations in the Eden in the west compared with the east (approximately 10.5 mg/l and 7.5 mg/l, respectively). All this confirms what was found in the interviews: that the intensive livestock producers and dairy farms are most likely responsible for diffuse agricultural pollution. They will also feel the impact of NVZ legislation the most.

Conclusions

The fact that none of the farmers interviewed in this study felt they were responsible for river and estuary pollution suggests there is a clear need to demonstrate that farming does produce nutrient surpluses. The findings from the NBs will be disseminated back to the farmers taking part in this study to gauge their reactions and to see if their views and intended management practices change as a result.

Nutrient budgeting is gathering increasing interest in the farming community in Europe. It is a simple but effective way of showing nutrient usage and loss on farms, however, its use in the farming community is not widespread. For example, only one farmer here had had a nutrient budget completed for his farm and this was only because the Scottish Agricultural College (SAC) had done it free of charge as part of a research study. Government farming agencies may need to consider more aggressively promoting the potential benefits of nutrient budgeting (cost savings associated with reduced mineral fertliser use) as part of whole farm management. By way of contrast, we can consider New Zealand's approach to this. The NZ Ministry of Agriculture and Forestry (MAF) has developed a quite sophisticated but fairly user-friendly nutrient budget software package which they will provide free to any farmer requesting it. Perhaps a similar strategy for NVZ designated areas would be appropriate in the UK. and in other EU States.

The farmers interviewed evidently do not readily perceive a connection with their farm management practices and the environment of the coastal zone. Basically, they continue to blame point sources for pollution to rivers and the estuary. The question associated with integrated farming also demonstrates that most lack what could be described as a 'systems thinking paradigm' about their farms and their place in the wider environment. They tend to focus only on what occurs within the boundaries of their farms; this is not to say that they lack concern for the environment but its value seems strongly associated with production motivations. What occurs off the farm seems over-looked or ignored partly at least because there is little incentive to minimise off-site impacts. In light of this it would appear that a legislative approach to encouraging sound environmental management might be the only serious way of ensuring compliance with sustainable practices. The responses of the farmers to the regulations associated with LERAPS regulations perhaps demonstrates well how restrictions can produce positive results – in short, most farmers find it so much trouble to manage pesticide and fertliser spraying close to water coursers that it is simply easier to install buffers. Importantly, few seem particularly troubled by the loss of production land and in some cases, strips were made wide enough (20 meters instead of the required 10 under LERAPS) so that they could continue to obtain an income from the land as set-aside.

One could question whether the current regulations associated with NVZ status go far enough. Arable producers certainly seemed the least concerned about restrictions over the quantity of fertlisers – all claimed they use less that that permissible under the prescribed limits. Certainly, given the results form the NBs, it seems these producers have less to be concerned about than others. Nevertheless, even this group is producing surpluses and it is clear from SEPA's continuing water quality tests of the Eden that a diffuse pollution problem still exists. Again, more widespread use of nutrient budgeting would help clarify what quantities of nutrients are escaping from which farms.

Livestock producers certainly do have some concerns with NVZ status. Their biggest concerns are those associated with the storage of silage and slurry. Many recognise that if they wish to maintain their herd sizes they may have to up-grade their tanks and pits. Most will also have to curb the quantities of fertilisers/farmyard manure they use. While most would probably qualify for a 40 percent grant to help improve their storage infrastructure, they maintain that there is no way to finance the remaining 60 percent; they comment, 'with commodity prices being so low, where am I to get the rest of the money?' Trading dung or slurry for straw with arable farmers seems like it may offer part of the solution in at least some cases but unfortunately, in an NVZ context, the farmyard manure rich farmers may have some difficulty identifying local arable producers willing to take part in such exchanges. One arable farmer in this study had already opted out of such an arrangement he had with a neighbour because of perceived possible NVZ conflicts. Exchanges of this kind may be possible wider afield, i.e. with arable farmers outside NVZs but even if this is possible the expense of transportation is likely to make it prohibitive.

This study did not set out to examine the social impact of CAP reforms but it is clear from comments received that the subsidies associated with it and environmental management are intertwined. For example, there is a strong dependence on income support generated through the AAPS scheme. The average arable or mixed farm in this study grew approximately 120 hectares of AAPS qualifying grains (typically wheat, barley and oil seed rape). This represents an annual income of about £29,000. If CAP reforms result in the loss of this it will surely have knock-on effects that will impact on the farmer's capacity to carry out or experiment with environmentally sensitive practices. Given the farmer's rather luke-warm attitudes towards alternate funding schemes, the shift from a production emphasis to one of environmental stewardship demanded by NVZ status and CAP reforms will certainly not be an easy one. Consequently, it may be some time before there are observable improvements in the quality of river and estuarine water.

References

Bergius, S., Parkes, P. and Stankevicius, R. 2002. Investigation of Carbon and Nitrogen Cycles in Pig Farming, Socrates European Common Curriculum (05 85 00) Spring semester 2002, Den Kongelige Veterinaer og Landbohojskole, Copenhagen.

Boesh, D. 2001. Agriculture and Coastal Eutrophication, [On-line], Available: http: //ca.umces.edu/ian/commonground/reports.htm [5 October 2001].

Agresearch 2002. Overseer User Manual: nutrient budget programs, Ministry of Agriculture and Forestry, New Zealand.

ECN (Environmental Change Network) 2002. ECN Freshwater Sites – River Eden, Fife Region, Scotland, [On-line], Available: http: //www.ecn.ac.uk/sites/eden2.html [19 April 2002].

Lanyon, L. and Beegle, D. 1989. The role of on-farm nutrient balance assessments in an integrated approach to nutrient management, Journal of Soil and Water Conservation **44**: 164-168.

LEAF (Linking Environment And Farming) 2001. What is Integrated Farm Management? [On-line], Available: http: //www.leafuk.org/LEAF/organisation/ifm.asp [15 June 2002].

Leeds-Harrison, P., Quinton, J., Walker, M., Harrison, K., Tyrrel, S., Morris, J. and Harrod, T., 1996. Buffer Zones in Headwater Catchments. Report on MAFF/English Nature Buffer Zone Project CSA 2285. Cranfield University, Silsoe, UK.

Mason, C., Underwood, G., Baker, N., Davey, P., Davidson, I., Hanlon, A., Long, S., Oxborough, K., Paterson, D. and Watson, A. 2003. The role of herbicides in the erosion of salt marshes in eastern England, Environmental Pollution **122**: 41-49.

Moore, Allen and Innocent 2003. A Guide to IACS and other Agricultural Subsidy Payments for 2003, [On-line], Available: http: //www.mooreallen.co.uk/agriculture/information-extra1.htm [10 June 2003].

Robbins, N. 1990. Hydrogeology in Scotland. London: HMSO for the British Geological Survey.

SAC (Scottish Agricultural College) 2002. The 4-point Plan: Straight forward guidance for livestock farmers to minimise pollution and benefit your business, [On-line], Available: http: //www.sac.ac.uk/corporate/publications.asp [7 July 2002].

Skinner, J.A., Lewis, K.A., Bardon, K.S., Tucker, P., Catt, J.A. and Chambers, B.J. 1997. An overview of the environmental impact of agriculture in the U.K. Journal of Environmental Management **50**: 111-128.

Scottish executive 2002. Fife farmers prepare to discuss NVZs. News Release: SE237/2002, [On-line], Available: http: //www.scotland.gov.uk/pages/news/2002/01/SE5237.aspx [28 January 2002].

Scottish executive 2003. Guidelines for Farmers in Nitrate Vulnerable Zones. The Scottish Executive.

SEPA (Scottish Environmental Protection Agency) 1996. State of the Environment Report 1996, SEPA Corporate Office, Stirling, Scotland.

Sheaves, J. 1999. FWAG's Nutrient Budget Handbook (unpublished), Farming and Wildlife Advisory Group, Prince of Wales Road, Dorchester, UK.

TRPB (Tay River Purification Board) 1994. A Catchment Study of the River Eden, Fife. Technical

Report TRPB 1/94. Tay River Purification Board, Fife.
UNWIN, R. 2001. 'New Initiatives to Control Soil Erosion in England', pages 426-430 in D. E. Stott, R. H. Mohtar and G. C. Steinhardt (eds). 2001. Sustaining the Global Farm. Selected papers from the 10th International Soil Conservation Organisation Meeting, May 24-29, 1999 at Purdue University.

Carsten Brockmann

Coastal management of estuarine environments: Application of remote sensing

Abstract

Proper management of estuarine environments requires appropriate management plans as well as tools to carry out these plans. This paper reviews the use of remote sensing as a valuable tool in monitoring of estuarine ecosystems. One problem with respect to monitoring of estuaries is that a small pixel size is required, but an overview of currently available space born sensors show that a number of satellite sensors provide the required spatial resolution. Another difficulty with the interpretation of optical remote sensing images is that they contain a mixture of signals originating from different origins, i.e. from mud, water, microphytobenthos etc. Decomposition of the composite signal into its components by Linear Spectral Unmixing provides good results for intertidal areas. It is concluded that the best way to monitor estuarine environment is by a combination of available methodologies, including remote sensing.

Introduction

The proper management of estuarine environments requires a) the appropriate policy directives and b) the necessary tools to carry out these directives. The increasing number of multilateral environmental treaties and EU Directives since the 1972 Stockholm Conference is an encouraging sign and testament to man's increasing concern and commitment to protecting particularly sensitive environments. However, because of this proliferation in the number of directives and treaties there is also a concomitant growing need for more information relating to the health of ecosystems, in order to better understand the biogeophysical processes involved and the socio-economic consequences which global, regional and national policies can have on these systems.

The growing acceptance of remote sensing data in the contribution to our understanding of coastal systems – in a visually compelling way – reflects the development which has taken place in the algorithm development necessary to properly process these data. A global assessment of ecosystems without such data is next to impossible and this is realized by scientists and policy makers alike. The use of remote sensing data to assist in environmental problems is mentioned in no less than 8 paragraphs and 11 sub-paragraphs in the recent Earth Summit on Sustainable Development - Plan of Implementation in Johannesburg in September, 2002. Most notably is the statement in Paragraph IV; 'Protecting and managing the natural resource base of economic and social development' subparagraph 35(c) which states:

> '*An integrated, multi-hazard, inclusive approach to address vulnerability, risk assessment and disaster management, including prevention, mitigation, preparedness, response and recovery, is an essential element of a safer world in the twenty-first century. Actions are required at all levels to: …………Strengthen, the institutional capacities of countries and promote international joint observation and research,* ***through improved surface-based monitoring and increased use of satellite data….***'

The following contribution will provide examples of how remotely sensed data can be used as an integral part of measurement, monitoring and management strategies for tidal flat areas. This will be illustrated using examples of results from two EU-funded projects, namely BIOPTIS (Assessing the Biological and Physical Dynamics of Intertidal Sediment Ecosystems: A Remote Sensing Approach) [BIOPTIS] and HIMOM (A System of Hierarchical Monitoring Methods for Assessing Changes in the Biological and Physical State of Intertidal Areas) [HIMOM].

Coastal management requirements

The management of estuarine and tidal flat environments requires, on the one hand, the monitoring of a multitude of parameters, on a multitude of spatial and temporal scales while on the other hand, the necessary tools to translate this scientific knowledge into facts which can be used in decision making. These tools help translate science into decisions, however, it should be clear to the scientists, who make the measurements, which parameters are of most importance in terms of the long-term sustainability of the system and how to structure these measurements to assist in long-term monitoring strategies. Tidal flat areas such as the Wadden Sea are a unique heritage which need careful management if they are to have long term sustainability. This is why the governments of Holland, Germany and Denmark place such emphasis on the advice and guidance of the Trilateral Monitoring and Assessment Group (mostly scientists) which laid down guidelines for what parameters are of importance to monitor to safeguard the future sustainability of this ecosystem [TMAP 2001].

A healthy ecosystem can be viewed as having the correct balance between productivity and biodiversity. The cornerstone of most aquatic ecosystems in terms of productivity has been shown to be the algae i.e. the phytoplankton and microphytobenthos. Berger *et al.* (1989) for instance estimated a two to five fold greater production in shelf areas compared to what occurs in open oceans. Postma (1982) concluded that the productivity of the phytoplankton in the Wadden Sea was about three times greater than that of the adjacent North Sea. This is a dramatic difference between a tidal flat area and the open ocean. Considering just the Wadden Sea alone, Asmus *et al* (1996) found that the productivity of the plants species (excluding macroalgae) can be divided between the phytoplankton contributing to 52% of the total, microphytobenthos to 45% and the seagrasses to 3% These results illustrated the enormous importance of the microphytobenthos to the overall productivity of tidal flat areas in the Wadden Sea.

However, the authors of the Trilateral Monitoring and Assessment Programme Evaluation Report (TMAP 2001) noted that the large majority of the parameter groups could be implemented but gaps still existed in the Common Package Parameters. It was noted that the TMAP provides no information for calculating the primary production of the benthic microalgae, which was considered to be a serious gap. It was noted that several gaps, which include information on the development of the coastline or mapping of habitat parameters of salt marsh areas, were not included in the Common Package (Figure 1) and that these gaps could be filled in using remote sensing techniques.

Solutions – current and future practices

Monitoring requirements of the various parameters, e.g. as in the TMAP Common Package, differ in their spatial and temporal resolution, which has a strong impact on the most suitable technique to use. Today most measurements are made by ground sampling with subsequent laboratory analysis. Only points are sampled and the selection of measurement points is biased by the accessibility of the area. Because of this time consuming procedure, large areas such as the Wadden Sea can only be sampled sparsely.

<table>
<tr><th colspan="3">Common Package of TMAP parameters</th></tr>
<tr>
<td>Chemical Parameters
• Nutrients
• Metals in sediment
• Contaminants in Blue Mussels, Flounders and birds eggs
• TBT in water and sediment</td>
<td rowspan="2">Biological Parameters
• Phytoplankton
• Macroalgae
• Eelgrass
• Macrozoobenthos
• Breeding birds
• Migratory birds
• Beached birds survey
• Common Seals
• Benthic microalgae*

*...addition after TMAP evaluation, Esjberg 2001

Italic = potentially measurable by remote sensing (direct of indirect)</td>
<td>Human Use Parameters
• Fisheries
• Recreational activities
• Agriculture
• Coastal Protection</td>
</tr>
<tr>
<td>Habitat Parameters
• Blue Mussel beds
• Salt marshes
• Beaches and Dunes</td>
<td>General Parameters
• Geomorphology
• Flooding
• Land use
• Weather conditions
• Hydrology</td>
</tr>
</table>

Figure 1. Parameters included in the TMAP Common Package. Parameters which can potentially be measured by remote sensing techniques are printed in italics.

Temporal monitoring requirements as laid down by the Water Framework Directive, the Habitat Directive or the TMAP, are ranging between daily to weekly measurements, e.g. for algal bloom monitoring (SISCAL), up to 5 year surveys for coastal morphology. The requirements for microphytobenthos, as formulated in the TMAP, are in the order of a seasonal assessment. In some cases the time for taking the measurement is event driven, i.e. triggered by natural circumstances, such as the evolution of a spring bloom. The spatial monitoring requirements vary between the decimetre/meter scale, e.g. for microphytobenthos, over a scale of 10-50 m, which is a typical resolution for a large scale assessment of primary productivity up to a scale of 1 km for monitoring water quality parameters. Today, several remote sensing systems which match these requirements are in operation (see Table 1).

Optical remote sensing instruments measure the sunlight backscattered from the sediment surface or from the euphotic zone of a water body. The light is spectrally resolved in several spectral bands. Methods are available today to qualitatively and quantitatively derive parameters from these measurements, which are requested by the monitoring programmes. The instruments can be operated from airplanes and are included on satellites. Satellites have the advantage of providing a larger overview and regular overpasses over an area of interest, while airplanes have the advantage of higher spatial resolution and are less dependent on cloud coverage. As an example, a Landsat 7 ETM image of the Sylt-Romo Bight in the North Frisian Wadden Sea is shown in Figure 2 as a black and white composite. In the Wadden Sea the vegetated areas are highlighted. The data have a spatial resolution of 30 m.

Every image pixel is the measurement of the sunlight reflected from an area at the surface covering 30 m x 30 m. This area is generally not uniform but a mixture of sediment of a certain grain size, vegetation of one or more species, and more or less water coverage. As a consequence, the signal measured at the sensor is a mixture of the reflectances of all different elementary surface types. Methods exists to decompose this measured signal into its components, one of these being the Linear Spectral Unmixing (LSU) (Adams *et al.*1989) which provides good results in intertidal areas (Stelzer *et al.* 2004). As a result, the portion of each elementary surface type in every image pixel is derived. The resulting image then shows, for example, the sand and mud fraction or the water coverage (Figure 3). This can be further calibrated with ground truth points in order to derive a quantitative image.

This is only one example of the contribution which remote sensing can provide to fulfil coastal monitoring requirements. On the smaller spatial scale, i.e. in the range of 1-5 m resolution, systems such as Ikonos have been available for quite a few years, which provide less spectral information but much higher spatial resolution compared to the Landsat ETM image described above. These data are combined with vector data in order to monitor coastal morphology, beach nourishment, area and position of dunes etc. Also a rough estimate of vegetation cover can be derived, but a sophisticated delineation of elementary surface types is not possible due to the lacking spectral information.

The monitoring of phytoplankton has been part of the TMAP from the beginning and is gaining more attention due to the EU Water Framework Directive. Remote sensing techniques to measure the chlorophyll-a concentration in the ocean have been

Table 1. Spatial and spectral characteristics of space borne remote sensing instruments

Sensor	Ikonos PAN	Quick Bird	IKONOS	SPOT	IRS	LANDSAT	MERIS FR	NOAA AVHRR	SeaWiFS	MERIS RR
Spatial resolution	1 m	2.4 m	4 m	20 m	23 m	30 m	300 m	1000 m	1100 m	1200 m
Swath width	11 km	16.5 km	11 km	60 km	140 km	183 km	1345 km	3000 km	2800 km	1345 km
Spectral resolution	Pan	4 (vis, nIR)	4 (vis, nIR)	4 (vis, nIR)	3 (vis, nIR, thIR)	7 (vis, nIR, mIR, thIR)	15 (vis, nIR)	4 (vis, nIR, mIR, thIR)	8 (vis, nIR)	25 (vis, nIR)
Temporal resolution	2.9 days		1.5 days	26 days	24 days	16 days	3 days	2-4 days	1 day	3 days
Price	25-98* $/ km2	25$/km2	18-98* $/ km^2	1900€ 53€/100km^2	2700 € 13€/100km^2	600-1500 € 2-5€/100km^2				

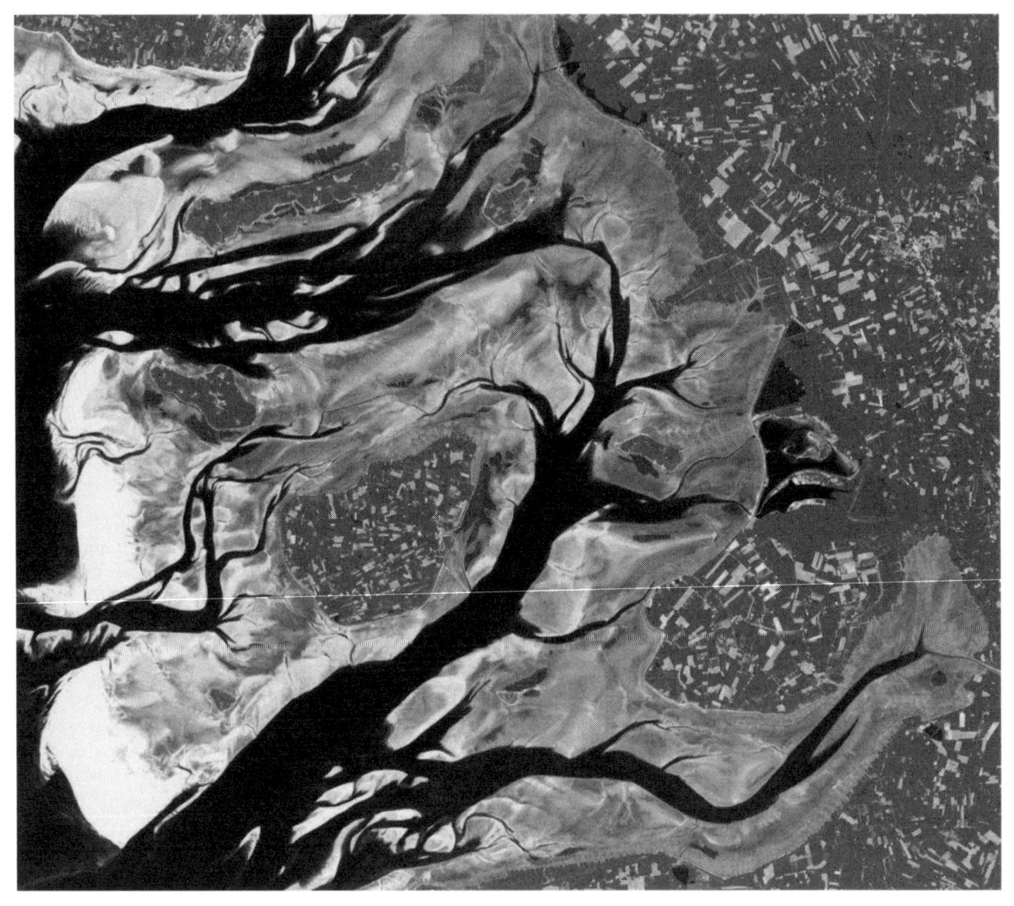

Figure 2. Example of a Landsat-7 ETM scene of the North Frisian Wadden Sea, 15.7.2002. Stelzer 2004. Landsat 7 ETM © Eurimage 2002

developed since the early 80s (Gordon and Morel 1983), and since the mid 90s space borne instruments dedicated to measuring the sea surface reflectance, such as SeaWiFS and MERIS, are in operation. In coastal waters, the retrieval of the chlorophyll-a concentration is particularly difficult due to the non correlated co-existence of suspended matter and yellow substance, which both have a strong influence on the measured signal. However, new retrieval methods are available (Doerffer and Schiller 1997) and weekly mean chlorophyll concentration maps of important European waters are now available on an operational basis (Marcoast).

An analysis carried out for the environmental ministries of Lower Saxony and Schleswig-Holstein in Germany (Stelzer 2004) has shown that the determination of the area and location of features and patterns of the following parameters, which are essential parts of the TMAP as shown in Figure 1, can be especially improved by using remote sensing techniques. These include:

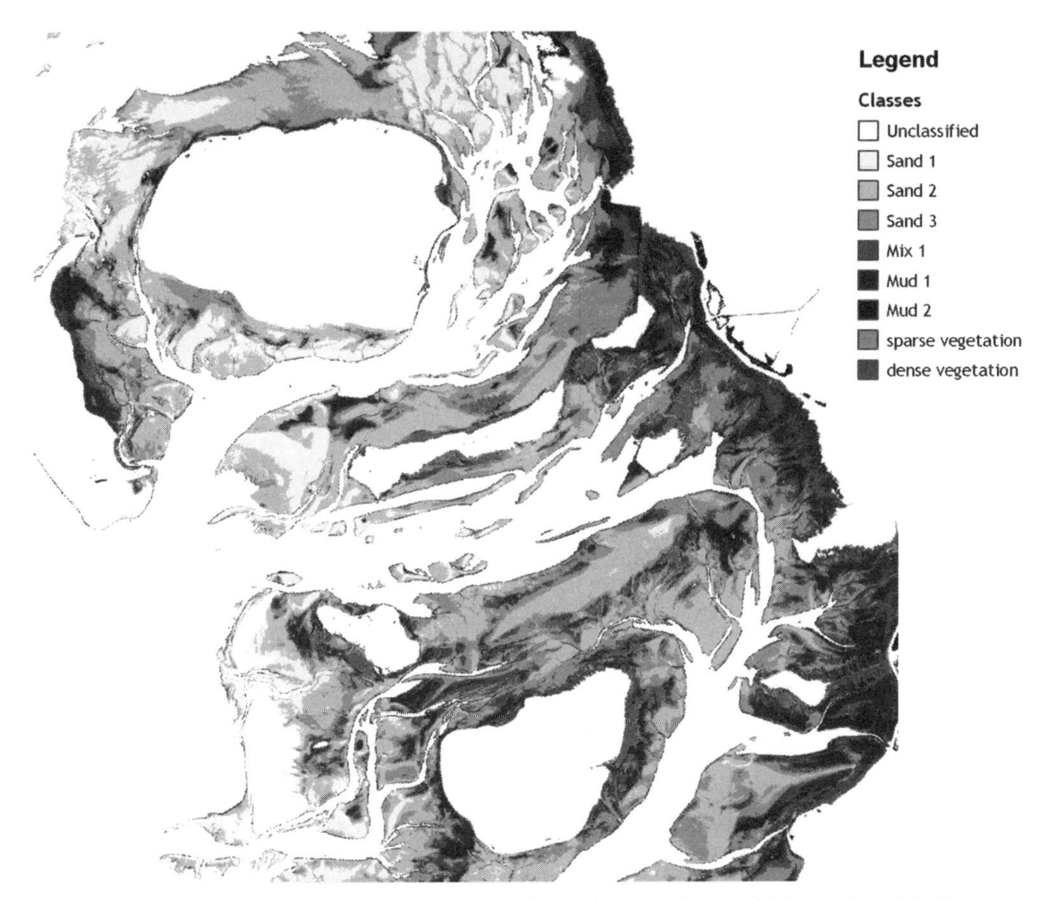

Figure 3. Result of the LSU (sediment classes) applied to the ETM image of North Frisian Wadden Sea. Stelzer 2004

- sediment type
- macrophytes
- phytoplankton
- blue mussel beds

Conclusion

The need for monitoring of coastal zones and in particular intertidal areas and coastal waters has become more urgent in the past decade due to European regulations, which themselves are a response on the high anthropogenic pressure being exerted on our coastal environments. Traditional techniques, namely the combination of intensive field survey and subsequent laboratory analysis, are high quality and well established methods, however, they are labour intensive and costly, and they cannot provide large scale assessments at reasonable frequencies. The need for new, more appropriate methods has been expressed for quite some time now, and in this respect, considerable expectations have been put on remote sensing.

Thanks to the availability of suitable instruments, both on air and space borne platforms, and due to the development of new evaluation procedures, it could be shown that some key parameters, such as sediment type or location and area of macrophytes, can be derived from these measurements. Satellite data are available at acceptable costs and are significantly cheaper compared to field surveys. However, the availability of space borne data is very much limited by cloud coverage in northern latitudes, and the spatial resolution, which is in the order to 30 m, is at the lower end of what is required by coastal managers. Airborne data overcome these limitations, however, they are much more expensive and the costs for one campaign can be compared with that for an extensive field survey.

The best way to proceed will be in using a combination of all available methods. Space borne remote sensing is beneficial for establishing references for key parameters and for monitoring large areas at a low frequency. Such data can also be used as a planning tool for airborne and field campaigns. Areas which are unambiguously mapped by space borne remote sensing can be left out, enabling more time to be spent, or an airborne campaign to be undertaken, in areas where for instance the macroalgae could not be differentiated from seagrass.

Acknowledgement

Part of this work was co-funded by the European Commission Research Directorate General XII, for the Project HIMOM (Contract No. EVK3-CT-2001-00052). We would like to thank the HIMOM project team, particularly Rodney Forster, for their support to the remote sensing work. We also thank Arfst Hinrichsen for the critical assessment of the study on high resolution imaging and provision of the vector data of the Amt für ländliche Räume.

References

BIOPTIS – Assessing the Biological and Physical Dynamics of Intertidal Sediment Systems: a Remote Sensing Approach. EU co-funded research project. Project reference MAS3970158. Homepage: http: //www.st-andrews.ac.uk/~serg/research/BIOPTIS/

HIMOM – A system of hierarchical monitoring methods for assessing changes in the biological and physical state of intertidal areas. EU co-funded research project. Project reference EVK3-CT-2001-00052. Homepage: http: //www.brockmann-consult.de/himom

Trilateral Monitoring and Assessment Group (2001). TMAP Evaluation Report. The Trilateral Monitoring and Assessment Program (TMAP). Common Wadden Sea Secretariat, Wilhelmshaven.

Berger, W.H., V.S. Smetacsk and G. Wefer. 1989. Productivity in the Ocean: Past and Present. Life Science Research Report, Vol 44, Chichester: John Wiley and Sons.

Postma, H. 1982. Hydrography of the Wadden Sea: Movements and properties of water and particulate matter. Report 2 of the Wadden Sea Working Group, 75 pp.

Asmus, R., M.H. Jensen, D. Murphy and R. Doerffer. 1998. Primary production of the microphytobenthos, phytoplankton and the annual yield of macrophytic biomass in the Sylt-Rømø Wadden Sea. (In German, English abstract) In C. Gätje and K. Reise (edts.) Ökosystem Wattenmeer – Austausch-, Transport- und Stoffumwandlungsprozesse. Springer-Verlag Berlin Heidelberg, pp. 367-391.

SISCAL – Satellite-based Information System on Coastal Areas and Lakes. EU co-funded research project. Project reference IST-2000-28187. Homepage www.siscal.net

ADAMS, J.B., SMITH, M.O, & GILLESPIE, A.R. 1989. Simple Models for Complex Natural Surfaces: A Strategy for the Hyperspectral Era of Remote Sensing. Proceedings IEEE Int. Geoscience and Remote Sensing Symposium, New York, 1, 16-21

STELZER K., BROCKMANN C. and MURPHY D. (2004). Using Remote Sensing Data for Coastal Zone Monitoring. RSPSoc 2004. Mapping and Resources Management. Aberdeen, 7.-10.9.2004. Proceedings

GORDON, H. R. and A. MOREL. 1983. Remote assessment of ocean color for interpretation of satellite visible imagery. A review. Lecture Notes on Coastal and Estuarine Study, Vol. 4, Springer Verlag.

DOERFFER, R., SCHILLER, H. 1997. Pigment Index, Sediment and Gelbstoff Retrieval from directional Water-leaving Reflectances using Inverse Modelling Technique. Algorithm Theoretical Basis Document. http: //envisat.esa.int/instruments/meris/pdf/atbd_2_12.pdf

COASTWATCH – a coastal information service for the European coastal environment. Homepage www.coastwatch.info

STELZER, K. 2004: Potenzial der Fernerkundung im Küstenraum für die Umsetzung der EG-Wasserrahmenrichtlinie, TMAP und FFH. www.brockmann-consult.de/english/projects/projects

List of contributors

List of contributors

(corresponding authors listed with e-mail address)

Dr. Linda K. Medlin
Alfred Wegener Institute
Am Handelshafen 12
D-27570 Bremerhaven Germany
lmedlin@AWI-Bremerhaven.DE

Dr. Jacco C. Kromkamp
Netherlands Institute of Ecology, Centre for Estuarine and Marine Ecology
PO Box 140, 4400 AC Yerseke
The Netherlands
j.kromkamp@nioo.knaw.nl

Dr. Rodney M. Forster
Netherlands Institute of Ecology, Centre for Estuarine and Marine Ecology
PO Box 140, 4400 AC Yerseke
The Netherlands
r.forster@nioo.knaw.nl

Dr. Ronnie N. Glud
University of Copenhagen
Marine Biological laboratory
Strandpromenaden 5
3000 Helsingør
Denmark
rnglud@zi.ku.dk

Dr. Jody F.C. de Brouwer
Netherlands Institute of Ecology, Centre for Estuarine and Coastal Ecology
P.O. Box 140
4400 AC Yerseke
The Netherlands
j.debrouwer@nioo.knaw.nl

Dr. Thomas R. Neu
UFZ Centre for Environmental Research Leipzig-Halle, Department of Inland Water Research,
Brückstrasse 3a
DE 39114 Magdeburg
Germany

Dr. Kevin R. Carman
Department of Biological Sciences
Louisiana State University
Baton Rouge, LA 70803
zocarm@lsu.edu

Dr. Padma Maddi
Department of Biological Sciences
Louisiana State University
Baton Rouge, LA 70803
USA

Dr. Brian Fry
Department of Oceanography and Coastal Science
Louisiana State University
Baton Rouge, LA 70803
USA

Dr. Bjoern Wissel
Department of Oceanography and Coastal Science
Louisiana State University
Baton Rouge, LA 70803
USA

Dr. Lucas J. Stal
Netherlands Institute of Ecology, Centre for Estuarine and Coastal Ecology
P.O. Box 140
4400 AC Yerseke
The Netherlands

Dr. Lawrence B. Cahoon
Department of Biological Sciences, University of North Carolina at Wilmington,
Wilmington, N.C. 28403 USA
cahoon@uncw.edu

Dr. Graham J. Underwood
Department of Biological Sciences
University of Essex
Wivenhoe Park
Colchester
Essex. CO4 3SQ
U.K.
gjcu@essex.ac.uk

Dr. Mark Barnett
Department of Biological Sciences
University of Essex
Wivenhoe Park
Colchester
Essex. CO4 3SQ
U.K.

Dr. Gérard F. Blanchard
University of La Rochelle, Laboratoire de Biologie et Environnement Marins
(LBEM, EA3168), Rue Enrico Fermi 17000 La Rochelle
France
gerard.blanchard@univ-lr.fr

Dr. T. Agion
University of La Rochelle, Laboratoire de Biologie et Environnement Marins
(LBEM, EA3168)
Rue Enrico Fermi
17000 La Rochelle
France

Dr. O. Herlory
University of La Rochelle, Laboratoire de Biologie et Environnement Marins
(LBEM, EA3168)
Rue Enrico Fermi
17000 La Rochelle
France

Dr. Jean-Marc Guarini
Laboratoire Arago
LOBB – UI 938 – UMR 76221
BP 44 – 66651 Banyuls sur Mer.
jean-marc.guarini@obs-banyuls.fr

Dr. P. Richard
CNRS-IFREMER, Centre de Recherche sur les Ecosystèmes Marins et Aquacoles
(CREMA, UMR10), B.P.5, 17137 L'Houmeau,
France

Dr. James L. Pinckney
Marine Science Program and
Department of Biological Science University of South Carolina
Columbia, SC 29208 (USA)
(803) 777-7133
jpinckney@biol.sc.edu

Dr. Laurent Seuront[1,2],
[1]Ecosystem Complexity Research Group, Station Marine de Wimereux CNRS UMR 8013 ELICO
Université des Sciences et Technologies de Lille
28 avenue Foch
F-62930 Wimereux, France
[2]School of Biological Sciences
The Flinders University of South Australia
GPO Box 2100
Adelaide 5001
South Australia
Laurent.Seuront@univ-lille1.fr / Laurent.Seuront@flinders.edu.au

Dr. Céline Leterme
Ecosystem Complexity Research Group
Station Marine de Wimereux
CNRS UMR 8013 ELICO
Université des Sciences et Technologies de Lille
28 avenue Foch
F-62930 Wimereux
France

Dr Colin J. Macgregor
School of Geography and Geosciences
University of St Andrews
St Andrews, Fife
Scotland
UK
cjm27@st-andrews.ac.uk

Dr. Carsten Brockmann
Brockmann Consult
Max-Planck-Str. 1
21502 Geesthacht
Germany
carsten.brockmann@brockmann-consult.de

